Dedication

To Sallie and Elaine

Brief Contents

Relations for Discrete Cash Flows with End-of-Period Compounding

Type	Find/Given	Factor Notation and Formula	Relation	Sample Cash Flow Diagram
Single Amount	F/P Compound amount	$(F/P,i,n) = (1 + i)^n$	$F = P(F/P,i,n)$	
	P/F Present worth	$(P/F,i,n) = \dfrac{1}{(1 + i)^n}$	$P = F(P/F,i,n)$	
Uniform Series	P/A Present worth	$(P/A,i,n) = \dfrac{(1 + i)^n - 1}{i(1 + i)^n}$	$P = A(P/A,i,n)$	
	A/P Capital recovery	$(A/P,i,n) = \dfrac{i(1 + i)^n}{(1 + i)^n - 1}$	$A = P(A/P,i,n)$	
	F/A Compound amount	$(F/A,i,n) = \dfrac{(1 + i)^n - 1}{i}$	$F = A(F/A,i,n)$	
	A/F Sinking fund	$(A/F,i,n) = \dfrac{i}{(1 + i)^n - 1}$	$A = F(A/F,i,n)$	
Arithmetic Gradient	P_G/G Present worth	$(P/G,i,n) = \dfrac{(1 + i)^n - in - 1}{i^2(1 + i)^n}$	$P_G = G(P/G,i,n)$	
	A_G/G Uniform series	$(A/G,i,n) = \dfrac{1}{i} - \dfrac{n}{(1 + i)^n - 1}$ (Gradient only)	$A_G = G(A/G,i,n)$	
Geometric Gradient	P_g/A_1 and g Present worth	$P_g = \begin{cases} \dfrac{A_1\left[1 - \left(\dfrac{1 + g}{1 + i}\right)^n\right]}{i - g} & g \neq i \\[4mm] A_1\dfrac{n}{1 + i} & g = i \end{cases}$ (Gradient and base)	$\begin{aligned} & g \neq i \\ & g = i \end{aligned}$	

F/P $F_g = A_1(F/A, g, i, n) = A_1\left[\dfrac{(1+i)^n - (1+g)^n}{i \pm g}\right] \quad g \neq i$

Basics of Engineering Economy

Leland Blank, P. E.

Dean Emeritus
American University of Sharjah
United Arab Emirates
and
Professor Emeritus
Industrial and Systems Engineering
Texas A&M University

Anthony Tarquin, P. E.

Professor
Civil Engineering
University of Texas—El Paso

McGraw-Hill
Higher Education

Boston   Burr Ridge, IL   Dubuque, IA   New York   San Francisco   St. Louis
Bangkok   Bogotá   Caracas   Kuala Lumpur   Lisbon   London   Madrid   Mexico City
Milan   Montreal   New Delhi   Santiago   Seoul   Singapore   Sydney   Taipei   Toronto

McGraw-Hill
Higher Education

BASICS OF ENGINEERING ECONOMY

Published by McGraw-Hill, a business unit of The McGraw-Hill Companies, Inc., 1221 Avenue of the Americas, New York, NY 10020. Copyright © 2008 by The McGraw-Hill Companies, Inc. All rights reserved. No part of this publication may be reproduced or distributed in any form or by any means, or stored in a database or retrieval system, without the prior written consent of The McGraw-Hill Companies, Inc., including, but not limited to, in any network or other electronic storage or transmission, or broadcast for distance learning.

Some ancillaries, including electronic and print components, may not be available to customers outside the United States.

This book is printed on acid-free paper.

1 2 3 4 5 6 7 8 9 0 DOC/DOC 0 9 8 7

ISBN 978–0–07–340129–4
MHID 0–07–340129–3

Global Publisher: *Raghothaman Srinivasan*
Executive Editor: *Michael Hackett*
Director of Development: *Kristine Tibbetts*
Developmental Editor: *Lorraine K. Buczek*
Executive Marketing Manager: *Michael Weitz*
Senior Project Manager: *Kay J. Brimeyer*
Senior Production Supervisor: *Sherry L. Kane*
Associate Design Coordinator: *Brenda A. Rolwes*
Cover/Interior Designer: *Studio Montage, St. Louis, Missouri*
Compositor: *Aptara*
Typeface: 10/12 *Times Roman*
Printer: *R. R. Donnelley Crawfordsville, IN*
(USE) Cover Image: *Transportation Market:* © Simon Fell/Getty Images; *Balancing Time and Money:* © Randy Allbritton/Getty Images; *Freeway Interchange:* © PhotoLink/Getty Images; *Construction:* © Digital Vision/PunchStock; *Offshore Oil Rig:* © Royalty-Free/CORBIS

Library of Congress Cataloging-in-Publication Data

Blank, Leland T.
  Basics of engineering economy / Leland Blank, Anthony Tarquin.—1st ed.
    p. cm.
  Includes indexes.
  ISBN 978–0–07?340129?4 ?ISBN 0–07?340129?3 (hard copy : alk. paper)
  1. Engineering economy. I.
Tarquin, Anthony J. II. Title.
    TA177.4.B565 2008
    658.15–dc22

                                                                    2007025300

Contents

Preface

This text covers the basic techniques and applications of engineering economy for all disciplines in the engineering profession. Its design and organization allow flexibility in topical coverage for any undergraduate curriculum, varying from ABET-accredited engineering and engineering technology programs to pre-engineering courses, to postgraduate and research-oriented programs. The text is adaptable to resident instruction and distance learning environments.

The writing style emphasizes brief, crisp coverage of the principle or technique discussed in order to reduce the time taken to present and grasp the essentials. A wide variety of examples, spreadsheet solutions, end-of-chapter problems, and test review/FE exam questions are included in every chapter.

OBJECTIVE AND USES OF TEXT

The objective of the text is to explain and demonstrate the principles and techniques of engineering economic analysis as applied in *different fields of engineering*. Interest factors and spreadsheet functions are used to perform equivalency calculations on estimated cash flows that account for the *time value of money* and *inflation*. Separate chapters discuss and illustrate all of the techniques used to evaluate a single project or to select from multiple alternatives. Mutually exclusive and independent projects are covered throughout the text.

Students should have attained a sophomore or higher level to thoroughly understand the engineering context of the techniques and problems addressed. A background in calculus is not necessary; however, a basic familiarity with engineering terminology in a student's own engineering discipline makes the material more meaningful and, therefore, easier to learn and apply.

The text may be used in a wide variety of ways in an undergraduate course—from a few weeks that introduce the basics of engineering economics, to a full two- or three-semester/quarter credit hour course. For senior students who have little or no background in engineering economic analysis in earlier courses, this text provides an excellent senior-level introduction as the *senior project* is designed and developed.

Engineering economy is one of the few engineering topics that is equally applicable to both individuals and corporate and government employees. It can analyze personal finances and investments in a fashion similar to corporate project finances. Students will find this text serves well as a reference throughout their courses and senior design projects, and especially after graduation as a reference source in engineering project work.

FEATURES

Each chapter includes a purpose statement followed by individual section *objectives*. Each section includes one or more examples and end-of-chapter problems that are keyed to the section. The last section in a chapter illustrates *spreadsheet usage* for techniques covered in the chapter. Spreadsheet images include *cell tags* that detail the Excel functions. Based on the hand-calculated examples and spreadsheet examples, a student should be able to solve problems manually or by spreadsheet, as requested by the instructor.

Each chapter includes *multiple-choice questions* that review the principles and calculations of the material. The questions are phrased in the same fashion as those on the *Fundamentals of Engineering (FE)* exam. However, when preparation for the FE exam (or any exam required for registration to practice engineering professionally) is not a requirement, these same questions will serve students as they prepare for exams in the course. Appendix D provides the answers to all review questions.

TOPICAL COVERAGE AND OPTIONAL MATERIAL FOR FLEXIBLE USAGE

Because various engineering curricula concentrate on different aspects of engineering economics, sections and chapters can be covered or skipped to tailor the text's usage. For example, cost estimation that is often of more importance to *chemical engineering* is concentrated in a special chapter. Public sector economics for *civil engineering* is discussed separately. After-tax analysis, cost of capital, and decision-making under risk are introduced for *industrial and systems engineering* and *engineering management* curricula that include a shortened course in engineering economy. Examples treat areas for *electrical, petroleum,* and *mechanical* and other engineering disciplines.

There are several appendices. One explains and illustrates the design of spreadsheets for efficient use and the development of *Excel functions*. A second appendix introduces *financial statements* and *business ratios* for students unfamiliar with accounting statements. Another appendix treats *multiple attribute evaluation* for instructors who want to include noneconomic dimensions in alternative evaluation. Additionally, a discussion of *risk* considerations introduces the fundamental elements of expected value, standard deviation, and probability distributions.

ACKNOWLEDGEMENTS

We would like to thank the following individuals who completed the content survey used in the development of this text:

     Larry Bland, *John Brown University*
     Frederick Bloetscher, *Florida Atlantic University*
     Jim Burati, *Clemson University*

Ronald A. Chadderton, *Villanova University*

Charles H. Gooding, *Clemson University*

David W. Gore, *Middle Tennessee State University*

Johnny R. Graham, *University of North Carolina at Charlotte*

Dr. Michael Hamid, *University of South Alabama*

Bruce Hartsough, *University of California-Davis*

Richard V. Helgason, *Southern Methodist University*

Krishna K. Krishnan, *Wichita State University*

Donald D. Liou, *University of North Carolina at Charlotte*

Daniel P. Loucks, *Cornell University*

Robert Lundquist, *Ohio State University*

Abu Masud, *Wichita State University*

Thomas J. Mclean, *University of Texas at El Paso*

James S. Noble, *University of Missouri at Columbia*

Surendra Singh, *University of Tulsa*

Gene Stuffle, *Idaho State University*

Meenakshi R. Sundaram, *Tennessee Tech University*

Janusz Supernak, *San Diego State University*

Dr. Mathias J. Sutton, *Purdue University*

Heng-Ming Tai, *University of Tulsa*

Lawrence E. Whitman, *Wichita State University*

We thank Jack Beltran for his diligence in accuracy checking the examples and problems. We thank Sallie Sheppard and Elaine Myers for their help in manuscript preparation and for their patience.

We welcome comments and corrections that will improve this text or its online learning material. Our e-mail addresses are lelandblank@yahoo.com and atarquin@utep.edu.

Lee Blank and *Tony Tarquin*

Foundations of Engineering Economy

Royalty-Free/CORBIS

The need for engineering economy is primarily motivated by the work that engineers do in performing analysis, synthesizing, and coming to a conclusion as they work on projects of all sizes. In other words, engineering economy is at the heart of *making decisions*. These decisions involve the fundamental elements of *cash flows of money, time,* and *interest rates*. This chapter introduces the basic concepts and terminology necessary for an engineer to combine these three essential elements in organized, mathematically correct ways to solve problems that will lead to better decisions.

Objectives

Purpose: Understand the fundamental concepts of engineering economy.

1. Determine the role of engineering economy in the decision-making process.

   Definition and role

2. Identify what is needed to successfully perform an engineering economy study.

   Study approach and terms

3. Perform calculations about interest rate and rate of return.

   Interest rate

4. Understand what equivalence means in economic terms.

   Equivalence

5. Calculate simple interest and compound interest for one or more interest periods.

   Simple and compound interest

6. Identify and use engineering economy terminology and symbols.

   Symbols

7. Understand cash flows, their estimation, and how to graphically represent them.

   Cash flows

8. Use the rule of 72 to estimate a compound interest rate or number of years for an amount of money to double in size.

   Doubling time

9. Formulate Excel© functions used in engineering economy.

   Spreadsheet functions

1.1 WHAT IS ENGINEERING ECONOMY?

Before we begin to develop the fundamental concepts upon which engineering economy is based, it would be appropriate to define what is meant by engineering economy. In the simplest of terms, *engineering economy* is a collection of techniques that simplify comparisons of alternatives on an *economic* basis. In defining what engineering economy is, it might also be helpful to define what it is not. Engineering economy is not a method or process for determining what the alternatives are. On the contrary, engineering economy begins only after the alternatives have been identified. If the best alternative is actually one that the engineer has not even recognized as an alternative, then all of the engineering economic analysis tools in this book will not result in its selection.

While economics will be the sole criterion for selecting the best alternatives in this book, real-world decisions usually include many other factors in the decision-making process. For example, in determining whether to build a nuclear-powered, gas-fired, or coal-fired power plant, factors such as safety, air pollution, public acceptance, water demand, waste disposal, global warming, and many others would be considered in identifying the best alternative. The inclusion of other factors (besides economics) in the decision-marking process is called multiple attribute analysis. This topic is introduced in Appendix C.

1.2 PERFORMING AN ENGINEERING ECONOMY STUDY

In order to apply economic analysis techniques, it is necessary to understand the basic terminology and fundamental concepts that form the foundation for engineering-economy studies. Some of these terms and concepts are described below.

1.2.1 Alternatives

An *alternative* is a stand-alone solution for a given situation. We are faced with alternatives in virtually everything we do, from selecting the method of transportation we use to get to work every day to deciding between buying a house or renting one. Similarly, in engineering practice, there are always several ways of accomplishing a given task, and it is necessary to be able to compare them in a rational manner so that the most economical alternative can be selected. The alternatives in engineering considerations usually involve such items as purchase cost (first cost), anticipated useful life, yearly costs of maintaining assets (annual maintenance and operating costs), anticipated resale value (salvage value), and the interest rate. After the facts and all the relevant estimates have been collected, an engineering economy analysis can be conducted to determine which is best from an economic point of view.

1.2.2 Cash Flows

The estimated inflows (revenues) and outflows (costs) of money are called cash flows. These estimates are truly the heart of an engineering economic analysis.

They also represent the weakest part of the analysis, because most of the numbers are judgments about what is going to happen in the *future*. After all, who can accurately predict the price of oil next week, much less next month, next year, or next decade? Thus, no matter how sophisticated the analysis technique, the end result is only as reliable as the data that it is based on.

1.2.3 Alternative Selection

Every situation has at least two alternatives. In addition to the one or more formulated alternatives, there is always the alternative of inaction, called the *do-nothing (DN)* alternative. This is the *as-is* or *status quo* condition. In any situation, when one consciously or subconsciously does not take any action, he or she is actually selecting the DN alternative. Of course, if the status quo alternative *is* selected, the decision-making process should indicate that doing nothing is the most favorable economic outcome at the time the evaluation is made. The procedures developed in this book will enable you to consciously identify the alternative(s) that is (are) economically the best.

1.2.4 Evaluation Criteria

Whether we are aware of it or not, we use criteria every day to choose between alternatives. For example, when you drive to campus, you decide to take the "best" route. But how did you define *best*? Was the best route the safest, shortest, fastest, cheapest, most scenic, or what? Obviously, depending upon which criterion or combination of criteria is used to identify the best, a different route might be selected each time. In economic analysis, *financial units* (dollars or other currency) are generally used as the tangible basis for evaluation. Thus, when there are several ways of accomplishing a stated objective, the alternative with the lowest overall cost or highest overall net income is selected.

1.2.5 Intangible Factors

In many cases, alternatives have noneconomic or intangible factors that are difficult to quantify. When the alternatives under consideration are hard to distinguish economically, intangible factors may tilt the decision in the direction of one of the alternatives. A few examples of noneconomic factors are goodwill, convenience, friendship, and morale.

1.2.6 Time Value of Money

It is often said that money makes money. The statement is indeed true, for if we elect to invest money today, we inherently expect to have more money in the future. If a person or company borrows money today, by tomorrow more than the original loan principal will be owed. This fact is also explained by the time value of money.

> **The change in the amount of money over a given time period is called the** *time value of money;* **it is the most important concept in engineering economy.**

The time value of money can be taken into account by several methods in an economy study, as we will learn. The method's final output is a *measure of worth,* for example, rate of return. This measure is used to accept/reject an alternative.

1.3 INTEREST RATE, RATE OF RETURN, AND MARR

Interest is the manifestation of the time value of money, and it essentially represents "rent" paid for use of the money. Computationally, interest is the difference between an ending amount of money and the beginning amount. If the difference is zero or negative, there is no interest. There are always two perspectives to an amount of interest—interest paid and interest earned. Interest is *paid* when a person or organization borrows money (obtains a loan) and repays a larger amount. Interest is *earned* when a person or organization saves, invests, or lends money and obtains a return of a larger amount. The computations and numerical values are essentially the same for both perspectives, but they are interpreted differently.

Interest paid or earned is determined by using the relation

$$\text{Interest} = \text{end amount} - \text{original amount} \qquad \textbf{[1.1]}$$

When interest over a *specific time unit* is expressed as a percentage of the original amount (principal), the result is called the *interest rate* or *rate of return (ROR).*

$$\textbf{Interest rate or rate of return} = \frac{\textbf{interest accrued per time unit}}{\textbf{original amount}} \times \textbf{100\%} \quad \textbf{[1.2]}$$

The time unit of the interest rate is called the *interest period.* By far the most common interest period used to state an interest rate is 1 year. Shorter time periods can be used, such as, 1% per month. Thus, the interest period of the interest rate should always be included. If only the rate is stated, for example, 8.5%, a 1-year interest period is assumed.

The term *return on investment (ROI)* is used equivalently with ROR in different industries and settings, especially where large capital funds are committed to engineering-oriented programs. The term *interest rate paid* is more appropriate for the borrower's perspective, while *rate of return earned* is better from the investor's perspective.

An employee at LaserKinetics.com borrows $10,000 on May 1 and must repay a total of $10,700 exactly 1 year later. Determine the interest amount and the interest rate paid.

EXAMPLE 1.1

Solution

The perspective here is that of the borrower since $10,700 repays a loan. Apply Equation [1.1] to determine the interest paid.

Interest paid = $10,700 − 10,000 = $700

Equation [1.2] determines the interest rate paid for 1 year.

$$\text{Percent interest rate} = \frac{\$700}{\$10,000} \times 100\% = 7\% \text{ per year}$$

EXAMPLE 1.2

a. Calculate the amount deposited 1 year ago to have $1000 now at an interest rate of 5% per year.
b. Calculate the amount of interest earned during this time period.

Solution

a. The total amount accrued ($1000) is the sum of the original deposit and the earned interest. If X is the original deposit,

Total accrued = original amount + original amount(interest rate)
$1000 = X + X(0.05) = X(1 + 0.05) = 1.05X$

The original deposit is

$$X = \frac{1000}{1.05} = \$952.38$$

b. Apply Equation [1.1] to determine interest earned.

Interest = $1000 − 952.38 = $47.62

In Examples 1.1 and 1.2 the interest period was 1 year, and the interest amount was calculated at the end of one period. When more than one interest period is involved (e.g., if we wanted the amount of interest owed after 3 years in Example 1.2), it is necessary to state whether the interest is accrued on a *simple* or *compound* basis from one period to the next. Simple and compound interest will be discussed in Section 1.5.

Engineering alternatives are evaluated upon the prognosis that a reasonable rate of return (ROR) can be realized. A reasonable rate must be established so that the accept/reject decision can be made. The reasonable rate, called the *minimum attractive rate of return* (MARR), must be higher than the cost of money used to finance the alternative, as well as higher than the rate that would be expected from a bank or safe (minimal risk) investment. Figure 1-1 indicates the relations between different rates of return. In the United States, the current U.S. Treasury bill rate of return is sometimes used as the benchmark safe rate.

For a corporation, the MARR is always set above its *cost of capital,* that is, the interest rate a company must pay for capital funds needed to finance projects. For exam-

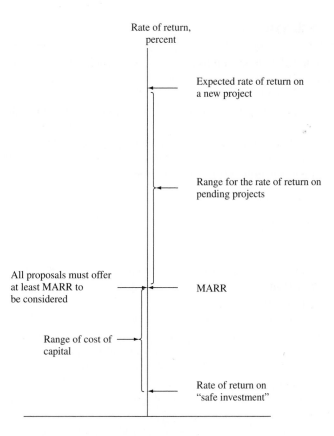

ple, if a corporation can borrow capital funds at an average of 5% per year and expects to clear at least 6% per year on a project, the minimum MARR will be 11% per year.

The MARR is also referred to as the *hurdle rate;* that is, a financially viable project's expected ROR must meet or exceed the hurdle rate. Note that the MARR is not a rate calculated like the ROR; MARR is established by financial managers and is used as a criterion for accept/reject decisions. The following inequality must be correct for any accepted project.

$$ROR \geq MARR > \text{cost of capital}$$

Descriptions and problems in the following chapters use stated MARR values with the assumption that they are set correctly relative to the cost of capital and the expected rate of return. If more understanding of capital funds and the establishment of the MARR is required, refer to Section 13.5 for further detail.

An additional economic consideration for any engineering economy study is *inflation*. In simple terms, bank interest rates reflect two things: a so-called real rate of return *plus* the expected inflation rate. The safest investments (such as U.S. government bonds) typically have a 3% to 4% real rate of return built into their overall interest rates. Thus, an interest rate of, say, 9% per year on a U.S. government bond means that investors expect the inflation rate to be in the range of 5% to 6% per year. Clearly, then, inflation causes interest rates to rise. Inflation is discussed in detail in Chapter 10.

1.4 EQUIVALENCE

Equivalent terms are used often in the transfer between scales and units. For example, 1000 meters is equal to (or equivalent to) 1 kilometer, 12 inches equals 1 foot, and 1 quart equals 2 pints or 0.946 liter.

In engineering economy, when considered together, the time value of money and the interest rate help develop the concept of *economic equivalence,* which means that different sums of money at different times would be equal in economic value. For example, if the interest rate is 6% per year, $100 today (present time) is equivalent to $106 one year from today.

$$\text{Amount in one year} = 100 + 100(0.06) = 100(1 + 0.06) = \$106$$

So, if someone offered you a gift of $100 today or $106 one year from today, it would make no difference which offer you accepted from an economic perspective. In either case you have $106 one year from today. However, the two sums of money are equivalent to each other *only* when the interest rate is 6% per year. At a higher or lower interest rate, $100 today is not equivalent to $106 one year from today.

In addition to future equivalence, we can apply the same logic to determine equivalence for previous years. A total of $100 now is equivalent to $100/1.06 = $94.34 one year ago at an interest rate of 6% per year. From these illustrations, we can state the following: $94.34 last year, $100 now, and $106 one year from now are equivalent at an interest rate of 6% per year. The fact that these sums are equivalent can be verified by computing the two interest rates for 1-year interest periods.

$$\frac{\$6}{\$100} \times 100\% = 6\% \text{ per year}$$

and

$$\frac{\$5.66}{\$94.34} \times 100\% = 6\% \text{ per year}$$

Figure 1.2 indicates the amount of interest each year necessary to make these three different amounts equivalent at 6% per year.

FIGURE 1.2
Equivalence of three amounts at a 6% per year interest rate.

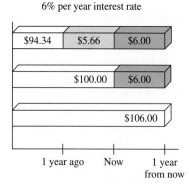

6% per year interest rate

$94.34 $5.66 $6.00

$100.00 $6.00

$106.00

1 year ago   Now   1 year from now

EXAMPLE **1.3**

AC-Delco makes auto batteries available to General Motors dealers through privately owned distributorships. In general, batteries are stored throughout the year, and a 5% cost increase is added each year to cover the inventory carrying charge for the distributorship owner. Assume you own the City Center Delco facility. Make the calculations necessary to show which of the following statements are true and which are false about battery costs.

a. The amount of $98 now is equivalent to a cost of $105.60 one year from now.
b. A truck battery cost of $200 one year ago is equivalent to $205 now.
c. A $38 cost now is equivalent to $39.90 one year from now.
d. A $3000 cost now is equivalent to $2887.14 one year ago.
e. The carrying charge accumulated in 1 year on an investment of $2000 worth of batteries is $100.

Solution

a. Total amount accrued = 98(1.05) = $102.90 ≠ $105.60; therefore, it is false. Another way to solve this is as follows: Required original cost is 105.60/1.05 = $100.57 ≠ $98.
b. Required old cost is 205.00/1.05 = $195.24 ≠ $200; therefore, it is false.
c. The cost 1 year from now is $38(1.05) = $39.90; true.
d. Cost now is 2887.14(1.05) = $3031.50 ≠ $3000; false.
e. The charge is 5% per year interest, or $2000(0.05) = $100; true.

1.5 SIMPLE AND COMPOUND INTEREST

The terms *interest, interest period,* and *interest rate* were introduced in Section 1.3 for calculating equivalent sums of money for one interest period in the past and one period in the future. However, for more than one interest period, the terms *simple interest* and *compound interest* become important.

Simple interest is calculated using the principal only, ignoring any interest accrued in preceding interest periods. The total simple interest over several periods is computed as

$$\text{Interest} = (\text{principal})(\text{number of periods})(\text{interest rate}) \qquad [1.3]$$

where the interest rate is expressed in decimal form.

EXAMPLE **1.4**

HP borrowed money to do rapid prototyping for a new ruggedized computer that targets desert oilfield conditions. The loan is $1 million for 3 years at 5% per year simple interest. How much money will HP repay at the end of 3 years? Tabulate the results in $1000 units.

Solution

The interest for each of the 3 years in $1000 units is

Interest per year = 1000(0.05) = $50

Total interest for 3 years from Equation [1.3] is

Total interest = 1000(3)(0.05) = $150

The amount due after 3 years in $1000 units is

Total due = $1000 + 150 = $1150

The $50,000 interest accrued in the first year and the $50,000 accrued in the second year do not earn interest. The interest due each year is calculated only on the $1,000,000 principal.

The details of this loan repayment are tabulated in Table 1.1 from the perspective of the borrower. The year zero represents the present, that is, when the money is borrowed. No payment is made until the end of year 3. The amount owed each year increases uniformly by $50,000, since simple interest is figured on only the loan principal.

TABLE 1.1   Simple Interest Computations (in $1000 units)

(1) End of Year	(2) Amount Borrowed	(3) Interest	(4) Amount Owed	(5) Amount Paid
0	$1000			
1	—	$50	$1050	$ 0
2	—	50	1100	0
3	—	50	1150	1150

For *compound interest,* the interest accrued for each interest period is calculated on the *principal plus* the *total amount of interest accumulated in all previous periods.* Thus, compound interest means interest on top of interest. Compound interest reflects the effect of the time value of money on the interest also. Now the interest for one period is calculated as

$$\textbf{Interest} = \textbf{(principal + all accrued interest)(interest rate)} \qquad [1.4]$$

per period

EXAMPLE 1.5   If HP borrows $1,000,000 from a different source at 5% per year compound interest, compute the total amount due after 3 years. Compare the results of this and the previous example.

TABLE 1.2   Compound Interest Computations (in $1000 units), Example 1.5

(1) End of Year	(2) Amount Borrowed	(3) Interest	(4) Amount Owed	(5) Amount Paid
0	$1000			
1	—	$50.00	$1050.00	$ 0
2	—	52.50	1102.50	0
3	—	55.13	1157.63	1157.63

Solution

The interest and total amount due each year are computed separately using Equation [1.4]. In $1000 units,

$$\text{Year 1 interest:} \quad \$1000(0.05) = \$50.00$$

$$\text{Total amount due after year 1:} \quad \$1000 + 50.00 = \$1050.00$$

$$\text{Year 2 interest:} \quad \$1050(0.05) = \$52.50$$

$$\text{Total amount due after year 2:} \quad \$1050 + 52.50 = \$1102.50$$

$$\text{Year 3 interest:} \quad \$1102.50(0.05) = \$55.13$$

$$\text{Total amount due after year 3:} \quad \$1102.50 + 55.13 = \$1157.63$$

The details are shown in Table 1.2. The repayment plan is the same as that for the simple interest example—no payment until the principal plus accrued interest is due at the end of year 3. An extra $1,157,630 − 1,150,000 = $7,630 of interest is paid compared to simple interest over the 3-year period.

Comment: The difference between simple and compound interest grows significantly each year. If the computations are continued for more years, for example, 10 years, the difference is $128,894; after 20 years compound interest is $653,298 more than simple interest.

Another and shorter way to calculate the total amount due after 3 years in Example 1.5 is to combine calculations rather than perform them on a year-by-year basis. The total due each year is as follows:

$$\text{Year 1:} \quad \$1000(1.05)^1 = \$1050.00$$
$$\text{Year 2:} \quad \$1000(1.05)^2 = \$1102.50$$
$$\text{Year 3:} \quad \$1000(1.05)^3 = \$1157.63$$

The year 3 total is calculated directly; it does not require the year 2 total. In general formula form,

Total due after a number of years = principal(1 + interest rate)$^{\text{number of years}}$

This fundamental relation is used many times in upcoming chapters.

We combine the concepts of interest rate, simple interest, compound interest, and equivalence to demonstrate that different loan repayment plans may be equivalent, but they may differ substantially in monetary amounts from one year to another. This also shows that there are many ways to take into account the time value of money. The following example illustrates equivalence for five different loan repayment plans.

EXAMPLE 1.6

a. Demonstrate the concept of equivalence using the different loan repayment plans described below. Each plan repays a $5000 loan in 5 years at 8% interest per year.

- **Plan 1: Simple interest, pay all at end.** No interest or principal is paid until the end of year 5. Interest accumulates each year on the *principal only*.
- **Plan 2: Compound interest, pay all at end.** No interest or principal is paid until the end of year 5. Interest accumulates each year on the total of principal *and* all accrued interest.
- **Plan 3: Simple interest paid annually, principal repaid at end.** The accrued interest is paid each year, and the entire principal is repaid at the end of year 5.
- **Plan 4: Compound interest and portion of principal repaid annually.** The accrued interest and one-fifth of the principal (or $1000) is repaid each year. The outstanding loan balance decreases each year, so the interest for each year decreases.
- **Plan 5: Equal payments of compound interest and principal made annually.** Equal payments are made each year with a portion going toward principal repayment and the remainder covering the accrued interest. Since the loan balance decreases at a rate slower than that in plan 4 due to the equal end-of-year payments, the interest decreases, but at a slower rate.

b. Make a statement about the equivalence of each plan at 8% simple or compound interest, as appropriate.

Solution

a. Table 1.3 presents the interest, payment amount, total owed at the end of each year, and total amount paid over the 5-year period (column 4 totals). The amounts of interest (column 2) are determined as follows:

Plan 1   Simple interest = (original principal)(0.08)
Plan 2   Compound interest = (total owed previous year)(0.08)
Plan 3   Simple interest = (original principal)(0.08)
Plan 4   Compound interest = (total owed previous year)(0.08)
Plan 5   Compound interest = (total owed previous year)(0.08)

Note that the amounts of the annual payments are different for each repayment schedule and that the total amounts repaid for most plans are different, even though each repayment plan requires exactly 5 years. The difference in the total amounts repaid can be explained (1) by the time value of money, (2) by simple or compound interest, and (3) by the partial repayment of principal prior to year 5.

TABLE 1.3 **Different Repayment Schedules Over 5 Years for $5000 at 8% Per Year Interest**

(1) End of Year	(2) Interest Owed for Year	(3) Total Owed at End of Year	(4) End-of-Year Payment	(5) Total Owed after Payment
Plan 1: Simple Interest, Pay All at End				
0				$5000.00
1	$400.00	$5400.00	—	5400.00
2	400.00	5800.00	—	5800.00
3	400.00	6200.00	—	6200.00
4	400.00	6600.00	—	6600.00
5	400.00	7000.00	$7000.00	
Totals			$7000.00	
Plan 2: Compound Interest, Pay All at End				
0				$5000.00
1	$400.00	$5400.00	—	5400.00
2	432.00	5832.00	—	5832.00
3	466.56	6298.56	—	6298.56
4	503.88	6802.44	—	6802.44
5	544.20	7346.64	$7346.64	
Totals			$7346.64	
Plan 3: Simple Interest Paid Annually; Principal Repaid at End				
0				$5000.00
1	$400.00	$5400.00	$ 400.00	5000.00
2	400.00	5400.00	400.00	5000.00
3	400.00	5400.00	400.00	5000.00
4	400.00	5400.00	400.00	5000.00
5	400.00	5400.00	5400.00	
Totals			$7000.00	
Plan 4: Compound Interest and Portion of Principal Repaid Annually				
0				$5000.00
1	$400.00	$5400.00	$1400.00	4000.00
2	320.00	4320.00	1320.00	3000.00
3	240.00	3240.00	1240.00	2000.00
4	160.00	2160.00	1160.00	1000.00
5	80.00	1080.00	1080.00	
Totals			$6200.00	

TABLE 1.3 (Continued)

(1) End of Year	(2) Interest Owed for Year	(3) Total Owed at End of Year	(4) End-of-Year Payment	(5) Total Owed after Payment
Plan 5: Equal Annual Payments of Compound Interest and Principal				
0				$5000.00
1	$400.00	$5400.00	$1252.28	4147.72
2	331.82	4479.54	1252.28	3227.25
3	258.18	3485.43	1252.28	2233.15
4	178.65	2411.80	1252.28	1159.52
5	92.76	1252.28	1252.28	
Totals			$6261.41	

b. Table 1.3 shows that $5000 at time 0 is equivalent to each of the following:

Plan 1 $7000 at the end of year 5 at 8% simple interest.

Plan 2 $7346.64 at the end of year 5 at 8% compound interest.

Plan 3 $400 per year for 4 years and $5400 at the end of year 5 at 8% simple interest.

Plan 4 Decreasing payments of interest and partial principal in years 1 ($1400) through 5 ($1080) at 8% compound interest.

Plan 5 $1252.28 per year for 5 years at 8% compound interest.

Beginning in Chapter 2, we will make many calculations like plan 5, where interest is compounded and a constant amount is paid each period. This amount covers accrued interest and a partial principal repayment.

1.6 TERMINOLOGY AND SYMBOLS

The equations and procedures of engineering economy utilize the following terms and symbols. Sample units are indicated.

P = value or amount of money at a time designated as the present or time 0. Also, P is referred to as present worth (PW), present value (PV), net present value (NPV), discounted cash flow (DCF), and capitalized cost (CC); dollars

F = value or amount of money at some future time. Also, F is called future worth (FW) and future value (FV); dollars

A = series of consecutive, equal, end-of-period amounts of money. Also, A is called the annual worth (AW) and equivalent uniform annual worth (EUAW); dollars per year, dollars per month

n = number of interest periods; years, months, days

i = interest rate or rate of return per time period; percent per year, percent per month, percent per day

t = time, stated in periods; years, months, days

The symbols P and F represent one-time occurrences: A occurs with the same value each interest period for a specified number of periods. It should be clear that a present value P represents a single sum of money at some time prior to a future value F or prior to the first occurrence of an equivalent series amount A.

It is important to note that the symbol A always represents a uniform amount (i.e., the same amount each period) that extends through *consecutive* interest periods. Both conditions must exist before the series can be represented by A.

The interest rate i is assumed to be a compound rate, unless specifically stated as simple interest. The rate i is expressed in percent per interest period, for example, 12% per year. Unless stated otherwise, assume that the rate applies throughout the entire n years or interest periods. The decimal equivalent for i is always used in engineering economy computations.

All engineering economy problems involve the element of time n and interest rate i. In general, every problem will involve at least four of the symbols P, F, A, n, and i, with at least three of them estimated or known.

EXAMPLE 1.7

A new college graduate has a job with Boeing Aerospace. She plans to borrow $10,000 now to help in buying a car. She has arranged to repay the entire principal plus 8% per year interest after 5 years. Identify the engineering economy symbols involved and their values for the total owed after 5 years.

Solution

In this case, P and F are involved, since all amounts are single payments, as well as n and i. Time is expressed in years.

$$P = \$10,000 \qquad i = 8\% \text{ per year} \qquad n = 5 \text{ years} \qquad F = ?$$

The future amount F is unknown.

EXAMPLE 1.8

Assume you borrow $2000 now at 7% per year for 10 years and must repay the loan in equal yearly payments. Determine the symbols involved and their values.

Solution

Time is in years.

$P = \$2000$

$A = ?$ per year for 5 years

$i = 7\%$ per year

$n = 10$ years

In Examples 1.7 and 1.8, the P value is a receipt *to* the borrower, and F or A is a disbursement *from* the borrower. It is equally correct to use these symbols in the reverse roles.

EXAMPLE 1.9   On July 1, 2008, your new employer Ford Motor Company deposits $5000 into your money market account, as part of your employment bonus. The account pays interest at 5% per year. You expect to withdraw an equal annual amount each year for the following 10 years. Identify the symbols and their values.

Solution
Time is in years.

$P = \$5000$
$A = ?$ per year
$i = 5\%$ per year
$n = 10$ years

EXAMPLE 1.10   You plan to make a lump-sum deposit of $5000 now into an investment account that pays 6% per year, and you plan to withdraw an equal end-of-year amount of $1000 for 5 years, starting next year. At the end of the sixth year, you plan to close your account by withdrawing the remaining money. Define the engineering economy symbols involved.

Solution
Time is expressed in years.

$P = \$5000$
$A = \$1000$ per year for 5 years
$F = ?$ at end of year 6
$i = 6\%$ per year
$n = 5$ years for the A series and 6 for the F value

1.7 CASH FLOWS: THEIR ESTIMATION AND DIAGRAMMING

Cash flows are inflows and outflows of money. These cash flows may be estimates or observed values. Every person or company has cash receipts—revenue and income (inflows); and cash disbursements—expenses, and costs (outflows). These receipts and disbursements are the cash flows, with a plus sign representing cash inflows and a minus sign representing cash outflows. Cash flows occur during specified periods of time, such as 1 month or 1 year.

Of all the elements of an engineering economy study, cash flow estimation is likely the most difficult and inexact. Cash flow estimates are just that—estimates about an uncertain future. Once estimated, the techniques of this book guide the decision-making process. But the time-proven accuracy of an alternative's estimated cash inflows and outflows clearly dictates the quality of the economic analysis and conclusion.

Cash inflows, or receipts, may be comprised of the following, depending upon the nature of the proposed activity and the type of business involved.

Samples of Cash Inflow Estimates

Revenues (from sales and contracts)

Operating cost reductions (resulting from an alternative)

Salvage value

Construction and facility cost savings

Receipt of loan principal

Income tax savings

Receipts from stock and bond sales

Cash outflows, or disbursements, may be comprised of the following, again depending upon the nature of the activity and type of business.

Samples of Cash Outflow Estimates

First cost of assets

Engineering design costs

Operating costs (annual and incremental)

Periodic maintenance and rebuild costs

Loan interest and principal payments

Major expected/unexpected upgrade costs

Income taxes

Background information for estimates may be available in departments such as accounting, finance, marketing, sales, engineering, design, manufacturing, production, field services, and computer services. The accuracy of estimates is largely dependent upon the experiences of the person making the estimate with similar situations. Usually *point estimates* are made; that is, a single-value estimate is developed for each economic element of an alternative. If a statistical approach to the engineering economy study is undertaken, a *range estimate* or *distribution estimate* may be developed. Though more involved computationally, a statistical study provides more complete results when key estimates are expected to vary widely. We will use point estimates throughout most of this book.

Once the cash inflow and outflow estimates are developed, the net cash flow can be determined.

$$\textbf{Net cash flow} = \textbf{receipts} - \textbf{disbursements}$$
$$= \textbf{cash inflows} - \textbf{cash outflows} \qquad \textbf{[1.5]}$$

Since cash flows normally take place at varying times within an interest period, a simplifying end-of-period assumption is made.

The *end-of-period convention* means that all cash flows are assumed to occur at the end of an interest period. When several receipts and disbursements occur within a given interest period, the *net* cash flow is assumed to occur at the *end* of the interest period.

FIGURE 1.3

A typical cash flow time scale for 5 years.

However, it should be understood that, although F or A amounts are located at the end of the interest period by convention, the end of the period is not necessarily December 31. In Example 1.9 the deposit took place on July 1, 2008, and the withdrawals will take place on July 1 of each succeeding year for 10 years. *Thus, end of the period means end of interest period, not end of calendar year.*

The *cash flow diagram* is a very important tool in an economic analysis, especially when the cash flow series is complex. It is a graphical representation of cash flows drawn on a time scale. The diagram includes what is known, what is estimated, and what is needed. That is, once the cash flow diagram is complete, another person should be able to work the problem by looking at the diagram.

Cash flow diagram time $t = 0$ is the present, and $t = 1$ is the end of time period 1. We assume that the periods are in years for now. The time scale of Figure 1.3 is set up for 5 years. Since the end-of-year convention places cash flows at the ends of years, the "1" marks the end of year 1.

While it is not necessary to use an exact scale on the cash flow diagram, you will probably avoid errors if you make a neat diagram to approximate scale for both time and relative cash flow magnitudes.

The direction of the arrows on the cash flow diagram is important. A vertical arrow pointing up indicates a positive cash flow. Conversely, an arrow pointing down indicates a negative cash flow. Figure 1.4 illustrates a receipt (cash inflow) at the end of year 1 and equal disbursements (cash outflows) at the end of years 2 and 3.

The perspective or vantage point must be determined prior to placing a sign on each cash flow and diagramming it. As an illustration, if you borrow $2500 to buy a $2000 used Harley-Davidson for cash, and you use the remaining $500 for a new paint job, there may be several different perspectives taken. Possible perspectives, cash flow signs, and amounts are as follows.

Perspéctive	Cash Flow, $
Credit union	−2500
You as borrower	+2500
You as purchaser,	−2000
and as paint customer	−500
Used cycle dealer	+2000
Paint shop owner	+500

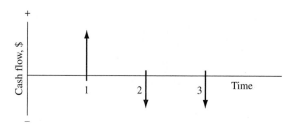

FIGURE 1.4
Example of positive
and negative cash
flows.

Reread Example 1.7, where $P = \$10,000$ is borrowed at 8% per year and F is   **EXAMPLE 1.11**
sought after 5 years. Construct the cash flow diagram.

Solution

Figure 1.5 presents the cash flow diagram from the vantage point of the borrower. The present sum P is a cash inflow of the loan principal at year 0, and
the future sum F is the cash outflow of the repayment at the end of year 5. The
interest rate should be indicated on the diagram.

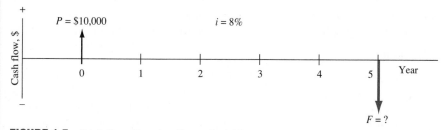

FIGURE 1.5   Cash flow diagram, Example 1.11.

Each year Exxon-Mobil expends large amounts of funds for mechanical safety   **EXAMPLE 1.12**
features throughout its worldwide operations. Carla Ramos, a lead engineer for
Mexico and Central American operations, plans expenditures of $1 million now
and each of the next 4 years just for the improvement of field-based pressure-
release valves. Construct the cash flow diagram to find the equivalent value of
these expenditures at the end of year 4, using a cost of capital estimate for
safety-related funds of 12% per year.

Solution

Figure 1.6 indicates the uniform and negative cash flow series (expenditures) for
five periods, and the unknown F value (positive cash flow equivalent) at exactly
the same time as the fifth expenditure. Since the expenditures start immediately,

the first $1 million is shown at time 0, not time 1. Therefore, the last negative cash flow occurs at the end of the fourth year, when F also occurs. To make this diagram appear similar to that of Figure 1.5 with a full 5 years on the time scale, the addition of the year -1 prior to year 0 completes the diagram for a full 5 years. This addition demonstrates that year 0 is the end-of-period point for the year -1.

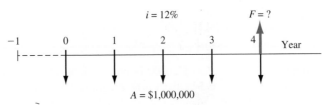

FIGURE 1.6   Cash flow diagram, Example 1.12.

EXAMPLE 1.13   A father wants to deposit an unknown lump-sum amount into an investment opportunity 2 years from now that is large enough to withdraw $4000 per year for state university tuition for 5 years starting 3 years from now. If the rate of return is estimated to be 15.5% per year, construct the cash flow diagram.

Solution

Figure 1.7 presents the cash flows from the father's perspective. The present value P is a cash outflow 2 years hence and is to be determined ($P = ?$). Note that this present value does not occur at time $t = 0$, but it does occur one period prior to the first A value of $4000, which is the cash inflow to the father.

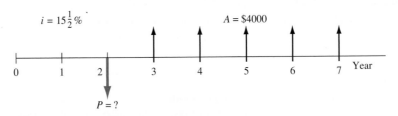

FIGURE 1.7   Cash flow diagram, Example 1.13.

1.8 THE RULE OF 72

The *rule of 72* can estimate either the number of years n or the compound interest rate (or rate of return) i required for a single amount of money to double in size ($2\times$). The rule is simple; the time required for doubling in size with a compound rate is approximately equal to 72 divided by the rate in percent.

$$\text{Approximate } n = 72/i$$ [1.6]

TABLE 1.4 **Number of Years Required for Money to Double**

	Time to Double, Years	
Compound Rate, % per year	Rule-of-72 Result	Actual
5	14.4	14.2
8	9.0	9.0
10	7.2	7.3
12	6.0	6.1
24	3.0	3.2

For example, at 5% per year, it takes approximately $72/5 = 14.4$ years for a current amount to double. (The actual time is 14.2 years.) Table 1.4 compares rule-of-72 results to the actual times required using time value of money formulas discussed in Chapter 2.

Solving Equation [1.6] for i approximates the compound rate per year for doubling in n years.

$$\text{Approximate } i = 72/n \qquad\qquad [1.7]$$

For example, if the cost of gasoline doubles in 6 years, the compound rate is approximately $72/6 = 12\%$ per year. (The exact rate is 12.25% per year.) The approximate number of years or compound rate for a current amount to quadruple $(4\times)$ is twice the answer obtained from the rule of 72 for doubling the amount.

For simple interest, the rule of 100 is used with the same equation formats as above, but the n or i value is exact. For example, money doubles in exactly 12 years at a simple rate of $100/12 = 8.33\%$ per year. And, at 10% simple interest, it takes exactly $100/10 = 10$ years to double.

1.9 INTRODUCTION TO USING SPREADSHEET FUNCTIONS

The functions on a computer spreadsheet can greatly reduce the amount of hand and calculator work for equivalency computations involving *compound interest* and the terms P, F, A, i, and n. Often a predefined function can be entered into one cell and we can obtain the final answer immediately. Any spreadsheet system can be used; Excel is used throughout this book because it is readily available and easy to use.

Appendix A is a primer on using spreadsheets and Excel. The functions used in engineering economy are described there in detail, with explanations of all the parameters (also called arguments) placed between parentheses after the function identifier. The Excel online help function provides similar information. Appendix A also includes a section on spreadsheet layout that is useful when the economic analysis is presented to someone else—a coworker, a boss, or a professor.

A total of seven Excel functions can perform most of the fundamental engineering economy calculations. However, these functions are no substitute for knowing how the time value of money and compound interest work. The functions are great supplemental tools, but they do not replace the understanding of engineering economy relations, assumptions, and techniques.

Using the symbols P, F, A, i, and n exactly as defined in Section 1.6, the Excel functions most used in engineering economic analysis are formulated as follows.

To find the present value P of an A series: $= \text{PV}(i\%,n,A,F)$

To find the future value F of an A series: $= \text{FV}(i\%,n,A,P)$

To find the equal, periodic value A: $= \text{PMT}(i\%,n,P,F)$

To find the number of periods n: $= \text{NPER}(i\%,A,P,F)$

To find the compound interest rate i: $= \text{RATE}(n,A,P,F)$

To find the compound interest rate i: $= \text{IRR}(\text{first\_cell:last\_cell})$

To find the present value P of any series: $= \text{NPV}(i\%, \text{second\_cell:last\_cell}) +$ first_cell

If some of the parameters don't apply to a particular problem, they can be omitted and zero is assumed. If the parameter omitted is an interior one, the comma must be entered. The last two functions require that a series of numbers be entered into contiguous spreadsheet cells, but the first five can be used with no supporting data. In all cases, the function must be preceded by an equals sign ($=$) in the cell where the answer is to be displayed.

Each of these functions will be introduced and illustrated at the point in this text where they are most useful. However, to get an idea of how they work, look back at Examples 1.7 and 1.8. In Example 1.7, the future amount F is unknown, as indicated by $F = ?$ in the solution. In Chapter 2, we will learn how the time value of money is used to find F, given P, i, and n. To find F in this example using a spreadsheet, simply enter the FV function preceded by an equals sign into any cell. The format is $=\text{FV}(i\%,n,,P)$ or $=\text{FV}(8\%,5,,10000)$. The comma is entered because there is no A involved. Figure 1.8a is a screen image of the Excel spreadsheet with the FV function entered into cell C4. The answer of $\$-14,693.28$ is displayed. The answer is a negative amount from the borrower's perspective to repay the loan after 5 years. The FV function is shown in the formula bar above the worksheet, and a cell tag shows the format of the FV function.

In Example 1.8, the uniform annual amount A is sought, and P, i, and n are known. Find A using the function $=\text{PMT}(7\%,10,2000)$. Figure 1.8b shows the result.

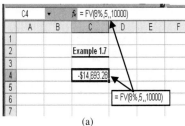

FIGURE 1.8
Use of spreadsheet functions for
(a) Example 1.7 and
(b) Example 1.8.

SUMMARY

Engineering economy is the application of economic factors to evaluate alternatives by considering the time value of money. The engineering economy study involves computing a specific economic measure of worth for estimated cash flows over a specific period of time.

The concept of *equivalence* helps in understanding how different sums of money at different times are equal in economic terms. The differences between simple interest (based on principal only) and compound interest (based on principal and interest upon interest) have been described in formulas, tables, and graphs.

The MARR is a reasonable rate of return established as a hurdle rate to determine if an alternative is economically viable. The MARR is always higher than the return from a safe investment and the corporation's cost of capital.

Also, this chapter introduced the estimation, conventions, and diagramming of cash flows.

PROBLEMS

Definitions and Basic Concepts

1.1 With respect to the selection of alternatives, state one thing that engineering economy will help you to do and one thing that it will not.

1.2 In economic analysis, revenues and costs are examples of what?

1.3 The analysis techniques that are used in engineering economic analysis are only as good as what?

1.4 What is meant by the term *evaluation criterion*?

1.5 What evaluation criterion is used in economic analysis?

1.6 What is meant by the term *intangible factors*?

1.7 Give three examples of intangible factors.

1.8 Interest is a manifestation of what general concept in engineering economy?

1.9 Of the fundamental dimensions length, mass, time, and electric charge, which one is the most important in economic analysis and why?

Interest Rate and Equivalence

1.10 The term that describes compensation for "renting" money is what?

1.11 When an interest rate, such as 3%, does not include the time period, the time period is assumed to be what?

1.12 The original amount of money in a loan transaction is known as what?

1.13 When the yield on a U.S government bond is 3% per year, investors are expecting the inflation rate to be approximately what?

1.14 In order to build a new warehouse facility, the regional distributor for Valco Multi-Position Valves borrowed $1.6 million at 10% per year interest. If the company repaid the loan in a lump sum amount after 2 years, what was (a) the amount of the payment, and (b) the amount of interest?

1.15 A sum of $2 million now is equivalent to $2.42 million 1 year from now at what interest rate?

1.16 In order to restructure some of its debt, General Motors decided to pay off one of its short-term loans. If the company borrowed the money 1 year ago at an interest rate of 8% per year and the total cost of repaying the loan was $82 million, what was the amount of the original loan?

1.17 A start-up company with multiple nano technology products established a goal of making a rate of return of at least 30% per year on its investments for the first 5 years. If the company acquired $200 million in venture capital, how much did it have to earn in the first year?

1.18 How many years would it take for an investment of $280,000 to accumulate to at least $425,000 at 15% per year interest?

1.19 Valley Rendering, Inc. is considering purchasing a new flotation system for recovering more grease. The company can finance a $150,000 system at 5% per year compound interest or 5.5% per year simple interest. If the total amount owed is due in a single payment at the end of 3 years, (a) which interest rate should the company select, and (b) how much is the difference in interest between the two schemes?

1.20 Valtro Electronic Systems, Inc. set aside a lump sum of money 4 years ago in order to finance a plant expansion now. If the money was invested in a 10% per year simple interest certificate of deposit, how much did the company set aside if the certificate is now worth $850,000?

Simple and Compound Interest

1.21 Two years ago, ASARCO, Inc. invested $580,000 in a certificate of deposit that paid *simple* interest of 9% per year. Now the company plans to invest the total amount accrued in another certificate that pays 9% per year *compound* interest. How much will the new certificate be worth 2 years from now?

1.22 A company that manufactures general-purpose transducers invested $2 million 4 years ago in high-yield junk bonds. If the bonds are now worth $2.8 million, what rate of return per year did the company make on the basis of (a) simple interest and (b) compound interest?

1.23 How many years would it take for money to triple in value at 20% per year simple interest?

1.24 If Farah Manufacturing wants its investments to double in value in 4 years, what rate of return would it have to make on the basis of (a) simple interest and (b) compound interest?

1.25 Companies frequently borrow money under an arrangement that requires them to make periodic payments of "interest only" and then pay the principal all at once. If Cisco International borrowed $500,000 (identified as loan A) at 10% per year simple interest and another $500,000 (identified as loan B) at 10% per year compound interest and paid *only the interest* at the end of each year for 3 years on both loans, (a) on which loan did the company pay more interest, and (b) what was the difference in interest paid between the two loans?

Symbols and Terminology

1.26 *All* engineering economy problems will involve which two of the following symbols: P, F, A, i, n?

1.27 Identify the symbols involved if a pharmaceutical company wants to have a liability fund worth $200 million 5 years from now. Assume the company will invest an equal amount of money *each year* beginning 1 year from now and that the investments will earn 20% per year.

1.28 Identify the following as cash inflows or outflows to Anderson and Dyess Design-Build Engineers: office supplies, GPS surveying equipment, auctioning of used earth-moving equipment, staff salaries, fees for services rendered, interest from bank deposits.

1.29 Vision Technologies, Inc. is a small company that uses ultra-wideband technology to develop devices that can detect objects (including people) inside buildings, behind walls, or below ground. The company expects to spend $100,000 per year for labor and $125,000 per year for supplies before a product can be marketed. If the company wants to know the total equivalent future amount of the company's expenses at the end of 3 years at 15% per year interest, identify the engineering economy symbols involved and the values for the ones that are given.

1.30 Corning Ceramics expects to spend $400,000 to upgrade certain equipment 2 years from now. If the company wants to know the equivalent value now of the planned expenditure, identify the symbols and their values, assuming Corning's minimum attractive rate of return is 20% per year.

1.31 Sensotech, Inc., a maker of microelectromechanical systems, believes it can reduce product recalls by 10% if it purchases new software for detecting faulty parts. The cost of the new software is $225,000. Identify the symbols involved and the values for the symbols that are given in determining how much the company would have to save each year to recover its investment in 4 years at a minimum attractive rate of return of 15% per year.

1.32 Atlantic Metals and Plastic uses austenitic nickel-chromium alloys to manufacture resistance heating wire. The company is considering a new annealing-drawing process to reduce costs. If the new process will cost $1.8 million dollars now, identify the symbols that are involved and the values of those that are given if the company wants to know how much it must save each year to recover the investment in 6 years at an interest rate of 12% per year.

1.33 Phelps-Dodge plans to expand capacity by purchasing equipment that will provide additional smelting capacity. The cost of the initial investment is expected to be $16 million. The company expects revenue to increase by $3.8 million per year after the expansion. If the company's MARR is 18% per year, identify the engineering economy symbols involved and their value.

Cash Flows

1.34 In the phrase "end-of-period convention," the word "period" refers to what?

1.35 The difference between cash inflows and cash outflows is known as what?

1.36 Construct a cash flow diagram for the following: $10,000 outflow at time zero, $3000 per year inflow in years 1 through 5 at an interest rate of 10% per year, and an unknown future amount in year 5.

1.37 Kennywood Amusement Park spends $75,000 each year in consulting services for ride inspection and maintenance recommendations. New actuator element technology enables engineers to simulate complex computer-controlled movements in any direction. Construct a cash flow diagram to determine how much the park could afford to spend now on the new technology, if the cost of annual consulting services will be reduced to $30,000 per year. Assume the park uses an interest rate of 15% per year and it wants to recover its investment in 5 years.

Spreadsheet Functions

1.38 Write the engineering economy symbol that corresponds to each of the following Excel functions.
  a. FV
  b. PMT
  c. NPER
  d. IRR
  e. PV

1.39 What are the values of the engineering economy symbols P, F, A, i, and n, in the following Excel functions? Use a "?" for the symbol that is to be determined.
  a. FV(8%,10,2000,10000)
  b. PMT(12%,30,16000)
  c. PV(9%,15,1000,700)

1.40 State the purpose for each of the following built-in Excel functions:
  a. FV($i\%,n,A,P$)
  b. IRR(first_cell:last_cell)
  c. PMT($i\%,n,P,F$)
  d. PV($i\%,n,A,F$)

PROBLEMS FOR TEST REVIEW AND FE EXAM PRACTICE

1.41 Of the five engineering economy symbols P, F, A, i, and n, every problem will *involve* at least how many of them?
  a. two
  b. three
  c. four
  d. all of them

1.42 Of the five engineering economy symbols P, F, A, i, and n, every problem will have at least how many of them *given*?
  a. one
  b. two
  c. three
  d. four

1.43 An example of an intangible factor is
  a. taxes.
  b. goodwill.
  c. labor costs.
  d. rent.

1.44 Amounts of $1000 1 year ago and $1345.60 1 year hence are equivalent at what compound interest rate per year?
  a. 12.5% per year
  b. 14.8% per year
  c. 17.2% per year
  d. None of the above

1.45 What simple interest rate per year would be required to accumulate the same amount of money in 2 years as 20% per year compound interest?
  a. 20.5%
  b. 21%
  c. 22%
  d. 23%

1.46 For the Excel built-in function of PV($i\%,n,A,F$), the only parameter that can be omitted is
  a. $i\%$
  b. n
  c. A
  d. F

Factors: How Time and Interest Affect Money

The McGraw-Hill Companies, Inc./John Flournoy

In Chapter 1 we learned the basic concepts of engineering economy and their role in decision making. The cash flow is fundamental to every economic study. Cash flows occur in many configurations and amounts—isolated single values, series that are uniform, and series that increase or decrease by constant amounts or constant percentages. This chapter develops the commonly used engineering economy factors that consider the time value of money for cash flows.

The application of factors is illustrated using their mathematical forms and a standard notation format. Spreadsheet functions are illustrated.

Objectives

Purpose: Use engineering economy factors to account for the time value of money.

1. Use the compound amount factor and present worth factor for single payments.

   F/P and *P/F* factors

2. Use the uniform series factors.

   P/A, A/P, F/A and *A/F* factors

3. Use the arithmetic gradient factors and the geometric gradient formula.

   Gradients

4. Use uniform series and gradient factors when cash flows are shifted.

   Shifted cash flows

5. Use a spreadsheet to make equivalency calculations.

   Spreadsheets

2.1 SINGLE-PAYMENT FORMULAS (*F/P* AND *P/F*)

The most fundamental equation in engineering economy is the one that determines the amount of money F accumulated after n years (or periods) from a *single* present worth P, with interest compounded one time per year (or period). Recall that compound interest refers to interest paid on top of interest. Therefore, if an amount P is invested at time $t = 0$, the amount F_1 accumulated 1 year hence at an interest rate of i percent per year will be

$$F_1 = P + Pi$$
$$= P(1 + i)$$

where the interest rate is expressed in decimal form. At the end of the second year, the amount accumulated F_2 is the amount after year 1 plus the interest from the end of year 1 to the end of year 2 on the entire F_1.

$$F_2 = F_1 + F_1 i$$
$$= P(1 + i) + P(1 + i)i$$

The amount F_2 can be expressed as

$$F_2 = P(1 + i + i + i^2)$$
$$= P(1 + 2i + i^2)$$
$$= P(1 + i)^2$$

Similarly, the amount of money accumulated at the end of year 3 will be

$$F_3 = P(1 + i)^3$$

By mathematical induction, the formula can be generalized for n years to

$$F = P(1 + i)^n \qquad \text{[2.1]}$$

The term $(1 + i)^n$ is called a factor and is known as the *single-payment compound amount factor* (SPCAF), but it is usually referred to as the *F/P factor*. This is the conversion factor that yields the future amount F of an initial amount P after n years at interest rate i. The cash flow diagram is seen in Figure 2.1*a*.

Reverse the situation to determine the P value for a stated amount F. Simply solve Equation [2.1] for P.

$$P = F \left[\frac{1}{(1 + i)^n} \right] \qquad \text{[2.2]}$$

The expression in brackets is known as the *single-payment present worth factor* (SPPWF), or the *P/F factor*. This expression determines the present worth P of a given future amount F after n years at interest rate i. The cash flow diagram is shown in Figure 2.1*b*.

Note that the two factors derived here are for *single payments;* that is, they are used to find the present or future amount when only one payment or receipt is involved.

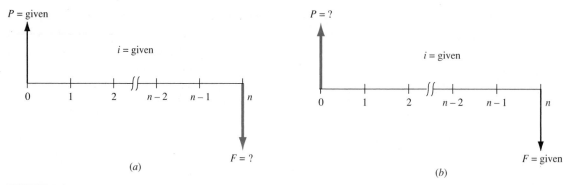

FIGURE 2.1   Cash flow diagrams for single-payment factors: (a) find *F* and (b) find *P*.

A standard notation has been adopted for all factors. It is always in the general form $(X/Y,i,n)$. The letter X represents what is sought, while the letter Y represents what is given. For example, F/P means *find F when given P*. The i is the interest rate in percent, and n represents the number of periods involved. Thus, $(F/P,6\%,20)$ represents the factor that is used to calculate the future amount F accumulated in 20 periods if the interest rate is 6% per period. The P is given. The standard notation, simpler to use than formulas and factor names, will be used hereafter. Table 2.1 summarizes the standard notation and equations for the F/P and P/F factors.

To simplify routine engineering economy calculations, tables of factor values have been prepared for a wide range of interest rates and time periods from 1 to large n values, depending on the i value. These tables are found at the end of this book following the Reference Materials.

For a given factor, interest rate, and time, the correct factor value is found at the intersection of the factor name and n. For example, the value of the factor $(P/F,5\%,10)$ is found in the P/F column of Table 10 at period 10 as 0.6139.

When it is necessary to locate a factor value for an i or n that is not in the interest tables, the desired value can be obtained in one of two ways: (1) by using the formulas derived in Sections 2.1 to 2.3 or (2) by linearly interpolating between the tabulated values.

For spreadsheet use, the F value is calculated by the FV function, and P is determined using the PV function. The formats are included in Table 2.1. Refer to Appendix A for more information on the FV and PV functions.

TABLE 2.1   *F/P and P/F* **Factors: Notation and Equation and Spreadsheet Function**

Factor			Standard Notation	Equation with	Excel
Notation	Name	Find/Given	Equation	Factor Formula	Functions
$(F/P,i,n)$	Single-payment compound amount	F/P	$F = P(F/P, i, n)$	$F = P(1 + i)^n$	$= \mathrm{FV}(i\%,n,,P)$
$(P/F,i,n)$	Single-payment present worth	P/F	$P = F(P/F, i, n)$	$P = F\left[1/(1 + i)^n\right]$	$= \mathrm{PV}(i\%,n,,F)$

An engineer received a bonus of $12,000 that he will invest now. He wants to calculate the equivalent value after 24 years, when he plans to use all the resulting money as the down payment on an island vacation home. Assume a rate of return of 8% per year for each of the 24 years. Find the amount he can pay down, using the tabulated factor, the factor formula, and a spreadsheet function.   **EXAMPLE 2.1**

Solution

The symbols and their values are

$$P = \$12,000 \qquad F = ? \qquad i = 8\% \text{ per year} \qquad n = 24 \text{ years}$$

The cash flow diagram is the same as that in Figure 2.1*a*.

Tabulated: Determine *F*, using the *F/P* factor for 8% and 24 years. Table 13 provides the factor value.

$$F = P(F/P,i,n) = 12,000(F/P,8\%,24)$$
$$= 12,000(6.3412)$$
$$= \$76,094.40$$

Formula: Apply Equation [2.1] to calculate the future worth *F*.

$$F = P(1 + i)^n = 12,000(1 + 0.08)^{24}$$
$$= 12,000(6.341181)$$
$$= \$76,094.17 \qquad = FV\left(8\%, 24, 12000\right)$$

Spreadsheet: Use the function $= FV(i\%,n,A,P)$. The cell entry is $= FV(8\%, 24,,12000)$. The *F* value displayed is ($76,094.17) in red or $-\$76,094.17$ in black to indicate a cash outflow.

The slight difference in answers is due to round-off error. An equivalence interpretation of this result is that $12,000 today is worth $76,094 after 24 years of growth at 8% per year compounded annually.

Hewlett-Packard has completed a study indicating that $50,000 in reduced maintenance this year (i.e., year zero) on one processing line resulted from improved wireless monitoring technology.   **EXAMPLE 2.2**

a. If Hewlett-Packard considers these types of savings worth 20% per year, find the equivalent value of this result after 5 years.

b. If the $50,000 maintenance savings occurs now, find its equivalent value 3 years earlier with interest at 20% per year.

Solution

a. The cash flow diagram appears as in Figure 2.1*a*. The symbols and their values are

$$P = \$50,000 \qquad F = ? \qquad i = 20\% \text{ per year} \qquad n = 5 \text{ years}$$

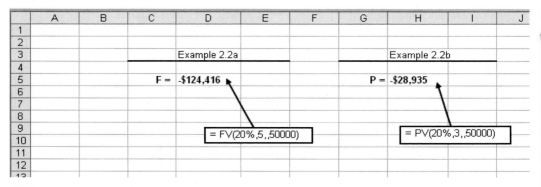

FIGURE 2.2   Use of single-cell spreadsheet functions to find *F* and *P* values, Example 2.2.

Use the F/P factor to determine F after 5 years.

$$F = P(F/P,i,n) = \$50,000(F/P,20\%,5)$$
$$= 50,000(2.4883)$$
$$= \$124,415.00$$

The function $=$ FV(20%,5,,50000) also provides the answer. See Figure 2.2.
b. The cash flow diagram appears as in Figure 2.1*b* with F placed at time $t = 0$ and the P value placed 3 years earlier at $t = -3$. The symbols and their values are

$$P = ? \qquad F = \$50,000 \qquad i = 20\% \text{ per year} \qquad n = 3 \text{ years}$$

Use the P/F factor to determine P three years earlier.

$$P = F(P/F,i,n) = \$50,000\,(P/F,20\%,3)$$
$$= 50,000(0.5787) = \$28,935.00$$

Use the PV function and omit the A value. Figure 2.2 shows the result of entering $=$ PV (20%,3,,50000) to be the same as using the P/F factor.

EXAMPLE 2.3   Jamie has become more conscientious about paying off his credit card bill promptly to reduce the amount of interest paid. He was surprised to learn that he paid $400 in interest in 2007 and the amounts shown in Figure 2.3 over the previous several years. If he made his payments to avoid interest charges, he would have these funds plus earned interest available in the future. What is the equivalent amount 5 years from now that Jamie could have available had he not paid the interest penalties? Let $i = 5\%$ per year.

Year	2002	2003	2004	2005	2006	2007
Interest paid, $	600	0	300	0	0	400

FIGURE 2.3　Credit card interest paid over the last 6 years, Example 2.3.

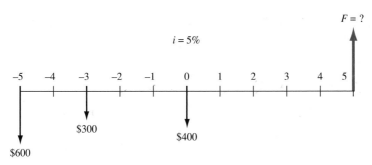

FIGURE 2.4　Cash flow diagram, Example 2.3.

Solution

Draw the cash flow diagram for the values $600, $300, and $400 from Jamie's perspective (Figure 2.4). Use *F/P* factors to find *F* in the year labeled 5, which is 10 years after the first cash flow.

$$F = 600(F/P,5\%,10) + 300(F/P,5\%,8) + 400(F/P,5\%,5)$$
$$= 600(1.6289) + 300(1.4775) + 400(1.2763)$$
$$= \$1931.11$$

The problem could also be solved by finding the present worth in year -5 of the $300 and $400 costs using the *P/F* factors and then finding the future worth of the total in 10 years.

$$P = 600 + 300(P/F,5\%,2) + 400(P/F,5\%,5)$$
$$= 600 + 300(0.9070) + 400(0.7835)$$
$$= \$1185.50$$
$$F = 1185.50(F/P,5\%,10) = 1185.50(1.6289)$$
$$= \$1931.06$$

Comment: It should be obvious that there are a number of ways the problem could be worked, since any year could be used to find the equivalent total of the costs before finding the future value in year 5. As an exercise, work the problem using year 0 for the equivalent total and then determine the final amount in year 5. All answers should be the same except for round-off error.

2.2 UNIFORM SERIES FORMULAS (*P/A, A/P, A/F, F/A*)

There are four *uniform series* formulas that involve *A*, where *A* means that:

1. The cash flow occurs in *consecutive interest periods*, and
2. The cash flow *amount* is the *same* in each period.

The formulas relate a present worth *P* or a future worth *F* to a uniform series amount *A*. The two equations that relate *P* and *A* are as follows. (See Figure 2.5 for cash flow diagrams.)

$$P = A\left[\frac{(1+i)^n - 1}{i(1+i)^n}\right]$$

$$A = P\left[\frac{i(1+i)^n}{(1+i)^n - 1}\right]$$

In standard factor notation, the equations are $P = A(P/A, i, n)$ and $A = P(A/P, i, n)$, respectively. It is important to remember that in these equations, the *P* and the first *A* value are separated by one interest period. That is, the present worth *P* is always located one interest period prior to the first *A* value. It is also important to remember that the *n* is always equal to the number of *A* values.

The factors and their use to find *P* and *A* are summarized in Table 2.2. The spreadsheet functions shown in Table 2.2 are capable of determining both

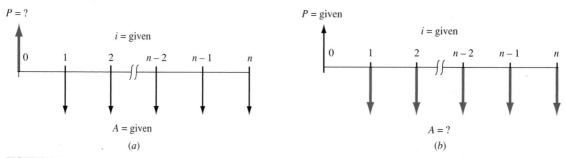

FIGURE 2.5   Cash flow diagrams used to determine (a) *P* of a uniform series and (b) *A* for a present worth.

TABLE 2.2 *P/A* and *A/P* Factors: Notation, Equation and Spreadsheet Function

Factor Notation	Factor Name	Find/Given	Factor Formula	Standard Notation Equation	Excel Function
(*P/A,i,n*)	Uniform-series present worth	*P/A*	$\dfrac{(1+i)^n - 1}{i(1+i)^n}$	$P = A(P/A,i,n)$	$= \mathrm{PV}(i\%,n,A,F)$
(*A/P,i,n*)	Capital recovery	*A/P*	$\dfrac{i(1+i)^n}{(1+i)^n - 1}$	$A = P(A/P,i,n)$	$= \mathrm{PMT}(i\%,n,P,F)$

P and *A* values in lieu of applying the *P/A* and *A/P* factors. The PV function calculates the *P* value for a given *A* over *n* years, and a separate *F* value in year *n*, if present. The format is

$$= \text{PV}(i\%,n,A,F)$$

Similarly, the *A* value is determined using the PMT function for a given *P* value in year 0 and a separate *F*, if present. The format is

$$= \text{PMT}(i\%,n,P,F)$$

How much money should you be willing to pay now for a guaranteed $600 per year for 9 years starting next year, at a rate of return of 16% per year?   **EXAMPLE 2.4**

Solution

The cash flow diagram (Figure 2.6) fits the *P/A* factor. The present worth is:

$$P = 600(P/A,16\%,9) = 600(4.6065) = \$2763.90$$

The PV function $= \text{PV}(16\%,9,600)$ entered into a single spreadsheet cell will display the answer $P = \$2763.93$.

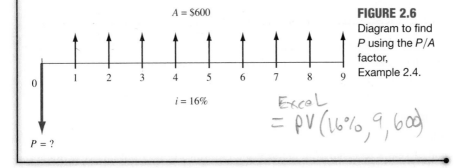

A = $600

i = 16%

Excel
= PV(16%, 9, 600)

P = ?

FIGURE 2.6
Diagram to find
P using the *P/A*
factor,
Example 2.4.

The uniform series formulas that relate *A* and *F* follow. See Figure 2.7 for cash flow diagrams.

$$A = F\left[\frac{i}{(1 + i)^n - 1}\right]$$

$$F = A\left[\frac{(1 + i)^n - 1}{i}\right]$$

It is important to remember that these equations are derived such that the last *A* value occurs in the *same* time period as the future worth *F*, and *n* is always equal to the number of *A* values.

Standard notation follows the same form as that of other factors. They are (*F/A,i,n*) and (*A/F,i,n*). Table 2.3 summarizes the notations and equations.

If *P* is not present for the PMT function, the comma must be entered to indicate that the last entry is an *F* value.

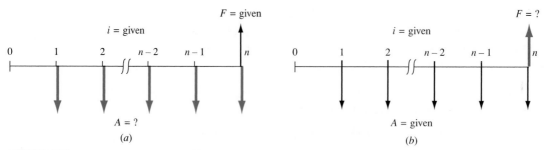

FIGURE 2.7   Cash flow diagrams to (a) find A, given F, and (b) find F, given A.

TABLE 2.3 *F/A and A/F Factors: Notation, Equation and Spreadsheet Function*

Factor			Factor	Standard	Excel
Notation	Name	Find/Given	Formula	Notation Equation	Function
(F/A,i,n)	Uniform-series compound amount	F/A	$\dfrac{(1 + i)^n - 1}{i}$	$F = A(F/A,i,n)$	$= FV(i\%,n,A,P)$
(A/F,i,n)	Sinking fund	A/F	$\dfrac{i}{(1 + i)^n - 1}$	$A = F(A/F,i,n)$	$= PMT(i\%,n,P,F)$

EXAMPLE 2.5   Formasa Plastics has major fabrication plants in Texas and Hong Kong. The president wants to know the equivalent future worth of $1 million capital investments each year for 8 years, starting 1 year from now. Formasa capital earns at a rate of 14% per year.

Solution

The cash flow diagram (Figure 2.8) shows the annual payments starting at the end of year 1 and ending in the year the future worth is desired. Cash flows are indicated in $1000 units. The F value in 8 years is

$$F = 1000(F/A,14\%,8) = 1000(13.2328) = \$13,232.80$$

The actual future worth is $13,232,800.

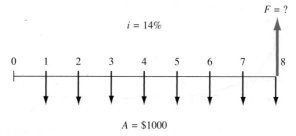

FIGURE 2.8   Diagram to find F for a uniform series, Example 2.5.

How much money must an electrical contractor deposit every year in her savings account starting 1 year from now at 5½% per year in order to accumulate $6000 seven years from now?

EXAMPLE 2.6

Solution

The cash flow diagram (Figure 2.9) fits the *A/F* factor.

$$A = \$6000(A/F,5.5\%,7) = 6000(0.12096) = \$725.76 \text{ per year}$$

The *A/F* factor value of 0.12096 was computed using the factor formula. Alternatively, use the spreadsheet function = PMT(5.5%,7,,6000) to obtain *A* = $725.79 per year.

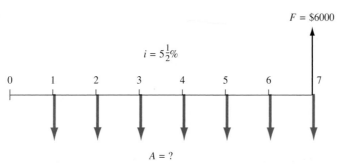

FIGURE 2.9 Cash flow diagram, Example 2.6. $= PMT(5.5\%, 7, 6000)$

When a problem involves finding *i* or *n* (instead of *P*, *F*, or *A*), the solution may require trial and error. Spreadsheet functions can be used to find *i* or *n* in most cases.

2.3 GRADIENT FORMULAS

The previous four equations involved cash flows of the *same magnitude A* in each interest period. Sometimes the cash flows that occur in consecutive interest periods are not the same amount (not an *A* value), but they do change in a predictable way. These cash flows are known as *gradients*, and there are two general types: arithmetic and geometric.

An *arithmetic gradient* is one wherein the cash flow changes (increases or decreases) by the same amount in each period. For example, if the cash flow in period 1 is $800 and in period 2 it is $900, with amounts increasing by $100 in each subsequent interest period, this is an arithmetic gradient *G*, with a value of $100.

The equation that represents the present worth of an arithmetic gradient series is:

$$P = \frac{G}{i}\left[\frac{(1+i)^n - 1}{i(1+i)^n} - \frac{n}{(1+i)^n}\right]$$

[2.3]

FIGURE 2.10
Conventional
arithmetic gradient
series without the
base amount.

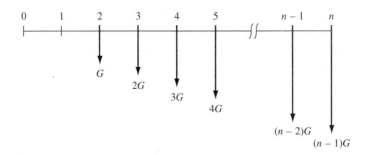

Equation [2.3] is derived from the cash flow diagram in Figure 2.10 by using the
P/F factor to find the equivalent *P* in year 0 of each cash flow. Standard factor
notation for the present worth of an arithmetic gradient is $P = G(P/G,i\%,n)$. This
equation finds the present worth of the *gradient only* (the $100 increases mentioned
earlier). It does *not* include the base amount of money that the gradient was built
upon ($800 in the example). The base amount in time period 1 must be accounted
for separately as a uniform cash flow series. Thus, the *general equation* to find the
present worth of an arithmetic gradient cash flow series is

P = Present worth of base amount + present worth of gradient amount   [2.4]
$$= A(P/A,i\%,n) + G(P/G,i\%,n)$$

where A = amount in *period* 1

G = amount of *change* in cash flow between periods 1 *and* 2

n = number of periods from 1 through *n* of gradient cash flow

i = interest rate per period

If the gradient cash flow *decreases* from one period to the next, the only change
in the general equation is that the plus sign becomes a minus sign. Since the gradient G begins between years 1 and 2, this is called a *conventional gradient*.

EXAMPLE 2.7

The Highway Department expects the cost of maintenance for a piece of heavy
construction equipment to be $5000 in year 1, to be $5500 in year 2, and to
increase annually by $500 through year 10. At an interest rate of 10% per year,
determine the present worth of 10 years of maintenance costs.

Solution

The cash flow includes an increasing gradient with $G = 500 and a base amount
of $5000 starting in year 1. Apply Equation [2.4].

$$P = 5000(P/A,10\%,10) + 500(P/G,10\%,10)$$
$$= 5000(6.1446) + 500(22.8913)$$
$$= \$42,169$$

In Example 2.7 an arithmetic gradient is converted to a P value using the P/G factor. If an equivalent A value for years 1 through n is needed, the A/G factor can be used directly to convert the gradient only. The equation follows with the $(A/G,i\%,n)$ factor formula included in the brackets.

$$A = G\left[\frac{1}{i} - \frac{n}{(1 + i)^n - 1}\right] \qquad [2.5]$$

As for the P/G factor, the A/G factor converts *only the gradient* into an A value. The base amount in year 1, A_1, must be added to the Equation [2.5] result to obtain the total annual worth A_T of the cash flows.

$$A_T = A_1 \pm A_G \qquad [2.6]$$

where A_1 = cash flow (base amount) in period 1

       A_G = annual worth of gradient

Alternatively, the annual worth of the cash flow could be obtained by first finding the total present worth P_T of the cash flows and then converting it into an A value using the relation $A_T = P_T(A/P,i,n)$.

EXAMPLE 2.8

The cash flow associated with a strip mining operation is expected to be $200,000 in year 1, $180,000 in year 2, and amounts decreasing by $20,000 annually through year 8. At an interest rate of 12% per year, calculate the equivalent annual cash flow.

Solution

Apply Equation [2.6] and the A/G factor.

$$
\begin{aligned}
A_T &= A_1 - A_G \\
&= 200{,}000 - 20{,}000(A/G,12\%,8) \\
&= 200{,}000 - 20{,}000(2.9131) \\
&= \$141{,}738
\end{aligned}
$$

The previous two gradient factors are for cash flows that change by a *constant amount* each period. Cash flows that change by a *constant percentage* each period are known as *geometric gradients*. The following equation is used to calculate the P value of a geometric gradient. The expression in brackets is called the $(P/A,g,i,n)$ factor.

$$P = A_1\left[\frac{1 - \left(\dfrac{1 + g}{1 + i}\right)^n}{i - g}\right] \qquad g \neq i \qquad [2.7]$$

where A_1 = total cash flow in period 1

       g = rate of change per period (decimal form)

       i = interest rate per period

This equation accounts for *all* of the cash flow, including the amount in period 1. For a decreasing geometric gradient, change the sign prior to both *g* values. When $g = i$, the *P* value is

$$P = A_1[n/(1 + i)] \qquad \text{[2.8]}$$

Geometric gradient factors are not tabulated; the equations are used. Spreadsheets are also an option.

EXAMPLE 2.9   A mechanical contractor has four employees whose combined salaries through the end of this year are $250,000. If he expects to give an average raise of 5% each year, calculate the present worth of the employees' salaries over the next 5 years. Let $i = 12\%$ per year.

Solution

The cash flow at the end of year 1 is $250,000, increasing by $g = 5\%$ per year (Figure 2.11). The present worth is found using Equation [2.7].

$$P = 250,000 \left[\frac{1 - \left(\dfrac{1.05}{1.12}\right)^5}{0.12 - 0.05} \right]$$

$$= 250,000(3.94005)$$

$$= \$985,013$$

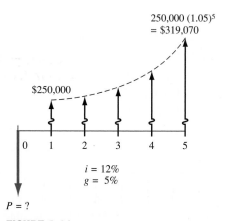

250,000 (1.05)^5 = $319,070

$250,000

0   1   2   3   4   5

$i = 12\%$
$g = 5\%$

$P = ?$

FIGURE 2.11   Cash flow with $g = 5\%$, Example 2.9

In summary, some basics for gradients are:

- Arithmetic gradients consist of two parts: a uniform series that has an *A* value equal to the amount of money in period 1, and a gradient that has a value equal to the change in cash flow between periods 1 and 2.
- For arithmetic gradients, the gradient factor is preceded by a plus sign for increasing gradients and a minus sign for decreasing gradients.

- Arithmetic and geometric cash flows start between periods 1 and 2, with the A value in each equation equal to the magnitude of the cash flow in period 1.
- Geometric gradients are handled with Equation [2.7] or [2.8], which yield the present worth of *all* the cash flows.

2.4 CALCULATIONS FOR CASH FLOWS THAT ARE SHIFTED

When a uniform series begins at a time other than at the end of period 1, it is called a *shifted series*. In this case several methods can be used to find the equivalent present worth P. For example, P of the uniform series shown in Figure 2.12 could be determined by any of the following methods:

- Use the P/F factor to find the present worth of each disbursement at year 0 and add them.
- Use the F/P factor to find the future worth of each disbursement in year 13, add them, and then find the present worth of the total using $P = F(P/F,i,13)$.
- Use the F/A factor to find the future amount $F = A(F/A,i,10)$, and then compute the present worth using $P = F(P/F,i,13)$.
- Use the P/A factor to compute the "present worth" (which will be located in year 3 not year 0), and then find the present worth in year 0 by using the $(P/F,i,3)$ factor. (Present worth is enclosed in quotation marks here only to represent the present worth as determined by the P/A factor in year 3, and to differentiate it from the present worth in year 0.)

Typically the last method is used. For Figure 2.12, the "present worth" obtained using the P/A factor is located in year 3. This is shown as P_3 in Figure 2.13.

Remember, the present worth is always located one period prior to the first uniform-series amount when using the P/A factor.

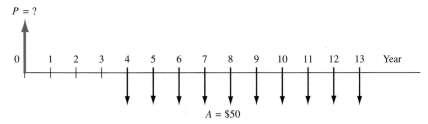

FIGURE 2.12
A uniform series that is shifted.

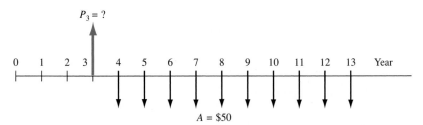

FIGURE 2.13
Location of present worth for the shifted uniform series in Figure 2.12.

FIGURE 2.14

Placement of *F* and renumbering for *n* for the shifted uniform series of Figure 2.12.

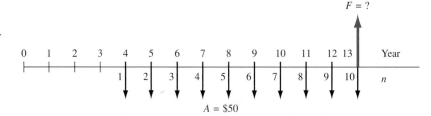

$A = \$50$

To determine a future worth or *F* value, recall that the *F/A* factor has the *F* located in the *same* period as the last uniform-series amount. Figure 2.14 shows the location of the future worth when *F/A* is used for Figure 2.12 cash flows.

Remember, the future worth is always located in the same period as the last uniform-series amount when using the *F/A* factor.

It is also important to remember that the number of periods *n* in the *P/A* or *F/A* factor is equal to the number of uniform-series values. It may be helpful to *renumber* the cash flow diagram to avoid errors in counting. Figure 2.14 shows Figure 2.12 renumbered to determine *n* = 10.

As stated above, there are several methods that can be used to solve problems containing a uniform series that is shifted. However, it is generally more convenient to use the uniform-series factors than the single-amount factors. There are specific steps that should be followed in order to avoid errors:

1. Draw a diagram of the positive and negative cash flows.
2. Locate the present worth or future worth of each series on the cash flow diagram.
3. Determine *n* for each series by renumbering the cash flow diagram.
4. Set up and solve the equations.

EXAMPLE 2.10   An engineering technology group just purchased new CAD software for $5000 now and annual payments of $500 per year for 6 years starting 3 years from now for annual upgrades. What is the present worth of the payments if the interest rate is 8% per year?

Solution

The cash flow diagram is shown in Figure 2.15. The symbol P_A is used throughout this chapter to represent the present worth of a uniform annual series *A*, and P'_A represents the present worth at a time other than period 0. Similarly, P_T represents the total present worth at time 0. The correct placement of P'_A and the diagram renumbering to obtain *n* are also indicated. Note that P'_A is located in actual year 2, not year 3. Also, *n* = 6, not 8, for the *P/A* factor. First find the value of P'_A of the shifted series.

$$P'_A = \$500(P/A, 8\%, 6)$$

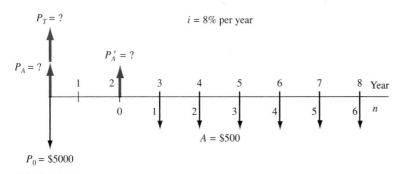

FIGURE 2.15
Cash flow diagram with placement of P values, Example 2.10.

Since P_A' is located in year 2, now find P_A in year 0.

$$P_A = P_A'(P/F,8\%,2)$$

The total present worth is determined by adding P_A and the initial payment P_0 in year 0.

$$
\begin{aligned}
P_T &= P_0 + P_A \\
&= 5000 + 500(P/A,8\%,6)(P/F,8\%,2) \\
&= 5000 + 500(4.6229)(0.8573) \\
&= \$6981.60
\end{aligned}
$$

To determine the present worth for a cash flow that includes both uniform series and single amounts at specific times, use the P/F factor for the single amounts and the P/A factor for the series. To calculate A for the cash flows, first convert everything to a P value in year 0, or an F value in the last year. Then obtain the A value using the A/P or A/F factor, where n is the total number of years over which the A is desired.

Many of the considerations that apply to shifted uniform series apply to gradient series as well. Recall that a conventional gradient series starts between periods 1 and 2 of the cash flow sequence. A gradient starting at any other time is called a *shifted gradient*. The n value in the P/G and A/G factors for the shifted gradient is determined by renumbering the time scale. The period in which the *gradient first appears is labeled period 2*. The n value for the factor is determined by the renumbered period where the last gradient increase occurs. The P/G factor values and placement of the gradient series present worth P_G for the shifted arithmetic gradients in Figure 2.16 are indicated.

It is important to note that the A/G factor *cannot* be used to find an equivalent A value in periods 1 through n for cash flows involving a shifted gradient. Consider the cash flow diagram of Figure 2.16b. To find the equivalent annual series in years 1 through 10 for the gradient series only, first find the present worth of the gradient in year 5, take this present worth back to year 0, and then annualize the present worth for 10 years with the A/P factor. If you apply the annual

FIGURE 2.16
Determination of G
and n values used in
factors for shifted
gradients.

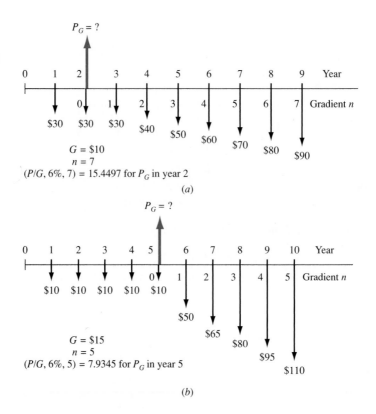

$P_G = ?$

(a)

$P_G = ?$

(b)

series gradient factor $(A/G,i,5)$ directly, the gradient is converted into an equivalent annual series over years 6 through 10 only.

> **Remember, to find the equivalent A series of a shifted gradient through all the periods, first find the present worth of the gradient at actual time 0, then apply the $(A/P,i,n)$ factor.**

If the cash flow series involves a *geometric gradient* and the gradient starts at a time other than between periods 1 and 2, it is a shifted gradient. The P_g is located in a manner similar to that for P_G above, and Equation [2.7] is the factor formula.

EXAMPLE 2.11   Chemical engineers at a Coleman Industries plant in the Midwest have determined that a small amount of a newly available chemical additive will increase the water repellency of Coleman's tent fabric by 20%. The plant superintendent has arranged to purchase the additive through a 5-year contract at $7000 per year, starting 1 year from now. He expects the annual price to increase by 12% per year starting in the sixth year and thereafter through year 13. Additionally, an initial investment of $35,000 was made now to prepare a site suitable for the contractor to deliver the additive. Use $i = 15\%$ per year to determine the equivalent total present worth for all these cash flows.

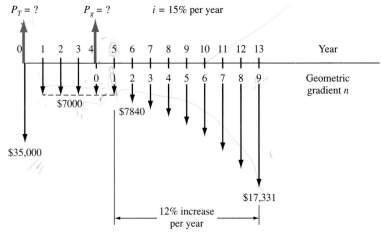

FIGURE 2.17 Cash flow diagram including a geometric gradient with $g = 12\%$, Example 2.11.

Solution

Figure 2.17 presents the cash flows. The total present worth P_T is found using $g = 0.12$ and $i = 0.15$. Equation [2.7] is used to determine the present worth P_g for the entire geometric series at actual year 4, which is moved to year 0 using $(P/F,15\%,4)$.

$$P_T = 35,000 + A(P/A,15\%,4) + A_1(P/A,12\%,15\%,9)(P/F,15\%,4)$$

$$= 35,000 + 7000(2.8550) + \left[7000\frac{1 - (1.12/1.15)^9}{0.15 - 0.12} \right](0.5718)$$

$$= 35,000 + 19,985 + 28,247$$

$$= \$83,232$$

Note that $n = 4$ in the $(P/A,15\%,4)$ factor because the $7000 in year 5 is the initial amount A_1 in Equation [2.7].

2.5 USING SPREADSHEETS FOR EQUIVALENCY COMPUTATIONS

The easiest single-cell Excel functions to apply to find P, F, or A require that the cash flows exactly fit the function format. Refer to Appendix A for more details. The functions apply the correct sign to the answer that would be on the cash flow diagram. That is, if cash flows are deposits (minus), the answer will have a plus sign. In order to retain the sign of the inputs, enter a minus sign prior to the function. Here is a summary and examples at 5% per year.

> **Present worth P:** Use the PV function $= PV(i\%,n,A,F)$ if A is exactly the same for each of n years; F can be present or not. For example, if $A = \$3000$ per year deposit for $n = 10$ years, the function $= PV(5\%, 10,-3000)$ will display $P = \$23,165$. This is the same as using the P/A factor to find $P = 3000(P/A,5\%,10) = 3000(7.7217) = \$23,165$.

Future worth F: Use the FV function $= \text{FV}(i\%,n,A,P)$ if A is exactly the same for each of n years; P can be present or not. For example, if $A = \$3000$ per year deposit for $n = 10$ years, the function $= \text{FV}(5\%, 10,-3000)$ will display $F = \$37{,}734$. This is the same as using the F/A factor to find $F = 3000(F/A,5\%,10) = 3000(12.5779) = \$37{,}734$.

Annual amount A: Use the PMT function $= \text{PMT}(i\%,n,P,F)$ when there is no A present, and either P or F or both are present. For example, for $P = \$-3000$ deposit now and $\text{F} = \$5000$ returned $n = 10$ years hence, the function $= \text{PMT}(5\%,10,-3000,5000)$ will display $A = -\$9$. This is the same as using the A/P and A/F factors to find the equivalent net $A = \$9$ per year between the deposit now and return 10 years later.

$$A = -3000\,(A/P,5\%,10) + 5000\,(A/F,5\%,10) = -389 + 398 = \$9$$

Number of periods n: Use the NPER functions $= \text{NPER}(i\%,A,P,F)$ if A is exactly the same for each of n years; either P or F can be omitted, but not both. For example, for $P = \$-25{,}000$ deposit now and $A = \$3000$ per year return, the function $= \text{NPER}(5\%,3000,-25000)$ will display $n = 11.05$ years to recover P at 5% per year. This is the same as using trail error to find n in the relation $0 = -25{,}000 + 3{,}000\,(P/A,5\%,n)$.

When cash flows vary in amount or timing, it is usually necessary to list them on a spreadsheet, including all zero amounts, and utilize other functions for P, F, or A values. All Excel functions allow another function to be embedded in them, thus reducing the time necessary to get final answers. Example 2.12 illustrates these functions and the embedding capability. Example 2.13 demonstrates how easily spreadsheets handle arithmetic and percentage gradients and how the Excel IRR (rate of return) function works.

EXAMPLE 2.12   Carol just entered college. Her grandparents have promised to give her $25,000 toward a new car if she graduates in 4 years. Alternatively, if she takes 5 years to graduate, they offered her $5000 each year starting after her second year is complete and an extra $5000 when she graduates. Draw the cash flow diagrams first. Then, use $i = 8\%$ per year to show Carol how to use spreadsheet functions to determine the following for each gift her grandparents offered:

a. Present worth P now
b. Future worth F five years from now
c. Equivalent annual amount A over a total of 5 years
d. Number of years it would take Carol to have $25,000 in hand for the new car if she were able to save $5000 each year starting next year.

Solution

The two cash flow series, labeled lump-sum gift and spread-out gift, are in Figure 2.18. The spreadsheet in Figure 2.19a lists the cash flows (don't forget to ⬇

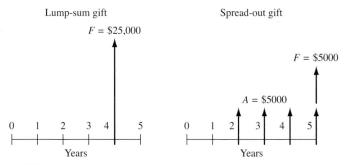

FIGURE 2.18 Cash flows for Carol's gift from her grandparents, Example 2.12.

enter the $0 cash flows so the NPV function can be used), and answers to each part using the PV, NPV, FV, or PMT functions are explained below. In some cases, there are alternative ways to obtain the answer.

Figure 2.19*b* shows the function formats with cell tag comments. Refer to Appendix A for a complete description of how each function operates. Remember that the PV, FV, and PMT functions will return an answer with the opposite sign from that of the cash flow entries. The same sign is maintained by entering a minus before the function name.

a. Rows 12 and 13: There are two ways to find P; either the PV or NPV function. NPV requires that the zeros be entered. (For lump-sum, omitting zeros in years 1, 2, and 3 will give the incorrect answer of $P = \$23,148$, because NPV assumes the $25,000 occurs in year 1 and discounts it only one year at 8%.) The single-cell PV is hard to use for the spread-out plan since cash flows do not start until year 2; using NPV is easier.

b. Rows 17 and 18: There are two ways to use the FV function to find F at the end of year 5. To develop FV correctly for the spread-out plan in a single cell without listing cash flows, add the extra $5000 in year 5 separate from the FV for the four $A = \$5000$ values. Alternatively, cell D18 incorporates the NPV function for the P value into the FV function. This is a very convenient way to combine functions.

c. Rows 21 and 22: There are two ways to use the PMT function to find A for 5 years; find P separately and use a cell reference, or imbed the NPV function into the PMT to find A in one operation.

d. Row 25: Finding the years to accumulate $25,000 by depositing $5000 each year using the NPER function is independent of either plan. The entry = NPER(8%,−5000,,25000) results in 4.3719 years. This can be confirmed by calculating $5000(F/A,8\%,4.3719) = 5000(5.0000) = \$25,000$ (The 4.37 years is about the time it will take Carol to finish college. Of course, this assumes she can actually save $5000 a year while working on the degree.)

	Year	Cash flows for gift offers	
		Lump sum gift	Spread out gift
	0		
	1	$0	$0
	2	$0	$ 5,000
	3	$0	$ 5,000
	4	$ 25,000	$ 5,000
	5	$0	$ 10,000
(a) Find P value now	Using single-cell PV function	$18,376	Harder to do in single cell
	Using NPV function	$18,376	$18,737
(b) Find F in year 5	Using single-cell FV function	$27,000	$27,531
	Using FV with imbedded NPV function	No need to imbed	$27,531
(c) Find A over 5 years	Using PMT with cell reference for P	$4,602	$4,693
	Using PMT with imbedded NPV function	$4,602	$4,693
(d) Years until F = $25,000	Using NPER function	4.37	4.37

(a)

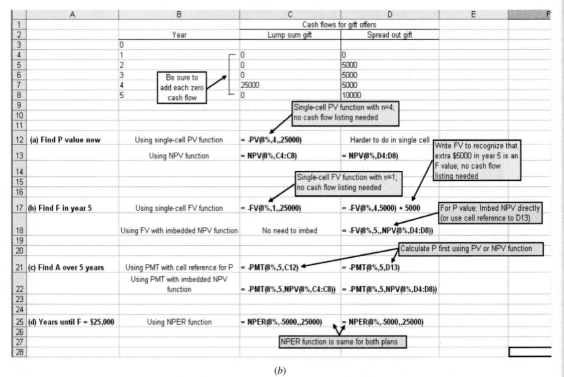

(b)

FIGURE 2.19 (a) Use of several Excel functions to find *P, F, A,* and *n* values, and (b) format of functions to obtain values, Example 2.12.

Bobby was desperate. He borrowed $600 from a pawn shop and understood he **EXAMPLE 2.13** was to repay the loan starting next month with $100, increasing by $10 per month for a total of 8 months. Actually, he misunderstood. The repayments increased by 10% each month after starting next month at $100. Use a spreadsheet to calculate the monthly interest rate that he thought he was to pay, and what he actually will pay.

Solution

Figure 2.20 lists the cash flows for the assumed arithmetic gradient $G = \$-10$ per month, and the actual percentage gradient $g = -10\%$ per month. Note the simple relations to construct the increasing cash flows for each type gradient. Apply the IRR function to each series using its format $= $ IRR (first_cell:last_cell). Bobby is paying an exorbitant rate per month (and year) at 14.9% *per month*, which is higher than he expected it to be at 13.8% per month. (Interest rates are covered in detail in Chapter 3.)

	A	B	C	D	E	F	G	
1		Cash flow			Cash flow			
2	Month	G = $10			g = 10%			
3	0	600.00			600.00			
4	1	-100.00			-100.00			
5	2	-110.00			-110.00			
6	3	-120.00	= B6-10		-121.00			
7	4	-130.00			-133.10			
8	5	-140.00			-146.41			
9	6	-150.00			-161.05			
10	7	-160.00	=SUM(B4:B11)		-177.16	= E10*(1.1)		
11	8	-170.00			-194.87			
12	Total paid back	-1080.00			-1143.59			
13	ROR per month using IRR function	**13.8%**			**14.9%**			
14								
15			= IRR(B3:B11)			= IRR(E3:E11)		
16								
17								

FIGURE 2.20 Use of a spreadsheet to generate arithmetic and percentage gradient cash flows and application of the IRR function, Example 2.13.

SUMMARY

In this chapter, we presented formulas that make it relatively easy to account for the time value of money. In order to use the formulas correctly, certain things must be remembered.

1. When using the P/A or A/P factors, the P and the first A value are separated by one interest period.

2. When using the F/A or A/F factors, the F and the last A value are in the *same* interest period.

3. The n in the uniform series formulas is equal to the number of A values involved.

4. Arithmetic gradients change by a uniform amount from one interest period to the next, and there are

two parts to the equation: a uniform series that has an A value equal to the magnitude of the cash flow in period 1 and the gradient that has the same n as the uniform series.

5. Geometric gradients involve cash flows that change by a uniform percentage from one period to the next, and the present worth of the entire cash flow sequence is determined from Equation [2.7] or [2.8].

6. For shifted gradients, the change equal to G or g occurs between periods 1 and 2. This requires renumbering the cash flows to properly identify which ones are accounted for in the gradient equations.

7. For decreasing arithmetic gradients, it is necessary to change the sign in front of the P/G or A/G factors from plus to minus. For decreasing geometric gradients, it is necessary to change the sign in front of both g's in Equation [2.7].

PROBLEMS

Use of Interest Tables

2.1 Find the correct numerical value for the following factors from the compound interest factor tables.
   a. $(F/P,10\%,20)$
   b. $(A/F,4\%,8)$
   c. $(P/A,8\%,20)$
   d. $(A/P,20\%,28)$
   e. $(F/A,30\%,15)$

Determination of P, F, A, and n

2.2 Beckton Steel Products, a company that specializes in crankshaft hardening, is investigating whether it should update certain equipment now or wait and do it later. If the cost now is $180,000, what would the equivalent amount be 3 years from now at an interest rate of 10% per year?

2.3 By filling carbon nanotubes with miniscule wires made of iron and iron carbide, incredibly thin nano wires could be extruded by blasting the carbon nanotubes with an electron beam. If Gentech Technologies spends $2.7 million *now* in developing the process, how much would the company have to receive in licensing fees 3 years from now to recover its investment at 20% per year interest?

2.4 Arctic and Antarctic regions are harsh environments in which to take data. A TempXZ 3000 portable temperature recorder can take and store 32,767 measurements at −40°C to 150°C. A research team from the University of Nova Scotia needs 20 of the recorders, and they are trying to decide whether they should buy them now at $649 each or purchase them 2 years from now, which is when they will be deployed. At an interest rate of 8% per year, how much will the 20 recorders have to cost in 2 years to render their decision indifferent?

2.5 Since many U.S. Navy aircraft are at or near their usual retirement age of 30 years, military officials want a precise system to assess when aircraft should be taken out of service. A computational method developed at Carnegie Mellon University maps in 3-D the microstructure of aircraft materials in their present state so that engineers can test them under different conditions of moisture, salt, dirt, etc. Military officials can then determine if an aircraft is fine, is in need of overhaul, or should be retired. If the 3-D system allows the Navy to use one airplane 2 years longer than it normally would have been used, thereby delaying the purchase of a $20 million aircraft for 2 years, what is the present worth of the assessment system at an interest rate of 15% per year?

2.6 GE Marine Systems is planning to supply a Japanese shipbuilder with aero-derivative gas turbines to power 11 DD-class destroyers for the Japanese Self-Defense Force. The buyer can pay the total contract price of $2,100,000 two years from now (when the turbines will be needed) or an equivalent amount now. At an interest rate of 15% per year, what is the equivalent amount now?

2.7 A maker of microelectromechanical systems can reduce product recalls by 10% if it purchases new packaging equipment. If the cost of the new equipment is expected to be $40,000 four years from now, how much could the company afford to spend now (instead of 4 years from now) if it uses a minimum attractive rate of return of 12% per year?

2.8 How much money could Tesla-Sino Inc., a maker of superconducting magnetic energy storage systems, afford to spend now on new equipment in lieu of spending $85,000 five years from now, if the company's rate of return is 18% per year?

2.9 French car maker Renault signed a $95 million contract with ABB of Zurich, Switzerland, for automated underbody assembly lines, body assembly workshops, and line control systems. If ABB will be paid in 3 years (when the systems are ready), what is the present worth of the contract at 12% per year interest?

2.10 What is the future worth 6 years from now of a present cost of $175,000 to Corning, Inc. at an interest rate of 10% per year?

2.11 A pulp and paper company is planning to set aside $150,000 now for possibly replacing its large synchronous refiner motors. If the replacement isn't needed for 8 years, how much will the company have in the account if it earns interest at a rate of 8% per year?

2.12 A mechanical consulting company is examining its cash flow requirements for the next 7 years. The company expects to replace office machines and computer equipment at various times over the 7-year planning period. Specifically, the company expects to spend $7000 two years from now, $9000 three years from now, and $5000 five years from now. What is the present worth of the planned expenditures at an interest rate of 10% per year?

2.13 A proximity sensor attached to the tip of an endoscope could reduce risks during eye surgery by alerting surgeons to the location of critical retinal tissue. If a certain eye surgeon expects that by using this technology, he will avoid lawsuits of $0.6 and $1.35 million 2 and 5 years from now, respectively, how much could he afford to spend now if his out-of-pocket costs for the lawsuits would be only 10% of the total amount of each suit? Use an interest rate of 10% per year.

2.14 Irvin Aerospace of Santa Ana, CA, was awarded a 5-year contract to develop an advanced space capsule airbag landing attenuation system for NASA's Langley Research Center. The company's computer system uses fluid structure interaction modeling to test and analyze each airbag design concept. What is the present worth of the contract at 10% per year interest if the annual cost (years 1 through 5) is $8 million per year?

2.15 Julong Petro Materials Inc. ordered $10 million worth of seamless tubes for its drill collars from the Timken Company of Canton, Ohio. (A drill collar is the heavy tubular connection between a drill pipe and drill bit.) At 10% per year interest, what is the

annual worth of the purchase over a 10-year amortization period?

2.16 Improvised explosive devices (IEDs) are responsible for many deaths in times of war. Unmanned ground vehicles (robots) can be used to disarm the IEDs and perform other tasks as well. If the robots cost $140,000 each and the U.S. Army signs a contract to purchase 4000 of them now, what is the equivalent annual cost of the contract if it is amortized over a 3-year period at 8% per year interest?

2.17 The U.S. Navy's robotics lab at Point Loma Naval Base in San Diego is developing robots that will follow a soldier's command or operate autonomously. If one robot would prevent injury to soldiers or loss of equipment valued at $1.5 million, how much could the military afford to spend on the robot and still recover its investment in 4 years at 8% per year?

2.18 PCM Thermal Products uses austenitic nickel-chromium alloys to manufacture resistance heating wire. The company is considering a new annealing-drawing process to reduce costs. If the new process will cost $2.55 million dollars now, how much must be saved each year to recover the investment in 6 years at an interest rate of 14% per year?

2.19 A green algae, chlamydomonas reinhardtii, can produce hydrogen when temporarily deprived of sulfur for up to 2 days at a time. How much could a small company afford to spend now to commercialize the process if the net value of the hydrogen produced is $280,000 per year? Assume the company wants to earn a rate of return of 18% per year and recover its investment in 8 years.

2.20 HydroKlean, LLC, an environmental soil cleaning company, borrowed $3.5 million to finance start-up costs for a site reclamation project. How much must the company receive each year in revenue to earn a rate of return of 20% per year for the 5-year project period?

2.21 A VMB pressure regulator allows gas suppliers and panel builders to provide compact gas handling equipment, thereby minimizing the space required in clean rooms. Veritech Micro Systems is planning to expand its clean room to accommodate a new product design team. The company estimates that it can reduce the space in the clean room by 7 square meters if it uses the compact equipment. If the cost of construction for a clean room is $5000 per square meter, what is the annual worth of the savings at 10% per year interest if the cost is amortized over 10 years?

2.22 New actuator element technology enables engineers to simulate complex computer-controlled movements in any direction. If the technology results in cost savings in the design of new roller coasters, determine the future worth in year 5 of savings of $70,000 now and $90,000 two years from now at an interest rate of 10% per year.

2.23 Under an agreement with the Internet Service Providers (ISPs) Association, ATT Communications reduced the price it charges ISPs to resell its high-speed digital subscriber line (DSL) service from $458 to $360 per year per customer line. A particular ISP, which has 20,000 customers, plans to pass 90% of the savings along to its customers. What is the total future worth of these savings in year 5 at an interest rate of 10% per year?

2.24 Southwestern Moving and Storage wants to have enough money to purchase a new tractor-trailer in 5 years at a cost of $290,000. If the company sets aside $100,000 in year 2 and $75,000 in year 3, how much will the company have to set aside in year 4 in order to have the money it needs if the money set aside earns 9% per year?

2.25 Vision Technologies, Inc., is a small company that uses ultra-wideband technology to develop devices that can detect objects (including people) inside buildings, behind walls, or below ground. The company expects to spend $100,000 per year for labor and $125,000 per year for supplies for three years before a product can be marketed. At an interest rate of 15% per year, what is the total equivalent present worth of the company's expenses?

2.26 How many years would it take for money to increase to four times the initial amount at an interest rate of 12% per year?

2.27 Acceleron, Inc., is planning to expand to new facilities in Indianapolis. The company will make the move when its real estate sinking fund has a total value of $1.2 million. If the fund currently has $400,000 and the company adds $50,000 per year, how many years will it take for the account to reach the desired value, if it earns interest at a rate of 10% per year?

2.28 The defined benefits pension fund of G-Tech Electronics has a net value of $2 billion. The company is switching to a defined contribution pension plan, but it guaranteed the current retirees that they will continue to receive their benefits as promised. If the withdrawal rate from the fund is $158 million per year starting 1 year from now, how many years will it take to completely deplete the fund if the conservatively managed fund grows at a rate of 7% per year?

Arithmetic and Geometric Gradients

2.29 Allen Bradley claims that its XM1Z1A and XM442 electronic overspeed detection relay modules provide customers a cost-effective monitoring and control system for turbo machinery. If the equipment provides more efficient turbine performance to the extent of $20,000 in year 1, $22,000 in year 2, and amounts increasing by $2000 per year, how much could Mountain Power and Light afford to spend now at 10% per year interest if it wanted to recover its investment in 5 years?

2.30 A low-cost noncontact temperature measuring tool may be able to identify railroad car wheels that are in need of repair long before a costly structural failure occurs. If the BNSF railroad saves $100,000 in year 1, $110,000 in year 2, and amounts increasing by $10,000 each year for five years, what is the equivalent annual worth of the savings at an interest rate of 10% per year?

2.31 Southwest Airlines hedged the cost of jet fuel by purchasing options that allowed the airline to purchase fuel at a fixed price for 5 years. If the market price of fuel was $0.50 per gallon higher than the option price in year 1, $0.60 per gallon higher in year 2, and amounts increasing by $0.10 per gallon higher through year 5, what was the present worth of SWA's savings per gallon if the interest rate was 10% per year?

2.32 NMTeX Oil company owns several gas wells in Carlsbad, NM. Income from the depleting wells has been decreasing according to an arithmetic gradient for the past five years. If the income in year 1 from well no. 24 was $390,000 and it decreased by $15,000 each year thereafter, (*a*) what was the income in year 3, and (*b*) what was the equivalent annual worth of the income through year 5 at an interest rate of 10% per year?

2.33 The present worth of income from an investment that follows an arithmetic gradient was projected to be $475,000. If the income in year one is expected to be $25.000, how much would the gradient have to be in each year through year 8 if the interest rate is 10% per year?

2.34 Very light jets (VLJs) are one-pilot, two-engine jets that weigh 10,000 pounds or less and have only five or six passenger seats. Since they cost half as much as the most inexpensive business jets, they are considered to be the wave of the future. MidAm Charter purchased 5 planes so that it can initiate service to small cities that have airports with short runways. MidAm expects revenue of $1 million in year 1, $1.2 million in year 2, and amounts increasing by $200,000 per year thereafter. If the company's MARR is 10% per year, what is the future worth of the revenue through the end of year 5?

2.35 Fomguard LLC of South Korea developed a high-tech fiber-optic fencing mesh (FOM) that contains embedded sensors that can differentiate between human and animal contact. In an effort to curtail illegal entry into the United States, a FOM fence has been proposed for certain portions of the U.S. border with Canada. The cost for erecting the fence in year 1 is expected to be $7 million, decreasing by $500,000 each year through year 5. At an interest rate of 10% per year, what is the equivalent uniform annual cost of the fence in years 1 to 5?

2.36 For the cash flows shown, determine the future worth in year 5 at an interest rate of 10% per year.

Year	1	2	3	4	5
Cash Flow	$300,000	$275,000	$250,000	$225,000	$200,000

2.37 Verizon Communications said it plans to spend $22.9 billion in expanding its fiber-optic Internet and television network through 2010 so that it can compete with cable TV providers like Comcast Corp. If the company gets 950,000 customers in year 1 and grows its customer base by 20% per year, what is the present worth of the subscription income through year 5 if income averages $600 per customer per year and the company uses a MARR of 10% per year?

2.38 A concept car that will get 100 miles per gallon and carry 4 persons would have a carbon-fiber and aluminum composite frame with a 900 cc three-cylinder turbodiesel/electric hybrid power plant. The extra cost of these technologies is estimated to be $11,000. If gasoline savings over a comparable conventional car would be $900 in year 1, increasing by 10% each year, what is the present worth of the savings over a 10-year period at an interest rate of 8% per year?

2.39 The National Institute on Drug Abuse has spent $15 million on clinical trials toward finding out whether two vaccines can end the bad habits of nicotine and cocaine addiction. A Switzerland-based company is now testing an obesity vaccine. If the vaccines are semi-successful such that treatment costs and medical bills are reduced by an average of $15,000 per person per year, what is the present worth of the vaccines if there are 10 million beneficiaries in year 1 and an additional 15% each year through year 5? Use an interest rate of 8% per year.

2.40 Find the future worth in year 10 of an investment that starts at $8000 in year 1 and increases by 10% each year. The interest rate is 10% per year.

2.41 The effort required to maintain a scanning electron microscope is known to increase by a fixed percentage each year. A high-tech equipment maintenance company has offered its services for a fee that includes automatic increases of 7% per year after year 1. A certain biotech company offered $75,000 as prepayment for a 3 year contract to take advantage of a temporary tax loophole. If the company used an interest rate of 12% per year in determining how much it should offer, what was the service fee amount that it assumed for year 1?

2.42 Hughes Cable Systems plans to offer its employees a salary enhancement package that has revenue sharing as its main component. Specifically, the company will set aside 1% of total sales for year-end bonuses for all its employees. The sales are expected to be $5 million the first year, $5.5 million the second year, and amounts increasing by 10% each year for the next 5 years. At an interest rate of 8% per year, what is the equivalent annual worth in years 1 through 5 of the bonus package?

2.43 Determine how much money would be in a savings account that started with a deposit of $2000 in year 1 with each succeeding amount increasing by 15% per year. Use an interest rate of 10% per year and a 7-year time period.

2.44 The future worth in year 10 of a *decreasing* geometric gradient series of cash flows was found to be $80,000. If the interest rate was 10% per year and the annual rate of decrease was 8% per year, what was the cash flow amount in year 1?

2.45 Altmax Ltd, a company that manufactures automobile wiring harnesses, has budgeted $P = \$400,000$ *now* to pay for a certain type of wire clip over the next 5 years. If the company expects the cost of the clips to increase by 4% each year, what is the expected cost in year 2 if the company uses and interest rate of 10% per year?

2.46 Thomasville Furniture Industries offers several types of high-performance fabrics that are capable of withstanding chemicals as harsh as chlorine. A certain midwestern manufacturing company that uses fabric in several products has a report showing that the present worth of fabric purchases over a specific 5-year period was $900,000. If the costs are known to have increased geometrically by 5% per year during that time and the company uses an interest rate of 15% per year for investments, what was the cost of the fabric in year 1?

2.47 Find the equivalent annual worth of a series of investments that starts at $1000 in year 1 and increases by 10% per year for 15 years. Assume the interest rate is 10% per year.

2.48 A northern California consulting firm wants to start saving money for replacement of network servers. If the company invests $5000 at the end of year 1 but decreases the amount invested by 5% each year, how much will be in the account 4 years from now, if it earns interest at a rate of 8% per year?

2.49 A company that manufactures purgable hydrogen sulfide monitors is planning to make deposits such that each one is 5% smaller than the preceding one. How large must the first deposit be (at the end of year 1) if the deposits extend through year 10 and the fourth deposit is $1250? Use an interest rate of 10% per year.

Shifted Cash Flows

2.50 Akron Coating and Adhesives (ACA) produces a hot melt adhesive that provides a strong bond between metals and thermoplastics used for weather stripping, seals, gaskets, hand grips for tools, appliances, etc. ACA claims that by eliminating a primer coat, manufacturers can cut costs and reduce scrap. If Porter Cable is able to save $60,000 now and $50,000 per year by switching to the new adhesive, what is the present worth of the savings for 3 years at an interest rate of 10% per year?

2.51 Attenuated Total Reflectance (ATR) is a method for looking at the surfaces of materials that are too opaque or too thick for standard transmission methods. A manufacturer of precision plastic parts

estimates that ATR spectroscopy can save the company $5000 per year by reducing returns of out-of-spec parts. What is the future worth of the savings if they start now and extend through year 5 at an interest rate of 10% per year?

2.52 To improve crack detection in aircraft, the U.S. Air Force combined ultrasonic inspection procedures with laser heating to identify fatigue cracks. Early detection of cracks may reduce repair costs by as much as $200,000 per year. If the savings start now and continue through year 5, what is the present worth of these savings at an interest rate of 10% per year?

2.53 Some studies have shown that taller men tend to earn higher salaries than equally qualified men of shorter stature. Homotrope growth hormone can increase a child's height at a cost of $50,000 per inch. Clyde's parents wanted him to be 3 inches taller than he was projected to be. They paid for the treatments for 3 years (at a cost of $50,000 per year) beginning on their son's 8th birthday. How much extra money would Clyde have to earn per year from his 26th through 60th birthdays (a total of 35 years) in order to justify their expenditure at an interest rate of 8% per year?

2.54 Calculate the annual worth (years 1 through 7) of Merchant Trucking Company's cash flow. Use an interest rate of 10% per year.

Year	0	1	2	3	4	5	6	7
Cash Flow, $ millions	450	−40	200	200	200	200	200	200

2.55 The by-product department of Iowa Packing utilizes a cooker that has the cost stream shown. Determine the present worth in year 0 of the costs at an interest rate of 10% per year.

Year	Cost ($1000)
0	850
1	300
2	400
3	400
4	400
5	500

2.56 An entrepreneurial electrical engineer approached a large water utility with a proposal that promises to reduce the utility's power bill by at least 15% through installation of patented surge protectors. The proposal states that the engineer will not be paid for the first year, but beginning in year 2, she will receive three equal, annual payments that are equivalent to 60% of the power savings achieved in year 1 due to the protectors. Assuming that the utility's power bill of $1 million per year is reduced by 15% after installation of the surge protectors, what will be the future worth in year 4 of the uniform payments to the engineer? Use an interest rate of 10% per year.

2.57 Metropolitan Water utility is planning to upgrade its SCADA system for controlling well pumps, booster pumps, and disinfection equipment so that everything can be controlled from one site. The first phase will reduce labor and travel costs by $31,000 per year. The second phase will reduce costs by $20,000 per year. If phase I will occur in years 1 through 3 and phase II in years 4 through 8, what is the equivalent annual worth (years 1 through 8) of the upgraded system at an interest rate of 8% per year?

2.58 Infrared thermometers by Delta Thermal Products are compatible with type K thermocouples and can provide rapid non contact measurement capabilities at a cost of $135 per unit. A small private electric utility company plans to purchase 100 of the thermometers now and 500 more 1 year from now if the anticipated savings in labor costs are realized. What would the future worth of the savings have to be 4 years from now in order to justify the equipment purchases at an interest rate of 12% per year?

2.59 Encon Systems, Inc., sales revenues for its main product line are as shown. Calculate the equivalent annual worth (years 1–7) using an interest rate of 10% per year.

Year	Revenue, $	Year	Revenue, $
0	4,000,000	4	5,000,000
1	4,000,000	5	5,000,000
2	4,000,000	6	5,000,000
3	4,000,000	7	5,000,000

2.60 Cisco's *gross revenue* (the percentage of revenue left after subtracting the cost of goods sold) was 70.1% of total revenue over a certain 4-year period. If the *total revenue* per year was $5.8 billion for the first 2 years and $6.2 billion per year for the last 2, what was the future worth of the *gross revenue* in year 4 at an interest rate of 14% per year?

2.61 Calculate the annual worth in years 1 to 8 of the following series of incomes and expenses at an interest rate of 10% per year.

Year	Income, $ per year	Expense, $ per year
0		20,000,000
1–5	8,000,000	1,000,000
6–8	9,000,000	2,000,000

2.62 A supplier of certain suspension system parts for General Motors wants to have a contingency fund that it can draw on during down periods of the economy. The company wants to have $15 million in the fund 5 years from now. If the company deposits $1.5 million now, what uniform amount must it add at the end of each of the next 5 years to reach its goal if the fund earns a rate of return of 10% per year?

2.63 A rural utility company provides standby power to pumping stations using diesel-powered generators. An alternative has arisen whereby the utility could use a combination of wind and solar power to run its generators, but it will be a few years before the alternative energy systems are available. The utility estimates that the new systems will result in savings of $15,000 per year for 3 years, starting *2 years from now,* and $25,000 per year for 4 more years after that (i.e., through year 8). At an interest rate of 8% per year, determine the equivalent annual worth (years 1–8) of the projected savings.

2.64 A design-construct-operate (DSO) company signed a contract to operate certain industrial wastewater treatment plants for 10 years. The contract will pay the company $2.5 million now and amounts increasing by $250,000 each year through year 10. At an interest rate of 10% per year, what is the present worth (year 0) of the contract to the DSO company?

2.65 Expenses associated with heating and cooling a large manufacturing facility owned by Nippon Steel are expected to increase according to an arithmetic gradient. If the cost is $550,000 now (i.e., year 0) and the company expects the cost to increase by $40,000 each year through year 12, what is the equivalent annual worth in years 1–12 of the energy costs at an interest rate of 10% per year?

2.66 Lifetime Savings Accounts, known as LSAs, allow people to invest after-tax money without being taxed on any of the gains. If an engineer invests $10,000 now and then increases his deposit by $1000 each year through year 20, how much will be in the account immediately after the last deposit, if the account grows by 12% per year?

2.67 A software company that installs systems for inventory control using RFID technology spent $600,000 per year for the past 3 years in developing its latest product. The company wants to recover its investment in 5 years beginning now. If the company signed a contract that will pay $250,000 now and amounts increasing by a uniform amount each year through year 5, how much must the increase be each year, if the company uses an interest rate of 15% per year?

2.68 The future worth in year 8 for the cash flows shown in $20,000. At an interest rate of 10% per year, what is the value x of the cash flow in year 4?

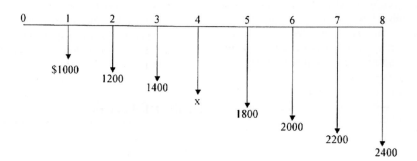

2.69 The annual worth (years 1 through 8) of the cash flows shown is $30,000. What is the amount x of the cash flow in year 3, if the interest rate is 10% per year?

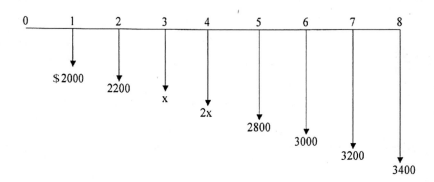

2.70 Levi Strauss has some of its jeans stone-washed under a contract with independent U.S. Garment Corp. If U.S. Garment's operating cost per machine is $22,000 per year for years 1 and 2 and then it increases by 8% per year through year 10, what is the equivalent uniform annual cost per machine (years 1–10) at an interest rate of 10% per year?

2.71 Burlington Northern is considering the elimination of a railroad grade crossing by constructing a dual-track overpass. The railroad subcontracts for maintenance of its crossing gates at $11,500 per year. Beginning 3 years from now, however, the costs are expected to increase by 10% per year into the foreseeable future (that is, $12,650 in year 3, $13,915 in year 4, etc.) If the railroad uses a 10-year study period and an interest rate of 15% per year, how much could the railroad afford to spend now on the overpass in lieu of the maintenance contracts?

2.72 Thunder Mountain Concrete and Building Materials is trying to bring the company-funded portion of its employee retirement fund into compliance with HB-301. The company has already deposited $200,000 in each of the last 5 years. If the company increases its deposits (beginning in year 6) by 15% per year through year 20, how much will be in the fund immediately after the last deposit, if the fund grows at a rate of 12% per year?

2.73 San Antonio is considering various options for providing water in its 50-year plan, including desalting. One brackish aquifer is expected to yield desalted water that will generate revenue of $4.1 million per year for the first 4 years, after which declining production will decrease revenue by 10% each year. If the aquifer will be totally depleted in 20 years, what is the present worth of the desalting option at an interest rate of 6% per year?

2.74 Revenue from gas wells that have been in production for at least 5 years tends to follow a decreasing geometric gradient. One particular rights holder received royalties of $4000 per year for years 1 through 6, but beginning in year 7, income decreased by 15% per year through year 14. What was the future value (year 14) of the income from the well, if all of the income was invested at 10% per year?

PROBLEMS FOR TEST REVIEW AND FE EXAM PRACTICE

2.75 An engineer conducting an economic analysis of wireless technology alternatives discovered that the A/F factor values were not in his copy of the tables. Therefore, he decided to create the $(A/F,i,n)$ factor values himself. He did so by
   a. multiplying the $(A/P,i,n)$ and $(P/F,i,n)$ values.
   b. dividing $(P/F,i,n)$ values by $(P/A,i,n)$ values.
   c. multiplying the $(P/A,i,n)$ and $(P/F,i,n)$ values.
   d. multiplying the $(F/A,i,n)$ and $(P/F,i,n)$ values.

2.76 The amount of money that would be accumulated in 12 years from an initial investment of $1000 at an interest rate of 8% per year is closest to
   a. $2376.
   b. $2490.
   c. $2518.
   d. $2643.

2.77 An executor for a wealthy person's estate discovered an uncollected note of $100,000 dated July 10, 1973. At an interest rate of 6% per year, the value of the note on July 10, 2008, would be closest to
   a. 684,060.
   b. $725,100.
   c. $768,610.
   d. $814,725.

2.78 An engineer planning for retirement decides that she wants to have income of $100,000 per year for 20 years with the first withdrawal beginning 30 years from now. If her retirement account earns interest at 8% per year, the annual amount she would have to deposit for 29 years beginning 1 year from now is closest to
   a. $7360.
   b. $8125.

   c. $8670.
   d. $9445.

2.79 A winner of the Texas State lottery was given two choices: receive a single lump sum payment *now* of $50 million *or* receive 21 uniform payments, with the first payment to be made *now,* and the rest to be made at the end of each of the next 20 years. At an interest rate of 4% per year, the amount of the 21 uniform payments that would be equivalent to the $50 million lump-sum payment would be closest to
   a. less than $3,400,000.
   b. $3,426,900.
   c. $3,623,600.
   d. $3,923,800.

2.80 A manufacturer of tygon tubing wants to have $3,000,000 available 10 years from now so that a new product line can be initiated. If the company plans to deposit money each year, starting *now,* the equation that represents how much it will have to deposit each year at 8% per year interest in order to have the $3,000,000 available immediately after the last deposit is made is
   a. $3,000,000(A/F,8\%,10)$
   b. $3,000,000(A/F,8\%,11)$
   c. $3,000,000 + 3,000,000(A/F,8\%,10)$
   d. $3,000,000 + 3,000,000(A/F,8\%,9)$

2.81 The maker of a motion-sensing towel dispenser is considering adding new products to enhance offerings in the area of touchless technology. If the company does not expand its product line now, it will definitely do so in 3 years. Assume the interest rate is 10% per year. The amount the company can afford to spend *now* if the cost 3 years from now is estimated to be $100,000 is

  a. $75,130.
  b. $82,640.
  c. $91,000.
  d. $93,280.

2.82 Assume you borrow $10,000 today and promise to repay the loan in two payments, one in year 2 and the other in year 5, with the one in year 5 being only half as large as the one in year 2. At an interest rate of 10% per year, the size of the payment in year 5 would be closest to

  a. less than $3900.
  b. $3975.
  c. $4398.
  d. $8796.

2.83 If you borrow $24,000 now at an interest rate of 10% per year and promise to repay the loan with payments of $3000 per year starting one year from now, the number of payments that you will have to make is closest to

  a. 8.
  b. 11.
  c. 14.
  d. 17.

2.84 You deposit $1000 now and you want the account to have a value as close to $8870 as possible in year 20. Assume the account earns interest at 10% per year. The year in which you must make another deposit of $1000 is

  a. 6.
  b. 8.
  c. 10.
  d. 12.

2.85 Levi Strauss has some of its jeans stone-washed under a contract with independent U.S. Garment Corp. U.S. Garment's operating cost per machine is $22,000 for year 1 and then it increases by $1000 per year through year 5. The equivalent uniform annual cost per machine (years 1–5) at an interest rate of 8% per year is closest to

  a. $23,850.
  b. $24,650.
  c. $25,930.
  d. over $26,000.

2.86 At an interest rate of 8% per year, the present worth in year 0 of a lease that requires a payment of $9,000 *now* and amounts increasing by 5% per year through year 7 is closest to

  a. $60,533.
  b. $65,376.
  c. $67,944.
  d. $69,328.

Nominal and Effective Interest Rates

Nick Koudis/Getty Images

In all engineering economy relations developed thus far, the interest rate has been a constant, annual value. For a substantial percentage of the projects evaluated by professional engineers in practice, the interest rate is compounded more frequently than once a year; frequencies such as semiannual, quarterly, and monthly are common. In fact, weekly, daily, and even continuous compounding may be experienced in some project evaluations. Also, in our own personal lives, many of the financial considerations we make—loans of all types (home mortgages, credit cards, automobiles, boats), checking and savings accounts, investments, stock option plans, etc.—have interest rates compounded for a time period shorter than 1 year. This requires the introduction of two new terms—nominal and effective interest rates. This chapter explains how to understand and use nominal and effective interest rates in engineering practice and in daily life situations.

Objectives

Purpose: Make economic calculations for interest rates and cash flows that occur on a time basis other than 1 year.

1. Understand nominal and effective interest rate statements.

2. Determine the effective interest rate for any time period.

3. Determine the correct *i* and *n* values for different payment and compounding periods.

4. Make equivalence calculations for various payment periods and compounding periods when only single amounts occur.

5. Make equivalence calculations when uniform or gradient series occur for payment periods equal to or longer than the compounding period.

6. Make equivalence calculations for payment periods shorter than the compounding period.

7. Use a spreadsheet to perform equivalency computations involving nominal and effective interest rates.

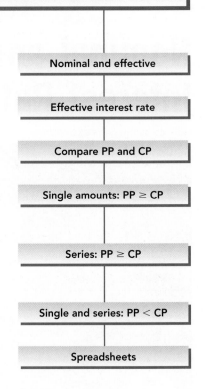

Nominal and effective

Effective interest rate

Compare PP and CP

Single amounts: PP $\geq$ CP

Series: PP $\geq$ CP

Single and series: PP < CP

Spreadsheets

3.1 NOMINAL AND EFFECTIVE INTEREST RATE STATEMENTS

In Chapter 1, we learned that the primary difference between simple interest and compound interest is that compound interest includes interest on the interest earned in the previous period, while simple interest does not. Here we discuss *nominal and effective interest rates,* which have the same basic relationship. The difference here is that the concepts of nominal and effective are used when interest is compounded more than once each year. For example, if an interest rate is expressed as 1% per month, the terms *nominal* and *effective* interest rates must be considered. Every nominal interest rate *must* be converted into an effective rate before it can be used in formulas, factor tables, or spreadsheet functions because they are all derived using effective rates.

Before discussing the conversion from nominal to effective rates, it is important to *identify* a stated rate as either nominal or effective. There are three general ways of expressing interest rates as shown by the three groups of statements in Table 3.1. The three statements in the top third of the table

TABLE 3.1   Various Interest Statements and Their Interpretations

(1) Interest Rate Statement	(2) Interpretation	(3) Comment
i = 12% per year	i = *effective* 12% per year compounded yearly	When no compounding period is given, interest rate is an effective rate, with compounding period assumed to be equal to stated time period.
i = 1% per month	i = *effective* 1% per month compounded monthly	
i = 3½% per quarter	i = *effective* 3½% per quarter compounded quarterly	
i = 8% per year, compounded monthly	i = *nominal* 8% per year compounded monthly	When compounding period is given without stating whether the interest rate is nominal or effective, it is assumed to be nominal. Compounding period is as stated.
i = 4% per quarter compounded monthly	i = *nominal* 4% per quarter compounded monthly	
i = 14% per year compounded semiannually	i = *nominal* 14% per year compounded semiannually	
i = effective 10% per year compounded monthly	i = *effective* 10% per year compounded monthly	If interest rate is stated as an effective rate, then it is an effective rate. If compounding period is not given, compounding period is assumed to coincide with stated time period.
i = effective 6% per quarter	i = *effective* 6% per quarter compounded quarterly	
i = effective 1% per month compounded daily	i = *effective* 1% per month compounded daily	

show that an interest rate can be stated over some designated time period without specifying the compounding period. Such interest rates are assumed to be effective rates with the *compounding period (CP)* the same as that of the stated interest rate.

For the interest statements presented in the middle of Table 3.1, three conditions prevail: (1) The compounding period is identified, (2) this compounding period is shorter than the time period over which the interest is stated, and (3) the interest rate is designated neither as nominal nor as effective. In such cases, the interest rate is assumed to be *nominal* and the compounding period is equal to that which is stated. (We learn how to get effective interest rates from these in the next section.)

For the third group of interest-rate statements in Table 3.1, the word *effective* precedes or follows the specified interest rate, and the compounding period is also given. These interest rates are obviously effective rates over the respective time periods stated.

The importance of being able to recognize whether a given interest rate is nominal or effective cannot be overstated with respect to the reader's understanding of the remainder of the material in this chapter and indeed the rest of the book. Table 3.2 contains a listing of several interest statements (column 1) along with their interpretations (columns 2 and 3).

TABLE 3.2   Specific Examples of Interest Statements and Interpretations

(1) Interest Rate Statement	(2) Nominal or Effective Interest	(3) Compounding Period
15% per year compounded monthly	Nominal	Monthly
15% per year	Effective	Yearly
Effective 15% per year compounded monthly	Effective	Monthly
20% per year compounded quarterly	Nominal	Quarterly
Nominal 2% per month compounded weekly	Nominal	Weekly
2% per month	Effective	Monthly
2% per month compounded monthly	Effective	Monthly
Effective 6% per quarter	Effective	Quarterly
Effective 2% per month compounded daily	Effective	Daily
1% per week compounded continuously	Nominal	Continuously

3.2 EFFECTIVE INTEREST RATE FORMULATION

Understanding effective interest rates requires a definition of a nominal interest rate r as the interest rate per period times the number of periods. In equation form,

$$r = \text{interest rate per period} \times \text{number of periods} \qquad \text{[3.1]}$$

A nominal interest rate can be found for any time period that is longer than the compounding period. For example, an interest rate of 1.5% per month can be expressed as a *nominal* 4.5% per quarter (1.5% per period $\times$ 3 periods), 9% per semiannual period, 18% per year, or 36% per 2 years. Nominal interest rates obviously neglect compounding.

The equation for converting a nominal interest rate into an effective interest rate is

$$i \text{ per period} = (1 + r/m)^m - 1 \qquad \text{[3.2]}$$

where i is the *effective* interest rate for a certain period, say six months, r is the *nominal* interest rate for that *period* (six months here), and m is the number of times interest is *compounded in that same period* (six months in this case). As was true for nominal interest rates, effective interest rates can be calculated for any time period longer than the compounding period of a given interest rate. The next example illustrates the use of Equations [3.1] and [3.2].

EXAMPLE 3.1

a. A Visa credit card issued through Chase Bank carries an interest rate of 1% per month on the unpaid balance. Calculate the effective rate per semiannual period.

b. If the card's interest rate is stated as 3.5% per quarter, find the effective semiannual and annual rates.

Solution

a. The compounding period is monthly. Since the effective interest rate per semiannual period is desired, the r in Equation [3.2] must be the nominal rate per 6 months.

$$r = 1\% \text{ per month} \times 6 \text{ months per semiannual period}$$
$$= 6\% \text{ per semiannual period}$$

The m in Equation [3.2] is equal to 6, since interest is compounded 6 times in 6 months. The effective semiannual rate is

$$i \text{ per 6 months} = \left(1 + \frac{0.06}{6}\right)^6 - 1$$
$$= 0.0615 \qquad (6.15\%)$$

b. For an interest rate of 3.5% per quarter, the compounding period is quarterly. In a semiannual period, $m = 2$ and $r = 7\%$.

$$i \text{ per 6 months} = \left(1 + \frac{0.07}{2}\right)^2 - 1$$

$$= 0.0712 \quad (7.12\%)$$

The effective interest rate per year is determined using $r = 14\%$ and $m = 4$.

$$i \text{ per year} = \left(1 + \frac{0.14}{4}\right)^4 - 1$$

$$= 0.1475 \quad (14.75\%)$$

Comment: Note that the term r/m in Equation [3.2] is always the effective interest rate per compounding period. In part (*a*) this is 1% per month, while in part (*b*) it is 3.5% per quarter.

If we allow compounding to occur more and more frequently, the compounding period becomes shorter and shorter. Then m, the number of compounding periods per payment period, increases. This situation occurs in businesses that have a very large number of cash flows every day, so it is correct to consider interest as compounded continuously. As m approaches infinity, the effective interest rate in Equation [3.2] reduces to

$$i = e^r - 1 \qquad\qquad [3.3]$$

Equation [3.3] is used to compute the *effective continuous interest rate*. The time periods on i and r must be the same. As an illustration, if the nominal annual $r = 15\%$ *per year,* the effective continuous rate *per year* is

$$i\% = e^{0.15} - 1 = 16.183\%$$

EXAMPLE 3.2   **a.** For an interest rate of 18% per year compounded continuously, calculate the effective monthly and annual interest rates.
   b. An investor requires an *effective* return of at least 15%. What is the minimum annual nominal rate that is acceptable for continuous compounding?

Solution

a. The nominal monthly rate is $r = 18\%/12 = 1.5\%$, or 0.015 per month. By Equation [3.3], the effective monthly rate is

$$i\% \text{ per month} = e^r - 1 = e^{0.015} - 1 = 1.511\%$$

Similarly, the effective annual rate using $r = 0.18$ per year is

$$i\% \text{ per year} = e^r - 1 = e^{0.18} - 1 = 19.72\%$$

TABLE 3.3 Effective Annual Interest Rates for Selected Nominal Rates

Nominal Rate r%	Semiannually ($m = 2$)	Quarterly ($m = 4$)	Monthly ($m = 12$)	Weekly ($m = 52$)	Daily ($m = 365$)	Continuously ($m = \infty; e^r - 1$)
1	1.003	1.004	1.005	1.005	1.005	1.005
2	2.010	2.015	2.018	2.020	2.020	2.020
3	3.023	3.034	3.042	3.044	3.045	3.046
4	4.040	4.060	4.074	4.079	4.081	4.081
5	5.063	5.095	5.116	5.124	5.126	5.127
6	6.090	6.136	6.168	6.180	6.180	6.184
7	7.123	7.186	7.229	7.246	7.247	7.251
8	8.160	8.243	8.300	8.322	8.328	8.329
9	9.203	9.308	9.381	9.409	9.417	9.417
10	10.250	10.381	10.471	10.506	10.516	10.517
12	12.360	12.551	12.683	12.734	12.745	12.750
15	15.563	15.865	16.076	16.158	16.177	16.183
18	18.810	19.252	19.562	19.684	19.714	19.722
20	21.000	21.551	21.939	22.093	22.132	22.140

b. Solve Equation [3.3] for r by taking the natural logarithm.

$$e^r - 1 = 0.15$$
$$e^r = 1.15$$
$$\ln e^r = \ln 1.15$$
$$r\% = 13.976\%$$

Therefore, a nominal rate of 13.976% per year compounded continuously will generate an effective 15% per year return.

Comment: The general formula to find the nominal rate, given the effective continuous rate i, is $r = \ln(1 + i)$

Table 3.3 summarizes the effective annual rates for frequently quoted nominal rates and various compounding frequencies.

3.3 RECONCILING COMPOUNDING PERIODS AND PAYMENT PERIODS

Now that the concepts of nominal and effective interest rates are introduced, in addition to considering the compounding period (which is also known as the interest period), it is necessary to consider the frequency of the payments of receipts

FIGURE 3.1 Cash-flow diagram for a monthly payment period (PP) and semiannual compounding period (CP).

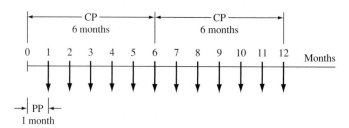

within the cash-flow time interval. For simplicity, the frequency of the payments or receipts is known as the *payment period (PP)*. It is important to distinguish between the compounding period (CP) and the payment period because in many instances the two do not coincide. For example, if a company deposited money each month into an account that pays a nominal interest rate of 6% per year compounded semi-annually, the payment period would be 1 month while the compounding period would be 6 months as shown in Figure 3.1. Similarly, if a person deposits money once each year into a savings account that compounds interest quarterly, the payment period is 1 year, while the compounding period is 3 months. Hereafter, for problems that involve either uniform-series or uniform-gradient cash-flow amounts, it will be necessary to determine the relationship between the compounding period and the payment period as a first step in the solution of the problem.

The next three sections describe procedures for determining the correct i and n values for use in formulas, factor tables, and spreadsheet functions. In general, there are three steps:

1. Compare the lengths of PP and CP.
2. Identify the cash-flow series as involving only single amounts (P and F) or series amounts (A, G, or g).
3. Select the proper i and n values.

3.4 EQUIVALENCE CALCULATIONS INVOLVING ONLY SINGLE-AMOUNT FACTORS

There are many correct combinations of i and n that can be used when only single-amount factors (F/P and P/F) are involved. This is because there are only two requirements: (1) An effective rate must be used for i, and (2) the time unit on n must be the same as that on i. In standard factor notation, the single-payment equations can be generalized.

$$P = F(P/F, \text{effective } i \text{ per period, number of periods})\qquad\text{[3.4]}$$

$$F = P(F/P, \text{effective } i \text{ per period, number of periods})\qquad\text{[3.5]}$$

Thus, for a nominal interest rate of 12% per year compounded monthly, any of the i and corresponding n values shown in Table 3.4 could be used (as well as many others not shown) in the factors. For example, if an effective quarterly

TABLE 3.4   **Various *i* and *n* Values for Single-Amount Equations Using *r* = 12% per Year, Compounded Monthly**

Effective Interest Rate, *i*	Units for *n*
1% per month	Months
3.03% per quarter	Quarters
6.15% per 6 months	Semiannual periods
12.68% per year	Years
26.97% per 2 years	2-year periods

interest rate is used for *i*, that is, $(1.01)^3 - 1 = 3.03\%$, then the *n* time unit is 4 quarters.

Alternatively, it is always correct to determine the effective *i* per payment period using Equation [3.2] and to use standard factor equations to calculate *P*, *F*, or *A*.

Sherry expects to deposit $1000 now, $3000 4 years from now, and $1500 6 years from now and earn at a rate of 12% per year compounded semiannually through a company-sponsored savings plan. What amount can she withdraw 10 years from now?   **EXAMPLE 3.3**

Solution

Only single-amount *P* and *F* values are involved (Figure 3.2). Since only effective rates can be present in the factors, use an effective rate of 6% per semiannual compounding period and semiannual payment periods. The future worth is calculated using Equation [3.5].

$$F = 1000(F/P,6\%,20) + 3000(F/P,6\%,12) + 1500(F/P,6\%,8)$$
$$= \$11.634$$

An alternative solution strategy is to find the effective annual rate by Equation [3.2] and express *n* in annual payment periods as stated in the problem.

$$i \text{ per year} = \left(1 + \frac{0.12}{2}\right)^2 - 1 = 0.1236 \qquad (12.36\%)$$

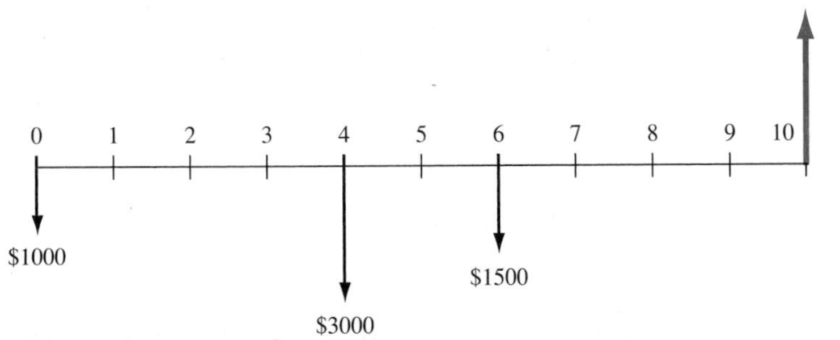

FIGURE 3.2   Cash flow diagram, Example 3.3

3.5 EQUIVALENCE CALCULATIONS INVOLVING SERIES WITH PP ≥ CP

When the cash flow of the problem dictates the use of one or more of the uniform-series or gradient factors, the relationship between the compounding period, CP, and payment period, PP, must be determined. The relationship will be one of the following three cases:

Type 1.   Payment period equals compounding period, PP = CP.
Type 2.   Payment period is longer than compounding period, PP > CP.
Type 3.   Payment period is shorter than compounding period, PP < CP.

The procedure for the first two cash flow types is the same. Type 3 problems are discussed in the following section. When PP = CP or PP > CP, the following procedure *always* applies:

Step 1.   Count the number of payments and use that number as *n*. For example, if payments are made quarterly for 5 years, *n* is 20.

Step 2.   Find the *effective* interest rate over the *same time period* as *n* in step 1. For example, if *n* is expressed in quarters, then the effective interest rate per quarter *must* be used.

Use these values of *n* and *i* (and only these!) in the standard factor notation or formulas. To illustrate, Table 3.5 shows the correct standard notation for sample cash-flow sequences and interest rates. Note in column 4 that *n* is always equal to the number of payments and *i* is an effective rate expressed over the same time period as *n*.

TABLE 3.5   Examples of *n* and *i* Values Where PP = CP or PP > CP

(1) Cash-flow Sequence	(2) Interest Rate	(3) What to Find; What is Given	(4) Standard Notation
$500 semiannually for 5 years	8% per year compounded semiannually	Find *P*; given *A*	$P = 500(P/A,4\%,10)$
$75 monthly for 3 years	12% per year compounded monthly	Find *F*; given *A*	$F = 75(F/A,1\%,36)$
$180 quarterly for 15 years	5% per quarter	Find *F*; given *A*	$F = 180(F/A,5\%,60)$
$25 per month increase for 4 years	1% per month	Find *P*; given *G*	$P = 25(P/G,1\%,48)$
$5000 per quarter for 6 years	1% per month	Find *A*; given *P*	$A = 5000(A/P,3.03\%,24)$

For the past 7 years, a quality manager has paid $500 every 6 months for the software maintenance contract of a LAN. What is the equivalent amount after the last payment, if these funds are taken from a pool that has been returning 10% per year compounded quarterly?

EXAMPLE 3.4

Pay 4 months
Comp Monthly
4/

Solution

The cash flow diagram is shown in Figure 3.3. The payment period (6 months) is longer than the compounding period (quarter); that is, PP > CP. Applying the guideline, determine an effective semiannual interest rate. Use Equation [3.2] or Table 3.3 with $r = 0.05$ per 6-month period and $m = 2$ quarters per semiannual period.

$$\text{Effective } i\% \text{ per 6-months} = \left(1 + \frac{0.05}{2}\right)^2 - 1 = 5.063$$

The value $i = 5.063\%$ is reasonable, since the effective rate should be slightly higher than the nominal rate of 5% per 6-month period. The number of semiannual periods is $n = 2(7) = 14$. The future worth is

$$F = A(F/A,5.063\%,14)$$
$$= 500(19.6845)$$
$$= \$9842$$

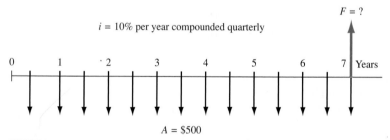

$i = 10\%$ per year compounded quarterly

$F = ?$

$A = \$500$

FIGURE 3.3 Diagram of semiannual payments used to determine F, Example 3.4.

Boeing has purchased composite wing fixtures for the assembly of its new Dreamliner Commercial airliner. Assume this system costs $3 million to install and an estimated $200,000 per year for all materials, operating, personnel, and maintenance costs. The expected life is 10 years. An engineer wants to estimate the total revenue requirement for each 6-month period that is necessary to recover the investment, interest, and annual costs. Find this semiannual A value if capital funds are evaluated at 8% per year compounded *semiannually*.

EXAMPLE 3.5

Solution

Figure 3.4 details the cash flows. There are several ways to solve this problem, but the most straightforward one is a two-stage approach. First, convert all cash

FIGURE 3.4

Cash flow diagram with different compounding and payment periods, Example 3.5.

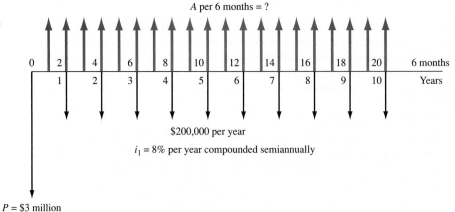

A per 6 months = ?

$200,000 per year

$i_1 = 8\%$ per year compounded semiannually

$P = \$3$ million

flows to a P at time 0, then find the A over the 20 semiannual periods. For stage 1, recognize that PP > CP, that is, 1 year > 6 months. According to the procedure for types 1 and 2 cash flows, $n = 10$, the number of annual payments. Now, find the effective i *per year* by Equation [3.2] or Table 3.3 and use it to find P.

$$i\% \text{ per year} = (1 + 0.08/2)^2 - 1 = 8.16\%$$

$$P = 3,000,000 + 200,000(P/A,8.16\%,10)$$
$$= 3,000,000 + 200,000(6.6620)$$
$$= \$4,332,400$$

For stage 2, P is converted to a semiannual A value. Now, PP = CP = 6 months, and $n = 20$ *semiannual* payments. The effective semiannual i for use in the A/P factor is determined directly from the problem statement using r/m.

$$i\% \text{ per 6 months} = 8\%/2 = 4\%$$

$$A = 4,332,400(A/P,4\%,20)$$
$$= \$318,778$$

In conclusion, \$318,778 every 6 months will repay the initial and annual costs, if money is worth 8% per year compounded semiannually.

3.6 EQUIVALENCE CALCULATIONS INVOLVING SERIES WITH PP < CP

If a person deposits money each *month* into a savings account where interest is compounded *quarterly,* do the so-called *interperiod deposits* earn interest? The usual answer is no. However, if a monthly payment on a \$10 million, quarterly compounded bank loan were made early by a large corporation, the corporate financial officer would likely insist that the bank reduce the amount of interest due, based on early payment. These two are type 3 examples of PP < CP. The timing of cash flow transactions between compounding points introduces the question of how *interperiod compounding* is handled. Fundamentally, there are two policies: interperiod cash flows earn *no interest*, or they earn *compound interest*. The only condition considered here is the first one (no interest), because most real-world transactions fall into this category.

For a no-interperiod-interest policy, deposits (negative cash flows) are all regarded as *deposited at the end of the compounding period,* and withdrawals are all regarded as *withdrawn at the beginning.* As an illustration, when interest is compounded quarterly, all monthly deposits are moved to the end of the quarter, and all withdrawals are moved to the beginning (no interest is paid for the entire quarter). This procedure can significantly alter the distribution of cash flows before the effective quarterly rate is applied to find P, F, or A. This effectively forces the cash flows into a PP = CP situation, as discussed in Section 3.5.

EXAMPLE 3.6

Rob is the on-site coordinating engineer for Alcoa Aluminum, where an under-renovation mine has new ore refining equipment being installed by a local contractor. Rob developed the cash flow diagram in Figure 3.5a in $1000 units from the project perspective. Included are payments to the contractor he has authorized for the current year and approved advances from Alcoa's home office. He knows that the interest rate on equipment "field projects" such as this is 12% per year compounded quarterly, and that Alcoa does not bother with interperiod compounding of interest. Will Rob's project finances be in the "red" or the "black" at the end of the year? By how much?

Solution

With no interperiod interest considered, Figure 3.5b reflects the moved cash flows. The future worth after four quarters requires an F at an effective rate per quarter such that PP = CP = 1 quarter, therefore, the effective $i = 12\%/4 = 3\%$. Figure 3.5b shows all negative cash flows (payments to contractor) moved to the end of the respective quarter, and all positive cash flows (receipts from home office) moved to the beginning of the respective quarter. Calculate the F value at 3%.

$$F = 1000[-150(F/P,3\%,4) - 200(F/P,3\%,3)$$
$$+ (-175 + 90)(F/P,3\%,2) + 165(F/P,3\%,1) - 50]$$
$$= \$-357,592$$

Rob can conclude that the on-site project finances will be in the red about $357,600 by the end of the year.

FIGURE 3.5
(*a*) Actual and
(*b*) moved cash
flows (in $1000)
for quarterly
compounding
periods using
no interperiod
interest,
Example 3.6.

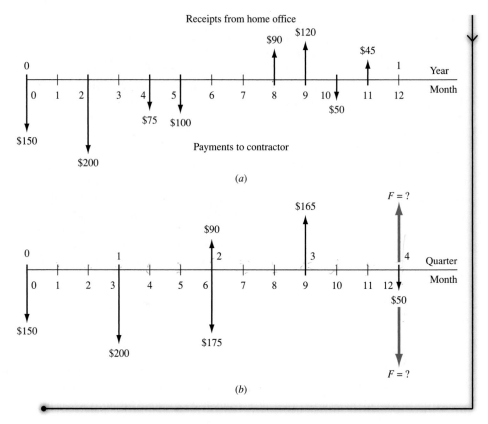

(*a*)

(*b*)

3.7 USING SPREADSHEETS FOR EFFECTIVE INTEREST RATE COMPUTATIONS

For hand calculations, converting between nominal and effective interest rates is accomplished with Equation [3.2]; for spreadsheets, the EFFECT or NOMINAL functions are applied as described below. Figure 3.6 gives two examples of each function.

Find effective rate: **EFFECT(nominal_rate, compounding frequency)**

As in Equation [3.2], the *nominal rate* is *r* and must be expressed over the same time period as that of the effective rate requested. The *compounding frequency* is *m*, which must equal the number of times interest is compounded for the period of time used in the effective rate. Therefore, in the second example of Figure 3.6 where the effective quarterly rate is requested, enter the nominal rate per quarter (3.75%) to get an effective rate per quarter, and enter *m* = 3, since monthly compounding occurs 3 times in a quarter.

Find nominal: **NOMINAL(effective_rate, compounding frequency per year)**

	A	B	C	D	E	F	G	H
1								
2	Example	What to find and what is given	Compounding frequency, m	Function	Function result			
3								
4	1	Find **effective** annual rate, given 15% nominal compounded quarterly	4	=EFFECT(15%,4)	15.87%		Must enter nominal rate	
5							per quarter, 15/4 = 3.75%	
6	2	Find **effective** quarterly rate, given 15% nominal compounded monthly	3	=EFFECT(3.75%,3)	3.80%			
7							Must divide result by 2	
8	3	Find **nominal** annual rate, given 15% effective compounded monthy	12	=NOMINAL(15%,12)	14.06%		to obtain semiannual	
9							nominal rate	
10	4	Find **nominal** semiannual rate, given 15% effective compounded monthly	12	=NOMINAL(15%,12)/2	7.03%			
11								
12								

FIGURE 3.6   Example uses of the EFFECT and NOMINAL functions to convert between nominal and effective interest rates.

This function always displays the *annual* nominal rate. Accordingly, the *m* entered must equal the number of times interest is compounded annually. If the nominal rate is needed for other than annually, use Equation [3.1] to calculate it. This is why the result of the NOMINAL function in Example 4 of Figure 3.6 is divided by 2.

If interest is compounded continuously, enter a very large value for the compounding frequency. A value of 10,000 or higher will provide sufficient accuracy. Say the nominal rate is 15% per year compounded continuously, Entering = EFFECT(15%,10000) displays 16.183%. The Equation [3.3] result is $e^{0.15} - 1 = 0.16183$, which is the same. The NOMINAL function is developed similarly to find the nominal rate per year, given the annual effective rate.

Once the effective interest rate is determined for the same timing as the cash flows, any function can be used, as illustrated in the next example.

Use a spreadsheet to find the semiannual cash flow requested in Example 3.5.   **EXAMPLE 3.7**

Solution

This problem is important to spreadsheet use because it involves annual and semiannual cash flows as well as nominal and effective interest rates. Review the solution approach of Example 3.5 before proceeding here. Since PP > CP, the *P* of annual cash flows requires an effective interest rate for the nominal 8% per year compounded semiannually. Use EFFECT to obtain 8.16% and the PV function to obtain $P = \$4,332,385$. The spreadsheet (left side) in Figure 3.7 shows these two results.

Now refer to the right side of Figure 3.7. Determine the requested semiannual repayment $A = \$-318,784$ based on the *P* in cell B5. For the PMT function, the semiannual rate 8/2 = 4% is entered. (Remember that effective rates must be used in all interest factors and spreadsheet functions. This 4% is the effective semiannual rate since interest is compounded semiannually.) Finally, note that the number of periods is 20 in the PMT function.

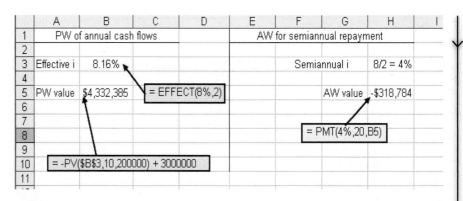

FIGURE 3.7   Use of EFFECT before applying spreadsheet functions to find *P* and *A* values over different time periods, Example 3.7.

SUMMARY

Since many real-world situations involve cash flow frequencies and compounding periods other than 1 year, it is necessary to use nominal and effective interest rates. When a nominal rate *r* is stated, the effective interest rate per payment period is determined by using the effective interest rate equation.

$$\text{Effective } i = \left(1 + \frac{r}{m}\right)^m - 1$$

The *m* is the number of compounding periods (CP) per payment period (PP). If interest compounding becomes more and more frequent, the length of the CP approaches zero, continuous compounding results, and the effective *i* is $e^r - 1$.

All engineering economy factors require the use of an effective interest rate. The *i* and *n* values placed in a factor depend upon the type of cash flow series. If only single amounts (*P* and *F*) are present, there are several ways to perform equivalence calculations using the factors. However, when series cash flows (*A*, *G*, and *g*) are present, only one combination of the effective rate *i* and number of periods *n* is correct for the factors. This requires that the relative lengths of PP and CP be considered as *i* and *n* are determined. *The interest rate and payment periods must have the same time unit for the factors to correctly account for the time value of money.*

PROBLEMS

Nominal and Effective Interest Rates

3.1 Simple interest is to nominal interest rate as compound interest is to what?

3.2 If a corporation deposits $20 million in a money market account for 1 year, what will be the difference in the total amount accumulated (i.e., *F*) at 18% per year *compound* interest versus 18% per year *simple* interest?

3.3 Identify the compounding period for the following interest statements: (*a*) 3% per quarter; (*b*) 10% per year compounded semiannually; (*c*) nominal 7.2% per year compounded monthly; (*d*) effective 3.4% per quarter compounded weekly; and (*e*) 2% per month compounded continuously.

3.4 Determine the number of times interest is compounded in 6 months for the following interest statements: (*a*) 3% per quarter; (*b*) 1% per month; and (*c*) 8% per year compounded semiannually.

3.5 For an interest rate of 12% per year compounded monthly, determine the number of times interest is compounded (*a*) per year; (*b*) per quarter; and (*c*) per 2 years.

3.6 For an interest rate of 4% per 6 months, determine the nominal interest rate per (*a*) semiannual period; (*b*) year; and (*c*) 2 years.

3.7 Identify the following interest rate statements as either nominal or effective: (*a*) 1.3% per month compounded weekly; (*b*) 0.5% per week compounded weekly; (*c*) effective 15% per year compounded monthly; (*d*) nominal 1.5% per month compounded daily; and (*e*) 15% per year compounded quarterly.

3.8 What effective interest rate per year is equivalent to 12% per year compounded quarterly?

3.9 An interest rate of 16% per year compounded quarterly is equivalent to what effective interest rate per year?

3.10 What nominal interest rate per year is equivalent to an effective 12% per year compounded quarterly?

3.11 What effective interest rate per year is equal to 1% per month compounded continuously?

3.12 What effective interest rate per quarter is equal to a nominal 2% per month compounded continuously?

3.13 What nominal rate per month is equivalent to an effective rate of 3.6% per quarter compounded continuously?

3.14 When interest is compounded monthly and single cash flows are separated by 3 years, what must the *time period* be on the interest rate if the value of *n* in the *P/F* or *F/P* equation is (*a*) *n* = 3, (*b*) *n* = 6, or (*c*) *n* = 12?

3.15 When interest is *compounded monthly* and a uniform series cash flow occurs over an *annual time period,* what time periods on *i* and *n* must be used?

3.16 When interest is *compounded quarterly* and an arithmetic gradient cash flow occurs over a *semiannual time period,* what time periods on *i* and *n* must be used?

Equivalence for Single Amounts

3.17 Beckton Steel Products, a company that specializes in crankshaft hardening, is investigating whether it should update equipment now or wait and do it later. If the cost 3 years from now is estimated to be $190,000, how much can the company afford to spend now if its minimum attractive rate of return is 2% per month?

3.18 Johnson Electronic Systems can reduce product recalls by 10% if it purchases new laser-based sensing equipment. If the cost of the new equipment is $250,000 now, how much can the company afford to spend in 2 years (instead of now) if it uses a minimum attractive rate of return of 12% per year compounded monthly?

3.19 Hydrex Mechanical Products is planning to set aside $160,000 now for possibly replacing its large synchronous refiner motors whenever it becomes necessary. If the replacement is expected to take place in 3-1/2 years, how much will the company have in its investment set-aside account. Assume a rate of return of 16% per year compounded quarterly?

3.20 What is the future worth 5 years from now of a present cost of $192,000 to Monsanto, Inc., at an interest rate of 1.5% per month?

3.21 Soil cleaning company Chemdex Partners plans to finance a site reclamation project that will require a 5-year cleanup period. If the company borrows $2.3 million now, how much will the company have to receive in a lump sum payment when the project is over in order to earn 20% per year compounded quarterly on its investment?

3.22 A present sum of $50,000 at an interest rate of 12% per year compounded quarterly is equivalent to how much money 8 years ago?

3.23 Maintenance costs for pollution control equipment on a pulverized coal cyclone furnace are expected to be $80,000 two years from now and another $90,000 four years from now. If Monongahela Power wants to set aside enough money now to cover these costs, how much must be invested at an interest rate of 12% per year compounded quarterly?

3.24 A plant expansion at a high-C steel company is expected to cost $14 million. How much money must the company set aside now in a lump sum investment in order to have the money in 2 years, if the investment earns interest at a rate of 14% per year compounded continuously?

3.25 Periodic outlays for inventory-control software at Baron Chemicals are expected to be $120,000 next year, $180,000 in 2 years, and $250,000 in 3 years. What is the present worth of the costs at an interest rate of 10% per year compounded continuously?

3.26 For the cash flow shown below, determine the future worth in year 5 at an interest rate of 10% per year compounded continuously.

Year	1	2	3	4	5
Cash Flow	$300,000	0	$250,000	0	$200,000

Equivalence when PP ≥ CP

3.27 A Pentsys lift table equipped with a powered deck extension can provide a safer and faster method for maintenance personnel to service large machinery. If labor savings are estimated to be $13,000 per 6 months, how much can Follansbee Steel afford to spend to purchase one, if the company uses a MARR of 1% per month and wants to recover its investment in 2-1/2 years?

3.28 Atlas Long-Haul Transportation is considering installing Valutemp temperature loggers in all of its refrigerated trucks for monitoring temperatures during transit. If the systems will reduce insurance claims by $40,000 per year for 5 years, how much should the company be willing to spend now if it uses an interest rate of 12% per year compounded quarterly?

3.29 According to U.S. Census Bureau data (2004), average monthly income for a person with a degree in engineering was $5296 versus $3443 for a degree in liberal arts. What is the future value in 40 years of the income difference at an interest rate of 6% per year compounded monthly?

3.30 Erbitux is a colorectal cancer treatment drug that is manufactured by ImClone Systems Inc. Assume treatment takes place over a 1 year period at a cost of $10,000 per month and the patient lives 5 years longer than he/she would have lived without the treatment. How much would that person have to earn each month beginning 1 month after treatment ends in order to earn an amount equivalent to the total treatment cost at an interest rate of 12% per year compounded monthly?

3.31 Plastics Engineering Inc., a developer of advanced belt drive components, recently began manufacturing sprockets made of polyacetal plastics that are approved for direct food contact. The company claims that high-precision manufacturing allows for greater belt life, eliminating costs associated with replacing belts. If Kraft Foods estimates that maintenance down time

costs the company an average of $12,000 per quarter, how much can the company afford to spend now if the new sprockets would cut their cost to $2000 per quarter? Assume Kraft Foods uses an interest rate of 12% per year compounded quarterly and a 2-year cost recovery period.

3.32 Linear actuators coupled to servo motors are expected to save a pneumatic valve manufacturer $19,000 per quarter through improved materials handling. If the savings begin *now,* what is the future worth of the savings through the end of year 3 if the company uses an interest rate of 12% per year compounded quarterly?

3.33 Entrex Exploration and Drilling bought $9 million worth of seamless tubes for its drill collars from Timken Roller Bearing, Inc. At an interest rate of 10% per year compounded semiannually, determine the annual worth of the purchase over a 10-year amortization period.

3.34 An environmental soil cleaning company got a contract to remove BTEX contamination from an oil company tank farm site. The contract required the soil cleaning company to provide quarterly invoices for materials and services provided. If the material costs were $140,000 per quarter and the service charges were calculated as 20% of the material costs, what is the present worth of the contract through the 3-year treatment period at an interest rate of 1% per month?

3.35 Redflex Traffic Systems manages red light camera systems that take photographs of vehicles that run red lights. Red light violations in El Paso, TX, result in fines of $75 per incident. A two-month trial period revealed that the police department could expect to issue 300 citations per month per intersection. If Redflex offered to install camera systems at 10 intersections, how much could the police department afford to spend on the project if it wanted to recover its investment in 2 years at an interest rate of 0.25% per month?

3.36 Setra's model 595 digitally compensated submersible pressure transducer can be used for many purposes, including flood warning. A lithium battery manufacturer is located in a low-lying area that is subject to flash flooding. An engineer for the company estimates that with early warning, equipment damage could be reduced by an average of $40,000. If a complete monitoring system (sensors, computers, installation, etc.) will cost

$25,000, how many months from now would the first flood have to occur in order to justify the system in one flood event? The company uses an interest rate of 1% per month.

3.37 Anderson-McKee Construction expects to invest $835,000 for heavy equipment replacement 2 years from now and another $1.1 million 4 years from now. How much must the company deposit into a sinking fund each month to provide for the purchases, if the fund earns a rate of return of 12% per year compounded monthly?

3.38 A Hilti PP11 pipe laser can be mounted inside or on top of a pipe, a tripod, or in a manhole to provide fast and accurate positioning and alignment of pipe sections. Jordon Construction is building a 23-mile pipeline through the desert to dispose of reverse osmosis concentrate by injection into a fractured-dolomite formation. The PP11 laser can reduce construction costs (through time savings) by $7500 per month. What will be the future worth of the savings for 2 years, if they start 1 month from now and the interest rate is 18% per year compounded monthly?

3.39 Under a program sponsored by the New Mexico Livestock Board, owners of horses and other livestock in Grant County can have chips inserted into their animals for $5 in vet fees and $30 for a permit. The state of New Mexico appropriated $840,000 to improve the program of animal inspection and movement control so that officials will be able to track and identify all animals and premises that have had contact with an animal disease within 48 hours of initial diagnosis. If the state receives the permit and vet fees from 500 animal owners per month, how long will it take for the state to recover its investment at an interest rate of 1.5% per quarter compounded monthly?

3.40 For the cash flow sequence described by $(10,000 - 100k)$, where k is in months, determine the annual worth of the cash flow in months 1 through 36 at an interest rate of 12% per year compounded continuously.

3.41 Magenn Power in Ottawa, Ontario, sells a rotating, bus-sized helium-filled electricity generator that floats 122 m above the ground. It sells for $10,000 (with helium) and will produce 4 kilowatts of electricity at its ground station. If the cost of electricity is $0.08 per kilowatt-hour, how many months must a purchaser use the system at full capacity to recover the initial investment at an interest rate

of 12% per year compounded monthly? Assume 30 days per month.

3.42 The cost of capturing and storing all CO_2 produced by coal-fired electricity generating plants during the next 200 years has been estimated to be $1.8 trillion. At an interest rate of 10% per year compounded monthly, what is the equivalent annual worth of such an undertaking?

3.43 For the cash flows shown, determine the present worth in year 0 at an interest rate of 12% per year compounded monthly.

Year	1	2	3	4	5
Cash Flow	$300,000	$275,000	$250,000	$225,000	$200,000

3.44 Income from sales of certain hardened steel connectors was $40,000 in the first quarter, $41,000 in the second, and amounts increasing by $1000 per quarter through year 4. What is the equivalent uniform amount per quarter if the interest rate is 12% per year compounded quarterly?

3.45 Atlas Moving and Storage wants to have enough money to purchase a new tractor-trailer in 4 years at a cost of $290,000. If the company sets aside $4000 in month 1 and plans to increase its set-aside by a uniform amount each month, how much must the increase be if the account earns 6% per year compounded monthly?

3.46 Frontier Airlines hedged the cost of jet fuel by purchasing options that allowed the airline to purchase fuel at a fixed price for 2 years. If the savings in fuel costs were $100,000 in month 1, $101,000 in month 2, and amounts increasing by 1% per month through the 2-year option period, what was the present worth of the savings at an interest rate of 18% per year compounded monthly?

3.47 Find the present worth of an investment that starts at $8000 in year 1 and increases by 10% each year through year 7, if the interest rate is 10% per year compounded continuously.

3.48 The initial cost of a pulverized coal cyclone furnace is $800,000. If the operating cost is $90,000 in year 1, $92,700 in year 2, and amounts increasing by 3% each year through a 10-year amortization period, what is the equivalent annual worth of the furnace at an interest rate of 1% per month?

3.49 Cheryl, a recent metallurgical engineering graduate, is happy because today she received notice of a raise of $100 per month. However, in the mail today she received her credit card bill, which shows a balance owed of $6697.44 at an interest rate of 13% per year compounded monthly (which equals the national average of an effective 13.8% per year). Cheryl's dad always told her that the effect of time and interest rate combined is something that can either work to your financial advantage or work against you very easily. Thinking about the longer term of the next 10 years, she has posed two options of what to do with the $100 per month.

Option 1: Use all $100 each month to pay off the credit card debt starting now.

Option 2: Start now to use $50 to pay off the debt and save $50 at the going rate of a nominal 6% per year compounded monthly.

Assuming no additional debt is accumulated on this credit card account, and that the interest rates are constant for all 10 years, determine the balances on the credit card and savings account for each option.

Equivalence When PP < CP

3.50 Copper Refiner Phelps Dodge purchased a model MTVS peristaltic pump for injecting antiscalant at its nanofiltration water conditioning plant. The cost of the pump was $950. If the chemical cost is $10 per day, determine the equivalent cost per month at an interest rate of 12% per year compounded monthly. Assume 30 days per month and a 3-year pump life.

3.51 A new way of leveraging intellectual property like patents and copyrights is to sell them at public auction. An offering that included patents in biotechnology, semiconductors, power solutions, and automotive systems resulted in 31 sales totaling $8.5 million. The cost of registering to bid was $1500. If a successful bidder paid $200,000 for a biotechnology patent, how much income must be generated *each month* to recover the investment in 3 years at an interest rate of 12% per year compounded semiannually?

3.52 How much money will be in a savings account at the end of 10 years from deposits of $1000 per month, if the account earns interest at a rate of 10% per year compounded semiannually?

3.53 Income from recycling the paper and cardboard generated in an office building has averaged $3000 per month for 3 years. What is the future worth of the income at an interest rate of 8% per year compounded quarterly?

3.54 Coal-fired power plants emit CO_2, which is one of the gases that is of concern with regard to global warming. A technique that power plants could adopt to keep most of the CO_2 from entering the air is called CO_2 capture and storage (CCS). If the incremental cost of the sequestration process is $0.019 per kilowatt-hour, what is the present worth of the extra cost over a 3-year period to a manufacturing plant that uses 100,000 kWh of power per month, if the interest rate is 12% per year compounded quarterly?

3.55 The Autocar E3 refuse truck has an energy recovery system developed by Parker Hannifin LLC that is expected to reduce fuel consumption by 50%. Pressurized fluid flows from carbon fiber–reinforced accumulator tanks to two hydrostatic motors that propel the vehicle forward. The truck recharges the accumulators when it brakes. If the fuel cost for a regular refuse truck is $800 per month, how much can a private waste hauling company afford to spend on the recovery system if it wants to recover its investment in 3 years at an interest rate of 12% per year?

PROBLEMS FOR TEST REVIEW AND FE EXAM PRACTICE

3.56 An interest rate obtained by multiplying an effective rate by a certain number of interest periods is the same as
   a. a simple rate.
   b. a nominal rate.
   c. an effective rate.
   d. either *a* or *b*.

3.57 An interest rate of a nominal 12% per year compounded weekly is
   a. an effective rate per year.
   b. an effective rate per week.
   c. a nominal rate per year.
   d. a nominal rate per week.

3.58 An interest rate of 1.5% per quarter is
 a. a nominal rate.
 b. a simple rate.
 c. an effective rate with an unknown compounding period.
 d. an effective rate with a known compounding period.

3.59 An interest rate of 1.5% per month compounded continuously is the same as
 a. an effective 1.5% per month.
 b. 4.5% per quarter compounded continuously.
 c. 6.0% per quarter compounded continuously.
 d. 9% per six months.

3.60 An effective 12.68% per year compounded monthly is closest to
 a. 12% per year.
 b. 12% per year compounded annually.
 c. 1% per month.
 d. 1% per month compounded annually.

3.61 In problems that involve an arithmetic gradient wherein the payment period is longer than the compounding period, the interest rate to use in the equations
 a. can be any effective rate, as long as the time units on i and n are the same.
 b. must be the interest rate that is exactly as stated in the problem.

 c. must be an effective interest rate that is expressed over a 1-year period.
 d. must be the effective interest rate that is expressed over the time period equal to the time where the first A value occurs.

3.62 An environmental testing company needs to purchase $40,000 worth of equipment 2 years from now. At an interest rate of 20% per year compounded quarterly, the present worth of the equipment is closest to
 a. $27,070.
 b. $27,800.
 c. $26,450.
 d. $28,220.

3.63 The cost of replacing part of a cell phone video chip production line in 6 years is estimated to be $500,000. At an interest rate of 14% per year compounded semiannually, the uniform amount that must be deposited into a sinking fund every 6 months is closest to
 a. $21,335.
 b. $24,825.
 c. $27,950.
 d. $97,995.

Present Worth Analysis

© Digital Vision/PunchStock

A future amount of money converted to its equivalent value now has a present worth (PW) that is always less than that of the actual cash flow, because for any interest rate greater than zero, all P/F factors have a value less than 1.0. For this reason, present worth values are often referred to as *discounted cash flows* (*DCF*). Similarly, the interest rate may be referred to as the *discount rate*. Besides PW, equivalent terms frequently used are present value (PV) and net present value (NPV). Up to this point, present worth computations have been made for one project. In this chapter, techniques for comparing two or more mutually exclusive alternatives by the present worth method are treated. Additionally, techniques that evaluate capitalized costs, life cycle costs, and independent projects are discussed.

Objectives

Purpose: Compare alternatives on a present worth basis.

Formulating alternatives	1. Identify mutually exclusive and independent projects, and define a revenue and a cost alternative.
PW of equal-life alternatives	2. Evaluate a single alternative and select the best of equal-life alternatives using present worth analysis.
PW of different-life alternatives	3. Select the best of different-life alternatives using present worth analysis.
Capitalized cost (CC)	4. Select the best alternative using capitalized cost calculations.
Independent projects	5. Select the best independent projects with and without a budget limit.
Spreadsheets	6. Use a spreadsheet to select an alternative by PW analysis.

4.1 FORMULATING ALTERNATIVES

Alternatives are developed from project proposals to accomplish a stated purpose. The logic of alternative formulation and evaluation is depicted in Figure 4.1. Some projects are economically and technologically viable, and others are not. Once the viable projects are defined, it is possible to formulate the alternatives.

Alternatives are one of two types: mutually exclusive or independent. Each type is evaluated differently.

- **Mutually exclusive (ME).** *Only one of the viable projects can be selected.* Each viable project *is* an alternative. If no alternative is economically justifiable, do nothing (DN) is the default selection.
- **Independent.** *More than one viable project may be selected* for investment. (There may be dependent projects requiring a particular project to be selected before another, and/or contingent projects where one project may be substituted for another.)

A mutually exclusive alternative selection is the most common type in engineering practice. It takes place, for example, when an engineer must select the one best diesel-powered engine from several competing models. Mutually exclusive alternatives are, therefore, the same as the viable projects; each one is evaluated, and the one best alternative is chosen. Mutually exclusive alternatives *compete with one another* in the evaluation. All the analysis techniques compare mutually exclusive alternatives. Present worth is discussed in the remainder of this chapter.

The *do-nothing (DN)* option is usually understood to be an alternative when the evaluation is performed. If it is absolutely required that one of the defined alternatives be selected, do nothing is not considered an option. (This may occur when a mandated function must be installed for safety, legal, or other purposes.) Selection of the DN alternative means that the current approach is maintained; no new costs, revenues, or savings are generated.

Independent projects are usually designed to accomplish different purposes, thus the possibility of selecting any number of the projects. These alternatives do not compete with one another; each project is evaluated separately, and the *comparison is* with the MARR. Independent project selection is treated in Section 4.5.

Finally, it is important to classify an *alternative's* cash flows as revenue-based or cost-based. All alternatives evaluated in one study must be of the same type.

- **Revenue.** *Each alternative generates cost and revenue cash flow estimates, and possibly savings,* which are treated like revenues. Revenues may be different for each alternative. These alternatives usually involve new systems, products, and services that require capital investment to generate revenues and/or savings. Purchasing new equipment to increase productivity and sales is a revenue alternative.
- **Cost.** *Each alternative has only cost cash flow estimates.* Revenues are assumed to be equal for all alternatives. These may be public sector (government) initiatives, or legally mandated or safety improvements.

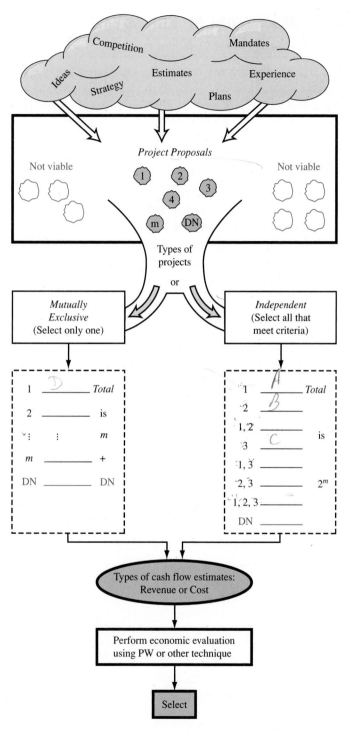

4.2 PRESENT WORTH ANALYSIS OF EQUAL-LIFE ALTERNATIVES

In present worth analysis, the *P* value, now called *PW*, is calculated at the MARR for each alternative. This converts all future cash flows into present dollar equivalents. This makes it easy to determine the economic advantage of one alternative over another.

The PW comparison of alternatives with equal lives is straightforward. If both alternatives are used in identical capacities for the same time period, they are termed *equal-service* alternatives.

For mutually exclusive alternatives the following guidelines are applied:

One alternative: Calculate PW at the MARR. If PW ≥ 0, the alternative is financially viable.

Two or more alternatives: Calculate the PW of each alternative at the MARR. *Select the alternative with the PW value that is numerically largest,* **that is, less negative or more positive.**

The second guideline uses the criterion of *numerically largest* to indicate a lower PW of costs only or larger PW of net cash flows. Numerically largest is *not the absolute value* because the sign matters here. The selections below correctly apply this guideline.

PW₁	PW₂	Selected Alternative
$-1500	$-500	2
-500	+1000	2
+2500	-500	1
+2500	+1500	1

EXAMPLE 4.1   Perform a present worth analysis of equal-service machines with the costs shown below, if the MARR is 10% per year. Revenues for all three alternatives are expected to be the same.

	Electric-Powered	Gas-Powered	Solar-Powered
First cost, $	-2500	-3500	-6000
Annual operating cost (AOC), $/year	-900	-700	-50
Salvage value, $	200	350	100
Life, years	5	5	5

Solution

These are cost alternatives. The salvage values are considered a "negative" cost, so a + sign precedes them. The PW of each machine is calculated at $i = 10\%$ for $n = 5$ years. Use subscripts E, G, and S.

$$PW_E = -2500 - 900(P/A,10\%,5) + 200(P/F,10\%,5) = \$-5788$$
$$PW_G = -3500 - 700(P/A,10\%,5) + 350(P/F,10\%,5) = \$-5936$$
$$PW_S = -6000 - 50(P/A,10\%,5) + 100(P/F,10\%,5) = \$-6127$$

The electric-powered machine is selected since the PW of its costs is the lowest; it has the numerically largest PW value.

Often a corporation or government obtains investment capital for projects by selling *bonds*. A good application of the PW method is the evaluation of a bond purchase alternative. If PW < 0 at the MARR, the do-nothing alternative is selected. A bond is like an IOU for time periods such as 5, 10, 20, or more years. Each bond has a *face value V* of \$100, \$1000, \$5000 or more that is fully returned to the purchaser when the bond maturity is reached. Additionally, bonds provide the purchaser with periodic *interest payments I* (also called bond dividends) using the *bond coupon* (or interest) *rate b*, and *c*, the number of payment periods per year.

$$I = \frac{(\text{bond face value})(\text{bond coupon rate})}{\text{number of payments per year}} = \frac{Vb}{c} \qquad \textbf{[4.1]}$$

$$y = \frac{V \cdot 4\%}{4}$$

At the time of purchase, the bond may sell for more or less than the face value, depending upon the financial reputation of the issuer. A purchase discount is more attractive financially to the purchaser; a premium is better for the issuer. For example, suppose a person is offered a 2% discount for an 8% \$10,000 20-year bond that pays the dividend quarterly. He will pay \$9800 now, and, according to Equation [4.1], he will receive quarterly dividends of $I = \$200$, plus the \$10,000 face value after 20 years.

To evaluate a proposed bond purchase, determine the PW at the MARR of all cash flows—initial payment and receipts of periodic dividends and the bond's face value at the maturity date. Then apply the guideline for one alternative, that is, if PW ≥ 0, the bond is financially viable. It is important to use the effective MARR rate in the PW relation that matches the time period of the payments. The simplest method is the procedure in Section 3.4 for PP = CP, as illustrated in the next example.

EXAMPLE 4.2

Marcie has some extra money that she wants to place into a relatively safe investment. Her employer is offering to employees a generous 5% discount for 10-year \$5,000 bonds that carry a coupon rate of 6% paid semiannually. The expectation is to match her return on other safe investments, which have averaged 6.7% per year compounded semiannually. (This is an effective rate of 6.81% per year.) Should she buy the bond?

Solution

Equation [4.1] results in a dividend of $I = (5000)(0.06)/2 = \$150$ every 6 months for a total of $n = 20$ dividend payments. The semiannual MARR is $6.7/2 = 3.35\%$, and the purchase price now is $-5000(0.95) = \$-4750$. Using PW evaluation,

$$PW = -4750 + 150\,(P/A,3.35\%,20) + 5000\,(P/F,3.35\%,20)$$
$$= \$-2.13$$

The effective rate is slightly less than 6.81% per year since PW < 0. If Marcie had to pay just \$2.13 less for the bond, she would meet her MARR goal. She should probably purchase the bond since the return is so close to her goal.

In order to speed up a PW analysis with Excel, the PV function is utilized. If all annual amounts for AOC are the same, the PW value for year 1 to n cash flows is found by entering the function $= P - \text{PV}(i\%, n, A, F)$. In Example 4.1, $PW_E = \$-5788$ is determined by entering $= -2500 - \text{PV}(10\%,5,-900,200)$ into any cell. Spreadsheet solutions are demonstrated in detail in Section 4.6.

4.3 PRESENT WORTH ANALYSIS OF DIFFERENT-LIFE ALTERNATIVES

Present worth analysis requires an *equal service* comparison of alternatives, that is, the number of years considered must be the same for all alternatives. *If equal service is not present, shorter-lived alternatives will be favored based on lower PW of total costs,* even though they may not be economically favorable. Fundamentally, there are two ways to use PW analysis to compare alternatives with unequal life estimates; evaluate over a specific study period (planning horizon), or use the least common multiple of lives for each pair of alternatives. In both cases, the PW is calculated at the MARR, and the selection guidelines of the previous section are applied.

4.3.1 Study Period

This is a commonly used approach. Once a study period is selected, only cash flows during this time frame are considered. If an expected life is *longer* than this period, the estimated market value of the alternative is used as a "salvage value" in the last year of the study period. If the expected life is *shorter* than the study period, cash flow estimates to continue equivalent service must be made for the time period between the end of the alternative's life and the end of the study period. In both cases, the result is an equal-service evaluation of the alternatives. As one example, assume a construction company wins a highway maintenance contract for 5 years, but plans to purchase specialized equipment expected to be operational for 10 years. For analysis purposes, the anticipated market value after 5 years is a salvage value in the PW equation, and any cash flows after year 5 are ignored. Example 4.3 *a* and *b* illustrate a study period analysis.

4.3.2 Least Common Multiple (LCM)

This approach can result in unrealistic assumptions since equal service comparison is achieved by assuming:

- The same service is needed for the LCM number of years. For example, the LCM of 5- and 9-year lives presumes the same need for 45 years!
- Cash flow estimates are initially expected to remain the same over each life cycle, which is correct *only* when changes in future cash flows exactly match the inflation or deflation rate.
- Each alternative is available for multiple life cycles, something that is usually not true.

Present worth analysis using the LCM method, as illustrated in Example 4.3, is correct, but it is not advocated. The same correct conclusion is easier to reach using each alternative's life and an annual worth (AW) computation as discussed in Chapter 5.

EXAMPLE 4.3

A project engineer with EnvironCare is assigned to start up a new office in a city where a 6-year contract has been finalized to collect and analyze ozone-level readings. Two lease options are available, each with a first cost, annual lease cost, and deposit-return estimates shown below. The MARR is 15% per year.

	Location A	Location B
First cost, $	−15,000	−18,000
Annual lease cost, $ per year	−3,500	−3,100
Deposit return, $	1,000	2,000
Lease term, years	6	9

a. EnvironCare has a practice of evaluating all projects over a 5-year period. If the deposit returns are not expected to change, which location should be selected?
b. Perform the analysis using an 8-year planning horizon.
c. Determine which lease option should be selected on the basis of a present worth comparison using the LCM.

Solution

a. For a 5-year study period, use the estimated deposit returns as positive cash flows in year 5.

$$PW_A = -15,000 - 3500(P/A,15\%,5) + 1000(P/F,15\%,5)$$
$$= \$-26,236$$
$$PW_B = -18,000 - 3100(P/A,15\%,5) + 2000(P/F,15\%,5)$$
$$= \$-27,397$$

Location A is the better economic choice.

b. For an 8-year study period, the deposit return for B remains at $2000 in year 8. For A, an estimate for equivalent service for the additional 2 years is needed. Assume this is expected to be relatively expensive at $6000 per year.

$$PW_A = -15,000 - 3500(P/A,15\%,6) + 1000(P/F,15\%,6)$$
$$- 6000(P/A,15\%,2)(P/F,15\%,6)$$
$$= \$-32,030$$
$$PW_B = -18,000 - 3100(P/A,15\%,8) + 2000(P/F,15\%,8)$$
$$= \$-31,257$$

Location B has an economic advantage for this longer study period.

c. Since the leases have different terms, compare them over the LCM of 18 years. For life cycles after the first, the first cost is repeated at the beginning (year 0) of each new cycle, which is the last year of the previous cycle. These are years 6 and 12 for location A and year 9 for B. The cash flow diagram is in Figure 4.2.

$$PW_A = -15,000 - 15,000(P/F,15\%,6) + 1000(P/F,15\%,6)$$
$$- 15,000(P/F,15\%,12) + 1000(P/F,15\%,12)$$
$$+ 1000(P/F,15\%,18) - 3500(P/A,15\%,18)$$
$$= \$-45,036$$
$$PW_B = -18,000 - 18,000(P/F,15\%,9) + 2000(P/F,15\%,9)$$
$$+ 2000(P/F,15\%,18) - 3100(P/A,15\%,18)$$
$$= \$-41,384$$

Location B is selected.

FIGURE 4.2

Cash flow diagram for different-life alternatives, Example 4.3c.

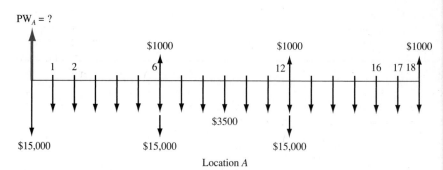

Location A

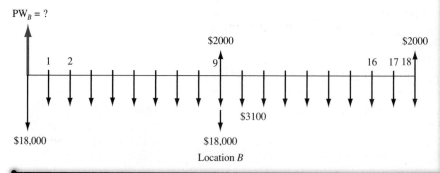

Location B

The future worth (FW) of an alternative may also be used to select an alternative. The FW is determined directly from the cash flows or by multiplying the PW value by the F/P factor at the established MARR. The n value in the F/P factor depends upon which time period has been used to determine PW—the LCM or a study period. Using FW values is especially applicable to large capital investment decisions when a prime goal is to maximize the *future wealth* of a corporation's stockholders. Alternatives such as electric generation facilities, toll roads, hotels, and the like can be analyzed using the FW value of investment commitments made during construction. Selection guidelines are the same as those for PW analysis.

Life-cycle cost (LCC) is another extension of present worth analysis. The LCC method, as its name implies, is commonly applied to alternatives with cost estimates over the entire *system life span*. This means that costs from the early stage of the project (needs assessment and design), through marketing, warranty, and operation phases and through the final stage (phaseout and disposal) are estimated. Typical applications for LCC are buildings (new construction or purchases), new product lines, manufacturing plants, commercial aircraft, new automobile models, defense systems, and the like.

A PW analysis with all definable costs (and possibly incomes) estimatable are considered in a LCC analysis. However, the broad definition of the LCC term *system life span* requires cost estimates not usually made for a regular PW analysis, such as design and development costs. *LCC is most effectively applied when a substantial percentage of the total costs over the system life span, relative to the initial investment, will be operating and maintenance costs* (postpurchase costs such as warranty, personnel, energy, upkeep, and materials). If Exxon-Mobil is evaluating the purchase of equipment for a large chemical processing plant for $150,000 with a 5-year life and annual costs of $15,000, LCC analysis is probably not justified. On the other hand, suppose Toyota is considering the design, construction, marketing, and after-delivery costs for a new automobile model. If the total startup cost is estimated at $125 million (over 3 years) and total annual costs are expected to be 20% of this figure to build, market, and service the cars for the next 15 years (estimated life span of the model), then the logic of LCC analysis will help the decision makers understand the profile of costs and their economic consequences using PW, FW, or AW analysis. LCC is required for most defense and aerospace industries, where the approach may be called Design to Cost (see Section 11.1). LCC is usually not applied to public sector projects, because the benefits and costs are difficult to estimate with much accuracy. Benefit/cost analysis is better applied here, as discussed in Chapter 7.

4.4 CAPITALIZED COST ANALYSIS

Capitalized cost (CC) is the present worth of an alternative that will last "forever." Public sector projects such as bridges, dams, irrigation systems, and railroads fall into this category, since they have useful lives of 30, 40, and more years. In addition, permanent and charitable organization endowments are evaluated using capitalized cost.

The formula to calculate CC is derived from the relation $PW = A(P/A,i,n)$, where $n = \infty$. The equation can be written

$$PW = A\left[\frac{1 - \dfrac{1}{(1 + i)^n}}{i}\right]$$

As n approaches ∞, the bracketed term becomes $1/i$. The symbol CC replaces PW, and AW (annual worth) replaces A to yield

$$CC = \frac{A}{i} = \frac{AW}{i} \qquad\qquad \textbf{[4.2]}$$

Equation [4.2] is illustrated by considering the time value of money. If $10,000 earns 10% per year, the interest earned at the end of every year for *eternity* is $1000. This leaves the original $10,000 intact to earn more next year. In general, the equivalent A value from Equation [4.2] for an infinite number of periods is

$$A = CC(i) \qquad\qquad \textbf{[4.3]}$$

The cash flows (costs or receipts) in a capitalized cost calculation are usually of two types: *recurring,* also called periodic, and *nonrecurring,* also called one-time. An annual operating cost of $50,000 and a rework cost estimated at $40,000 every 12 years are examples of recurring cash flows. Examples of nonrecurring cash flows are the initial investment amount in year 0 and one-time cash flow estimates, for example, $500,000 in royalty fees 2 years hence. The following procedure assists in calculating the CC for an infinite sequence of cash flows.

1. Draw a cash flow diagram showing all nonrecurring and at least two cycles of all recurring cash flows. (Drawing the cash flow diagram is more important in CC calculations than elsewhere.)
2. Find the present worth of all nonrecurring amounts. This is their CC value.
3. Find the equivalent uniform annual worth (A value) through *one life cycle* of all recurring amounts. This is the same value in all succeeding life cycles. Add this to all other uniform amounts occurring in years 1 through infinity and the result is the total equivalent uniform annual worth (AW).
4. Divide the AW obtained in step 3 by the interest rate i to obtain a CC value. This is an application of Equation [4.2].
5. Add the CC values obtained in steps 2 and 4.

EXAMPLE 4.4   The property appraisal district for Marin County has just installed new software to track residential market values for property tax computations. The manager wants to know the total equivalent cost of all future costs incurred when the ⌄

three county judges agreed to purchase the software system. If the new system will be used for the indefinite future, find the equivalent value (*a*) now and (*b*) for each year hereafter.

The system has an installed cost of $150,000 and an additional cost of $50,000 after 10 years. The annual software maintenance contract cost is $5000 for the first 4 years and $8000 thereafter. In addition, there is expected to be a recurring major upgrade cost of $15,000 every 13 years. Assume that $i = 5\%$ per year for county funds.

Solution

a. The detailed procedure is applied.

1. Draw a cash flow diagram for two cycles (Figure 4.3).
2. Find the present worth of the nonrecurring costs of $150,000 now and $50,000 in year 10 at $i = 5\%$. Label this CC_1.

$$CC_1 = -150,000 - 50,000(P/F,5\%,10) = \$-180,695$$

3. Convert the recurring cost of $15,000 every 13 years into an annual worth A_1 for the first 13 years.

$$A_1 = -15,000(A/F,5\%,13) = \$-847$$

The same value, $A_1 = \$-847$, applies to all the other 13-year periods as well.

4. The capitalized cost for the two annual maintenance cost series may be determined in either of two ways: (1) consider a series of $-5000 from now to infinity plus a series of $-3000 from year 5 on; or (2) a series of $-5000 for 4 years followed by a series of $-8000 from year 5 to infinity. Using the first method, the annual cost (A_2) is $-5000 forever.

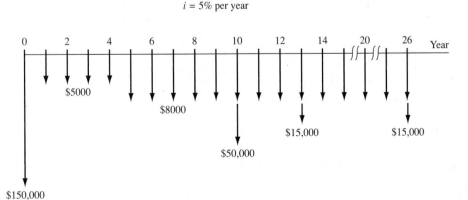

$i = 5\%$ per year

FIGURE 4.3 Cash flows for two cycles of recurring costs and all nonrecurring amounts, Example 4.4.

The capitalized cost CC_2 of $\$-3000$ from year 5 to infinity is found using Equation [4.2] times the P/F factor.

$$CC_2 = \frac{-3000}{0.05}(P/F,5\%,4) = \$-49,362$$

The CC_2 value is calculated using $n = 4$ because the present worth of the annual \$3000 cost is located in year 4, one period ahead of the first A. The two annual cost series are converted into a capitalized cost CC_3.

$$CC_3 = \frac{A_1 + A_2}{i} = \frac{-847 + (-5000)}{0.05} = \$-116,940$$

5. The total capitalized cost CC_T is obtained by adding the three CC values.

$$CC_T = -180,695 - 49,362 - 116,940 = \$-346,997$$

b. Equation [4.3] determines the A value forever.

$$A = CC_T(i) = \$-346,997(0.05) = \$-17,350$$

Correctly interpreted, this means Marin County officials have committed the equivalent of \$17,350 forever to operate and maintain the property appraisal software.

The CC evaluation of two or more alternatives compares them for the same number of years—infinity. The alternative with the smaller capitalized cost is the more economical one.

EXAMPLE 4.5 Two sites are currently under consideration for a bridge over a small river. The north site requires a suspension bridge. The south site has a much shorter span, allowing for a truss bridge, but it would require new road construction.

The suspension bridge will cost \$500 million with annual inspection and maintenance costs of \$350,000. In addition, the concrete deck would have to be resurfaced every 10 years at a cost of \$1,000,000. The truss bridge and approach roads are expected to cost \$250 million and have annual maintenance costs of \$200,000. This bridge would have to be painted every 3 years at a cost of \$400,000. In addition, the bridge would have to be sandblasted every 10 years at a cost of \$1,900,000. The cost of purchasing right-of-way is expected to be \$20 million for the suspension bridge and \$150 million for the truss bridge. Compare the alternatives on the basis of their capitalized cost if the interest rate is 6% per year.

Solution

Construct the cash flow diagrams over two cycles (20 years).
Capitalized cost of suspension bridge (CC_S):

CC_1 = capitalized cost of initial cost
= $-500 - 20 = \$-520$ million

The recurring operating cost is $A_1 = \$-350,000$, and the annual equivalent of the resurface cost is

$$A_2 = -1,000,000 \, (A/F, 6\%, 10) = \$-75,870$$

$$CC_2 = \text{capitalized cost of recurring costs} = \frac{A_1 + A_2}{i}$$

$$= \frac{-350,000 + (-75,870)}{0.06} = \$-7,097,833$$

The total capitalized cost is

$$CC_S = CC_1 + CC_2 = \$-527.1 \text{ million}$$

Capitalized cost of truss bridge (CC_T):

$CC_1 = -250 + (-150) = \$-400$ million
$A_1 = \$-200,000$
A_2 = annual cost of painting = $-400,000(A/F,6\%,3) = \$-125,644$
A_3 = annual cost of sandblasting = $-1,900,000(A/F,6\%,10) = \$-144,153$

$$CC_2 = \frac{A_1 + A_2 + A_3}{i} = \frac{\$-469,797}{0.06} = \$-7,829,950$$

$$CC_T = CC_1 + CC_2 = \$-407.83 \text{ million}$$

Conclusion: Build the truss bridge, since its capitalized cost is lower by $119 million.

If a finite-life alternative (e.g., 5 years) is compared to one with an indefinite or very long life, capitalized costs can be used. To determine CC for the finite life alternative, calculate the A value for *one life cycle* and divide by the interest rate (Equation [4.2]).

EXAMPLE 4.6

APSco, a large electronics subcontractor for the Air Force, needs to immediately acquire 10 soldering machines with specially prepared jigs for assembling components onto circuit boards. More machines may be needed in the future. The lead production engineer has outlined two simplified, but viable, alternatives. The company's MARR is 15% per year and capitalized cost is the evaluation technique.

Alternative LT (long-term). For $8 million now, a contractor will provide the necessary number of machines (up to a maximum of 20), now and in the future, for as long as APSco needs them. The annual contract fee is a total of $25,000 with no additional per-machine annual cost. There is no time limit placed on the contract, and the costs do not escalate.

Alternative ST (short-term). APSco buys its own machines for $275,000 each and expends an estimated $12,000 per machine in annual operating cost (AOC). The useful life of a soldering system is 5 years.

Solution

For the LT alternative, find the CC of the AOC using Equation [4.2]. Add this amount to the initial contract fee, which is already a capitalized cost.

$$CC_{LT} = CC \text{ of contract fee} + CC \text{ of AOC}$$
$$= -8 \text{ million} - 25{,}000/0.15 = \$-8{,}166{,}667$$

For the ST alternative, first calculate the equivalent annual amount for the purchase cost over the 5-year life, and add the AOC values for all 10 machines. Then determine the total CC using Equation [4.2].

$$AW_{ST} = AW \text{ for purchase} + AOC$$
$$= -2.75 \text{ million}(A/P,15\%,5) - 120{,}000 = \$-940{,}380$$
$$CC_{ST} = -940{,}380/0.15 = \$-6{,}269{,}200$$

The ST alternative has a lower capitalized cost by approximately $1.9 million present value dollars.

4.5 EVALUATION OF INDEPENDENT PROJECTS

Consider a biomedical company that has a new genetics engineering product that it can market in three different countries (S, U, and R), including any combination of the three. The do nothing (DN) alternative is also an option. All possible options are: S, U, R, SU, SR, UR, SUR, and DN. In general, for m independent projects, there are 2^m alternatives to evaluate. Selection from independent projects uses a fundamentally different approach from that for mutually exclusive (ME) alternatives. When selecting independent projects, each project's PW is compared with the MARR. (In ME alternative evaluation, the projects compete with each other, and only one is selected.) The selection rule is quite simple for one or more *independent* projects:

Select all projects that have PW ≥ 0 at the MARR.

All projects must be developed to have revenue cash flows (not costs only) so that projects earning more than the MARR have positive PW values.

Unlike ME alternative evaluation, which assumes the need for the service over multiple life cycles, independent projects are considered one-time investments. This means the PW analysis is performed over the respective life of each project and the assumption is made that any leftover cash flows earn at the MARR when the project ends. As a result, the equal service requirement does not impose the use of a specified study period or the LCM method. The implied study period is that of the longest lived project.

There are two types of selection environments—unlimited and budget constrained.

- **Unlimited.** All projects that make or exceed the MARR are selected. Selection is made using the PW $\geq$ 0 guideline.
- **Budget constrained.** No more than a specified amount, b, of funds can be invested in all of the selected projects, and each project must make or exceed the MARR. Now the solution methodology is slightly more complex in that *bundles* of projects that do not exceed the investment limit b are the only ones evaluated using PW values. The procedure is:
  1. Determine all bundles that have total initial investments no more than b. (This limit usually applies in year 0 to get the project started).
  2. Find the PW value at the MARR for all projects contained in the bundles.
  3. Total the PW values for each bundle in (1).
  4. Select the bundle with the largest PW value.

EXAMPLE 4.7

Marshall Aqua Technologies has four separate projects it can pursue over the next several years. The required amounts to start each project (initial investments) now and the anticipated cash flows over the expected lives are estimated by the Project Engineering Department. At MARR = 15% per year, determine which projects should be pursued if initial funding is (*a*) not limited, and (*b*) limited to no more than $15,000.

Project	Initial Investment	Annual Net Cash Flow	Life, Years
F	$ −8,000	$3870	6
G	−15,000	2930	9
H	−6,000	2080	5
J	−10,000	5060	3

Solution

a. Determine the PW value for each independent project and select all with PW $\geq$ 0 at 15%.

$$PW_F = -8000 + 3870(P/A,15\%,6) = \$6646$$
$$PW_G = -15,000 + 2930(P/A,15\%,9) = \$-1019$$
$$PW_H = -6000 + 2080(P/A,15\%,5) = \$973$$
$$PW_J = -10,000 + 5060(P/A,15\%,3) = \$1553$$

Select the three projects F, H, and J for a total investment of $24,000.

TABLE 4.1   **Present Worth Analysis of Independent Projects with Investment Limited to $15,000, Example 4.7**

Bundle	Projects	Total Initial Investment	PW of Bundle at 15%
1	F	$ −8,000	$ 6646
2	G	−15,000	−1019
3	H	−6,000	973
4	J	−10,000	1553
5	FH	−14,000	7619
6	Do-nothing	0	0

b. Use the steps for a budget-constrained selection with b = $15,000.

1 and **2.** Of the $2^4 = 16$ possible bundles, Table 4.1 indicates that 6 are acceptable. These bundles involve all four projects plus do-nothing with $PW_{DN} = \$0$.

3. The PW value for a bundle is obtained by adding the respective project PW values. For example, $PW_5 = 6646 + 973 = \$7619$.

4. Select projects F and H, since their PW is the largest and both projects exceed the MARR, as indicated by PW > 0 at $i = 15\%$.

Comment: Budget-constrained selection from independent projects is commonly called the capital rationing or capital budgeting problem. It may be worked efficiently using a variety of techniques, one being the integer linear programming technique. Excel and its optimizing tool SOLVER handle this type of problem rather nicely.

4.6 USING SPREADSHEETS FOR PW ANALYSIS

Spreadsheet-based evaluation of equal-life, mutually exclusive alternatives can be performed using the single-cell PV function when the annual amount A is the same. The general format to determine the PW is

$$= P - PV(i\%,n,A,F) \qquad \textbf{[4.4]}$$

It is important to pay attention to the sign placed on the PV function in order to get the correct answer for the alternative's PW value. Excel returns the opposite sign of the A series because it interprets cash flows in the manner explained in Chapter 1, that is, costs are negative and the PV function value is a positive equivalent at time 0. Therefore, to retain the negative sense of a cost series A, place a minus sign immediately in front of the PV function. This is illustrated in the next example.

Cesar, a petroleum engineer, has identified two equivalent diesel-powered genera- **EXAMPLE 4.8**
tors to be purchased for an offshore platform. Use $i = 12\%$ per year to determine
which is the more economic.

	Generator 1	Generator 2
P, $	−80,000	−120,000
S, $	15,000	40,000
n, years	3	3
AOC, $/year	−30,000	−8,000

Solution

Follow the format in Equation [4.4] in a single cell for each alternative. Figure
4.4 cell tags show the details. Note the use of minus signs on P, the PV function, and AOC value. Generator 2 is selected with the smaller PW of costs
(numerically larger value).

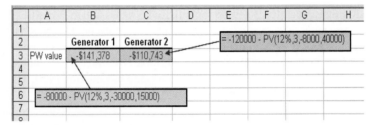

FIGURE 4.4   Equal-life alternatives evaluated using the PV function,
Example 4.8.

When different-life alternatives are evaluated, using the LCM basis, it is necessary to input all the cash flows for the LCM of the lives to ensure an equal-service
evaluation. Develop the NPV function to find PW. If cash flow is identified by CF,
the general format is

$$= P + \text{NPV}(i\%,\text{year\_1\_CF\_cell:last\_year\_CF\_cell}) \qquad [4.5]$$

It is very important that the *initial cost P not be included* in the cash flow series
identified in the NPV function. Unlike the PV function, the NPV function returns
the correct sign for the PW value.

Continuing with the previous example, once Cesar had selected generator 2 to **EXAMPLE 4.9**
purchase, he approached the manufacturer with the concerns that the first cost
was too high and the expected life was too short. He was offered a lease arrangement for 6 years with a $20,000 annual cost and an extra $20,000 payment in
the first and last years to cover installation and removal costs. Determine if
generator 2 or the lease arrangement is better at 12% per year.

	A	B	C	D	E	F	G
1							
2	Year	Generator 2	Lease				
3	0	-120,000	-40,000				
4	1	-8,000	-20,000				
5	2	-8,000	-20,000				
6	3	-88,000	-20,000				
7	4	-8,000	-20,000				
8	5	-8,000	-20,000				
9	6	32,000	-40,000				
10	PW value	-$189,568	-$132,361				
11							
12							
13							
14							
15							

= -40000 + NPV(12%,C4:C9)

= 40000 - 8000 - 120000

= -120000 + NPV(12%,B4:B9)

FIGURE 4.5   Different-life alternatives evaluated using the NPV function, Example 4.9.

Solution

Assuming that generator 2 can be repurchased 3 years hence and all estimates remain the same, PW evaluation over 6 years is correct. Figure 4.5 details the cash flows and NPV functions. The year 3 cash flow for generator 2 is $S - AOC - P = \$-88,000$. Note that the first costs are not included in the NPV function but are listed separately, as indicated in Equation [4.5]. The lease option is the clear winner for the next 6 years.

When evaluating alternatives in which the annual cash flows do not form an A series, the individual amounts must be entered on the spreadsheet and Equation [4.5] is used to find PW values. Also, remember that any zero-cash-flow year must be entered as 0 to ensure that the NPV function correctly tracks the years.

SUMMARY

This chapter explained the difference between mutually exclusive and independent alternatives, as well as revenue and cost cash flows. It discussed the use of present worth (PW) analysis to select the economically best alternative. In general, always choose an alternative with the largest PW value calculated at the MARR.

Important points to remember for mutually exclusive alternatives selection are:

1. Compare equal service alternatives over the same number of years using a specified study period. Alternatives with lives shortened by the study period have as a "salvage value" the estimated market value. For longer lives, an equivalent-

service cost must be estimated through the end of the study period.

2. PW evaluation over the least common multiple (LCM) of lives can be used to obtain equal service.

3. Capitalized cost (CC) analysis, an application of PW analysis with $n = \infty$, compares alternatives with indefinite or very long lives. In short, the CC value is determined by dividing the equivalent A value of all cash flows by the interest rate i.

For independent projects, select all with PW ≥ 0 at the MARR if there is no budget limitation. When funds are limited, form bundles of projects that do not exceed the limit and select the bundle that maximizes PW.

PROBLEMS

Alternative Formulation

4.1 Why is it necessary to consider do-nothing as a viable alternative when evaluating mutually exclusive alternatives or independent projects?

4.2 A biomedical engineer with Johnston Implants just received estimates for replacement equipment to deliver online selected diagnostic results to doctors performing surgery who need immediate information on the patient's condition. The cost is $200,000, the annual maintenance contract costs $5000, and the useful life (technologically) is 5 years.

a. What is the alternative if this equipment is not selected? What other information is necessary to perform an economic evaluation of the two?

b. What type of cash flow series will these estimates form?

c. What additional information is needed to convert the cash flow estimates to the other type?

4.3 When performing an engineering economy evaluation, only one of several mutually exclusive alternatives is selected, whereas any number of independent projects can be selected. Explain the fundamental difference between mutually exclusive and independent projects that makes these selection rules correct.

4.4 The lead engineer at Bell Aerospace has $1 million in research funds to commit this year. She is considering five separate R&D projects, identified as A through E. Upon examination, she determines that three of these projects (A, B, and C) accomplish exactly the same objective using different techniques.

a. Identify each project as mutually exclusive or independent.

b. If the selected alternative between A, B, and C is labeled X, list all viable options (bundles) for the five projects.

4.5 List all possible bundles for the four independent projects 1, 2, 3, and 4. Projects 3 and 4 cannot both be included in the same bundle.

Alternative Evaluation—Equal Lives

4.6 Johnston Implants is planning new online patient diagnostics for surgeons while they operate. The new system will cost $200,000 to install in an operating room, $5000 annually for maintenance, and have an expected life of 5 years. The revenue per system is estimated to be $40,000 in year 1 and to increase by $10,000 per year through year 5. Determine if the project is economically justified using PW analysis and an MARR of 10% per year.

4.7 An engineer and her husband operate a pet sitting service on the side. They want to add a daily service of a photo placed online for pet owners who are traveling. The estimates are: equipment and setup cost $600, and net monthly income over costs $30. For a period of 2 years, will the service make at least 12% per year compounded monthly?

4.8 The CFO of Marta Aaraña Cement Industries knows that many of the diesel-fueled systems in its quarries must be replaced at an estimated cost of $20 million 10 years from now. A fund for these replacements has been established with the commitment of $1 million at the end of next year (year 1) with 10% increases through the 10th year. If the fund earns at 5.25% per year, will the company have enough to pay for the replacements?

4.9 Burling Water Cooperative currently contracts the removal of small amounts of hydrogen sulfide from its well water using manganese dioxide filtration prior to the addition of chlorine and fluoride. Contract renewal for 5 years will cost $75,000 annually for the next 3 years and $100,000 in years 4 and 5. Assume payment is made at the end of each contract year. Burling Coop can install the filtration equipment for $150,000 and perform the process for $60,000 per year. At a discount rate of 6% per year, does the contract service still save money?

4.10 Jamal bought a 5% $1000 20-year bond for $825. He received a semiannual dividend for 8 years, then sold it immediately after the 16th dividend for $800. Did Jamal make the return of 5% per year compounded semiannually that he wanted?

4.11 Atari needs $4.5 million in new investment capital to develop and market downloadable games software for its new GPS2-ZX system. The plan is to sell $10,000 face-value corporate bonds at a discount of $9000 now. A bond pays a dividend each 6 months based on a bond interest rate of 5% per year with the $10,000 face value returned after 20 years. Will a purchase make at least 6% per year compounded semiannually?

4.12 Fairchild Industries issued 4% bonds some years ago. Carla's grandfather told her he purchased one at a 5% discount when they were issued, and it has paid him $1000 each 3 months for the last 15 years. He told her that he just received the 60th dividend and he plans to sell the bond tomorrow for the face value and give her this money to help with college tuition and expenses.

a. Determine the amount Carla will receive.

b. Use the PW method to determine the number of years to maturity printed on the bond certificate at issue time.

c. Calculate the effective annual rate of return on this investment using a PW relation.

4.13 Nicole and Amy share an apartment and work at Pure H$_2$O Engineering, and they have decided to drink more water each day. The recommended amount is eight 8-ounce glasses, or 64 ounces, per day. This is a total of one gallon per day for the two of them. Using this 1 gallon (or 3.78 liters equivalent) basis for three possible sources of drinking water, determine the monthly cost and present worth value for one year of use. For each source, the average cost and amount of source water estimated to be used to get a gallon of drinking water is tabulated. Let $i = 6\%$ per year compounded monthly and assume 30 days per month.

Method	Average Cost	Gallons used per Gallon Drank
Bottles from store shelf	$1 per gallon	1
In-home filtration system	27¢ per gallon	2
Tap water	$2.75 per 1000 gallons	5

4.14 The Briggs and Stratton Commercial Division designs and manufactures small engines for golf turf maintenance equipment. A robotics-based testing system will ensure that their new signature guarantee program entitled "Always Insta-Start" does indeed work for every engine produced. Compare the two systems at MARR = 10% per year.

	Pull System	Push System
Robot and support equipment first cost, $	−1,500,000	−2,250,000
Annual maintenance and operating cost, $ per year	−700,000	−600,000
Rebuild cost in year 3, $	0	−500,000
Salvage value, $	100,000	50,000
Estimated life, years	8	8

4.15 The Bureau of Indian Affairs provides various services to American Indians and Alaskan Natives. The Director of Indian Health Services is working with chief physicians at some of the 230 clinics nationwide to select the better of two medical X-ray system alternatives to be located at secondary-level clinics. At 5% per year, select the more economical system.

	Del Medical	Siemens
First cost, $	−250,000	−224,000
Annual operating cost, $ per year	−231,000	−235,000
Overhaul in year 3, $	–	−26,000
Overhaul in year 4, $	−140,000	–
Salvage value, $	50,000	10,000
Expected life, years	6	6

4.16 Chevron Corporation has a capital and exploratory budget for oil and gas production of $19.6 billion in one year. The Upstream Division has a project in Angola for which three offshore platform equipment alternatives are identified. Use the present worth method to select the best alternative at 12% per year.

	A	B	C
First cost, $ million	−200	−350	−475
Annual cost, $ million per year	−450	−275	−400
Salvage value, $ million	75	50	90
Estimated life, years	20	20	20

4.17 The TechEdge Corporation offers two forms of 4-year service contracts on its closed-loop water purification system used in the manufacture of semiconductor packages for microwave and high-speed digital devices. The Professional Plan has an initial fee of $52,000 with annual fees starting at $1000 in contract year 1 and increasing by $500 each year. Alternatively, the Executive Plan costs $62,000 up front with annual fees starting at $5000 in contract year 1 and decreasing by $500 each year. The initial charge is considered a setup cost for which there is no salvage value expected. Evaluate the plans at a MARR of 9% per year.

4.18 Harold and Gwendelyn are engineers at Raytheon. Each has presented a proposal to track fatigue development in composites materials installed on special-purpose aircraft. Which is the better plan economically if $i = 12\%$ per year compounded monthly?

	Harold's Plan	Gwendelyn's Plan
Initial cost, $	−40,000	−60,000
Monthly M&O costs, $ per month	−5,000	−
Semiannual M&O cost, $ per 6-month	−	−13,000
Salvage value, $	10,000	8,000
Life, years	5	5

Alternative Evaluation—Unequal Lives

4.19 An environmental engineer must recommend one of two methods for monitoring high colony counts of E. coli and other bacteria in watershed area "hot spots." Estimates are tabulated, and the MARR is 10% per year.

	Method A	Method B
Initial cost, $	−100,000	−250,000
Annual operating cost, $ per year	−30,000 in year 1 increasing by 5000 each year	−20,000
Life, years	3	6

a. Use present worth analysis to select the better method.

b. For a study period of 3 years, use PW analysis to select the better method.

4.20 Use the LCM method and PW analysis to solve problem 4.19a. The engineer has updated the estimates for repurchasing A. The purchase 3 years hence is estimated to cost $150,000, and annual costs will likely increase by $10,000, making them $40,000, $45,000, and $50,000.

4.21 Allen Auto Group owns corner property that can be a parking lot for customers or sold for retail sales space. The parking lot option can use concrete or asphalt. Concrete will cost $375,000 initially, last for 20 years, and have an estimated annual maintenance cost of $200 starting at the end of the eighth year. Asphalt is cheaper to install at $250,000, but it will last 10 years and cost $2500 per year to maintain starting at the end of the second year. If asphalt is replaced after 10 years, the $2500 maintenance cost will be ex-

pended in its last year. There are no salvage values to be considered. Use $i = 8\%$ per year and PW analysis to select the more economic surface, provided the property is (a) used as a parking lot for 20 years, and (b) sold after 5 years and the parking lot is completely removed.

4.22 The manager of engineering at the 900-megawatt Hamilton Nuclear Power Plant has three options to supply personal safety equipment to employees. Two are vendors who sell the items, and a third will rent the equipment for $50,000 per year, but for no more than 3 years per contract. These items have relatively short lives due to constant use. The MARR is 10% per year.

	Vendor R	Vendor T	Rental
Initial cost, $	−75,000	−125,000	0
Annual upkeep cost, $/year	−27,000	−12,000	0
Annual rental cost, $ per year	0	0	−50,000
Salvage value, $	0	30,000	0
Estimated life, years	2	3	Maximum of 3

a. Select from the two sales vendors using the LCM and PW analysis.

b. Determine which of the three options is cheaper over a study period of 3 years.

4.23 Akash Uni-Safe in Chennai, India, makes Terminater fire extinguishers. It needs replacement equipment to form the neck at the top of each extinguisher during production. Select between two metal-constricting systems. Use the corporate MARR of 15% per year with (a) present worth analysis, and (b) future worth analysis.

	Machine D	Machine E
First cost, $	−62,000	−77,000
Annual operating cost, $ per year	−15,000	−21,000
Salvage value, $	8,000	10,000
Life, years	4	6

4.24 The Engineering Director at Akash Uni-Safe (Problem 4.23) wants to perform the evaluation over a 5-year study period. When asked how he would resolve the dilemma that neither machine has a 5-year life, he provided the following estimates:

Machine D: Rework the machine at an additional cost of $30,000 in year 4, plan for the AOC of $15,000 in year 5, and expect zero salvage when it is discarded.

Machine E: Sell it prematurely for the $10,000 salvage value.

Evaluate the alternatives over 5 years. If you worked problem 4.23, is the selection the same?

4.25 If energy usage guidelines expect future increases in deep freeze efficiency, which one of two energy efficiency improvements is more economical based on future worth at an interest rate of 10% per year? A 20% increase is expected to add $150 to the current price of a freezer, while a 35% increase will add $340 to the price. The estimated cost for energy is $115 per year with the 20% increase in efficiency and $80 per year with the 35% increase. Assume a 15-year life for all freezer models.

4.26 HJ Heinz Corporation is constructing a distribution facility in Italy for products such as Heinz Ketchup, Jack Daniel's Sauces, HP steak sauce, and Lea & Perrins Worcestershire sauce. A 15-year life is expected for the structure. The exterior of the building is not yet selected. One alternative is to use the concrete walls as the facade. This will require painting now and every 5 years at a cost of $80,000 each time. Another alternative is an anodized metal exterior attached to the concrete wall. This will cost $200,000 now and require only minimal maintenance of $500 every 3 years. A metal exterior is more attractive and will have a market value of an estimated $25,000 more than concrete 15 years from now. Assume painting (concrete) or maintenance (metal) will be performed in the last year of ownership to promote selling the property. Use future worth analysis and $i = 12\%$ per year to select the exterior finish.

Life Cycle Cost

4.27 Emerald International Airlines plans to develop and install a new passenger ticketing system. A life-cycle cost approach has been used to categorize costs into four categories: development, programming and testing (P/T), operating, and support. There are three alternatives under consideration, identified as T (tailored system), A (adapted system), and C (current system). Use a life-cycle cost analysis to identify the best alternative at 8% per year. All estimates are in $ millions.

Alternative	Category	Estimated Cost
T	Development	$250 now, $150 for years 1–3
	P/T	$45 now, $35 for years 1, 2
	Operation	$50 for years 1 through 10
	Support	$30 for years 1 through 4
A	Development	$10 now
	P/T	$45 year 0, $30 for years 1–3
	Operation	$100 for years 1 through 10
	Support	$40 for years 1 through 10
C	Operation	$190 for years 1 through 10

4.28 Gatorade Endurance Formula, introduced in 2004, contains more electrolytes (such as calcium and magnesium) than the original sports drink formula, thus causing Endurance to taste saltier to some. It is important that the amount of electrolytes be precisely balanced in the manufacturing process. The currently installed system (called EMOST) can be upgraded to monitor the amount more precisely. It costs $12,000 per year for equipment maintenance and $45,000 per year for labor now, and the upgrade will cost $25,000 now. This can serve for 10 more years, the expected remaining time the product will be financially successful. A new system (UPMOST) will also serve for the 10 years and have the following estimated costs. All costs are per year for the indicated time periods.

Equipment: $150,000; years 0 and 1
Development: $120,000; years 1 and 2
Maintain and phaseout EMOST: $20,000; years 1, 2, and 3
Maintain hardware and software: $10,000; years 3 through 10
Personnel costs: $90,000; years 3 through 10
Scrapped formula: $30,000; years 3 through 10

Sales of Gatorade Endurance are expected to be improved by $150,000 per year beginning in year 3 and increase by $50,000 per year through year 10. Use LCC analysis at an MARR of 20% per year to select the better electrolyte monitoring system.

Capitalized Cost

4.29 Suppose that at 35 years old you are a well-paid project leader with Electrolux USA. You plan to work for this company for 20 more years and retire from it with a retirement income of $5000 per month for the rest of your life. Assume an earning rate of 6% per year compounded monthly on your investments. (*a*) How much must you place into your retirement plan monthly starting at the end of next month (month 1) and ending one month prior to the commencement of the retirement benefit? (*b*) Alternatively, assume you decided to deposit the required amount once at the end of each year. Determine the annual retirement savings needed to reach your goal.

4.30 The president of GE Healthcare is considering an external long-term contract offer that will significantly improve the energy efficiency of their imaging systems. The payment schedule has two large payments in the first years with continuing payments thereafter. The proposed schedule is $200,000 now, $300,000 four years from now, $50,000 every 5 years, and an annual amount of $8000 beginning 15 years from now. Determine the capitalized cost at 6% per year.

4.31 The Rustin Transportation Planning Board estimates the cost of upgrading a 4-mile section of 4-lane highway from public use to toll road to be $85 million now. Resurfacing and other maintenance will cost $550,000 every 3 years. Annual toll revenue is expected to average $18.5 million for the foreseeable future. If $i = 8\%$ per year, what is (*a*) the capitalized cost now, and (*b*) the equivalent *A* value of this capitalized cost? Explain the meaning of the *A* value just calculated.

4.32 A specially built computer system has just been purchased by Progress Greenhouse Products to monitor moisture level and to control drip irrigation in hydroponics beds that grow cluster "tradiro" tomatoes. The system's first cost was $97,000 with an AOC of $10,000 and a salvage value on the international market of $20,000 after 4 years, when Progress expects to sell the system. Determine the capitalized cost at 12% per year and explain its meaning.

4.33 A rich graduate of your university wants to set up a scholarship fund for economically disadvantaged high school seniors who have a mental aptitude to become good engineers. The scholarships will be awarded starting now and continue forever to several individuals for a total of $100,000 per year. If investments earn at 5% per year, how much must the alumnus contribute immediately?

4.34 Tom, a new honors graduate, just received a Visa credit card with a high spending limit and an 18% per year compounded monthly interest rate. He plans to use the card starting next month to reward himself by purchasing things that he has wanted for some time at a rate of $1000 per month and pay none of the balance or accrued interest. One month after he stops spending, Tom is willing to then start payments of interest only on the total credit card balance at the rate of $500 per month *for the rest of his life*. Use the capitalized cost method to determine how many months Tom can spend before he must start making the $500 per month payments. Assume the payments can be made for many years, i.e., forever. (Hint: Draw the cash flow diagram first.)

4.35 A pipeline engineer working in Kuwait for BP (formerly British Petroleum) wants to perform a capitalized cost analysis on alternative pipeline routings—one predominately by land and the second primarily undersea. The undersea route is more expensive initially due to extra corrosion protection and installation costs, but cheaper security and maintenance reduce annual costs. Select the better routing at 10% per year.

	Land	Undersea
Installation cost, $ million	−225	−350
Pumping, operating, security, $ million per year	−20	−2
Replacement of deteriorated pipe, $ million each 40 years	−50	—
Expected life, years	∞	∞

4.36 A Dade County project engineer has received draft cost and revenue estimates for a new exhibit and convention center. She has asked you to perform a capitalized cost analysis at 6% per year. Which plan is more economical?

Plan 1: Initial costs of $40 million with an expansion costing $8 million 10 years from now. AOC is $250,000 per year. Net revenue expectation: $190,000 the first year increasing by $20,000 per year for 4 additional years and then leveling off until year 10; $350,000 in year 11 and thereafter.

Plan 2: Initial cost of $42 million with an AOC of $300,000 per year. Net revenue expectation: $260,000 the first year increasing by $30,000 per year for 6 additional years and then leveling off at $440,000 in year 8 and thereafter.

4.37 UPS Freight plans to spend $100 million on new long-haul tractor-trailers. Some of these vehicles will include a new shelving design with adjustable shelves to transport irregularly sized freight that requires special handling during loading and unloading. Though the life is relatively short, the director wants a capitalized cost analysis performed on the two final designs. Compare the alternatives at the MARR of 10% per year.

	Design A: Movable Shelves	Design B: Adaptable Frames
First cost, $	−2,500,000	−1,100,000
AOC, $ per year	−130,000	−65,000
Annual revenue, $ per year	800,000	625,000
Salvage value, $	50,000	20,000
Life, years	6	4

4.38 A water supply cooperative plans to increase its water supply by 8.5 million gallons per day to meet increasing demand. One alternative is to spend $10 million to increase the size of an existing reservoir in an environmentally acceptable way. Added annual upkeep will be $25,000 for this option. A second option is to drill new wells and provide added pipelines for transportation to treatment facilities at an initial cost of $1.5 million and annual cost of $120,000. The reservoir is expected to last indefinitely, but the productive well life is 10 years. Compare the alternatives at 5% per year.

4.39 Three alternatives to incorporate improved techniques to manufacture computer drives to play HD DVD optical disc formats have been developed and costed. Compare the alternatives below using capitalized cost and an interest rate of 12% per year compounded quarterly.

	Alternative E	Alternative F	Alternative G
First cost, $	−2,000,000	−3,000,000	−10,000,000
Net income, $ per quarter	300,000	100,000	400,000
Salvage value, $	50,000	70,000	—
Life, years	4	8	∞

Independent Projects

4.40 State the primary difference between mutually exclusive alternatives and independent projects for the equal-service requirement when economic evaluation is performed.

4.41 Carlotta, the general manager for Woodsome appliance company Plant #A14 in Mexico City has 4 independent projects she can fund this year to improve surface durability on stainless steel products. The project costs and 12%-per-year PW values are available. What projects are acceptable if the budget limit is (a) no limit, and (b) $60,000?

Project	Initial Investment, $	Life, Years	PW at 12% per Year, $
1	−15,000	3	−400
2	−25,000	3	8500
3	−20,000	2	500
4	−40,000	5	7600

4.42 Dwayne has four independent vendor proposals to contract the nationwide oil recycling services for the Ford Corporation manufacturing plants. All combinations are acceptable, except that vendors B and C cannot both be chosen. Revenue sharing of recycled oil sales with Ford is a part of the requirement. Develop all possible mutually exclusive bundles under the additional following restrictions and select the best projects. The corporate MARR is 10% per year.
a. A maximum of $4 million can be spent.
b. A larger budget of $5.5 million is allowed, but no more than two vendors can be selected.
c. There is no limit on spending.

Vendor	Initial Investment, $	Life, Years	Annual Net Revenue, $ per Year
A	−1.5 million	8	360,000
B	−3.0 million	10	600,000
C	−1.8 million	5	620,000
D	−2.0 million	4	630,000

4.43 Use the PW method at 8% per year to select up to 3 projects from the 4 available ones if no more than $20,000 can be invested. Estimated lives and annual net cash flows vary.

Project	Initial Investment, $	Net Cash Flow, $ per Year					
		1	2	3	4	5	6
W	−5,000	1000	1700	1800	2,500	2,000	
X	−8,000	900	950	1000	1,050	10,500	
Y	−8,000	4000	3000	1000	500	500	2000
Z	−10,000	0	0	0	17,000		

4.44 Chloe has $7,000 to spend on as many as 3 electronic features to enhance revenue from the telemarketing business she operates on weekends. Use the PW method to determine which of these independent investments are financially acceptable at 6% per year compounded monthly. All are expected to last 3 years.

Feature	Installed Cost, $	Estimated Added Revenue, $ per Month
Auto-caller	−4500	220
Web interface	−3000	200
Fast search software	−2200	140

PROBLEMS FOR TEST REVIEW AND FE EXAM PRACTICE

Problems 4.45 through 4.48 utilize the following estimates. Costs are shown with a minus sign, and the MARR is 12% per year.

	Alternative 1	Alternative 2
First cost, $	−40,000	−65,000
Annual cost, $ per year	−20,000	−15,000
Salvage value, $	10,000	25,000
Life, years	3	4

4.45 The relation that correctly calculates the present worth of alternative 2 over its life is
a. $-65,000 - 15,000(P/A,6\%,8) + 25,000(P/F,6\%,8)$
b. $-65,000 - 15,000(P/A,12\%,4) + 25,000(P/F,12\%,4)$
c. $-65,000 - 15,000(P/A,12\%,12) + 25,000(P/F,12\%,12)$
d. $-65,000[1 + (P/F,12\%,4) + (P/F,12\%,8)] - 15,000(P/A,12\%,12) + 10,000(P/F,12\%,4)$

4.46 The number of life cycles for each alternative when performing a present worth evaluation based on equal service is
a. 2 for each alternative.
b. 1 for each alternative.
c. 4 for alternative 1; 3 for alternative 2.
d. 3 for alternative 1; 4 for alternative 2.

4.47 If the evaluation of the two alternatives is performed on a present worth basis, the PW value for alternative 2 is closest to
a. $−193,075.
b. $−219,060.
c. $−189,225.
d. $−172,388.

4.48 The FW value of alternative 2 when performing a future worth evaluation over a study period of 8 years is closest to
a. $−383,375.
b. $−408,375.
c. $−320,433.
d. $−603,245.

4.49 The correct relation to calculate the capitalized cost of a project with the following estimates at 8% per year is

$P = \$-100,000 \qquad A = \$-24,000$ per year
$n = 6$ years

a. $[-100,000 - 24,000(P/A,8\%,6)] \times (0.08)$
b. $-100,000 - [24,000(P/A,8\%,6)] \div (0.08)$
c. $-100,000 - [24,000(P/A,8\%,6)] \times (0.08)$
d. $[-100,000(A/P,8\%,6) - 24,000] \div (0.08)$

4.50 Three mutually exclusive cost alternatives (only costs are estimated) are to be evaluated on a PW basis at 15% per year. All have a life of 5 years. An engineer working for you provides the PW values below with a recommendation to select K. The first thing you should do is

$$PW_G = \$-34{,}500$$
$$PW_H = \$-18{,}900$$
$$PW_K = \$+4{,}500$$

a. recommend H as the cheapest alternative.
b. question the positive PW_K value since these are cost alternatives.
c. accept the recommendation.
d. question the negative PW_G and PW_H values since these are cost alternatives.

4.51 The PW method requires evaluation of two mutually exclusive alternatives over the least common multiple (LCM) of their lives. This is required
   a. because the study period is always the LCM of the lives.
   b. to maximize the number of calculations to find PW.
   c. to ensure that the equal service assumption is not violated.
   d. to compare them over a period equal to the life of the longer-lived alternative.

4.52 In a present worth or future worth evaluation over a specified study period that is shorter than the life of an alternative, the alternative's salvage value used in the PW relation is
   a. an estimated market value at the end of the study period.
   b. the same as the estimated salvage value at the end of the alternative's life.
   c. not included in the evaluation.
   d. equal to an amount that is a linear interpolation between the first cost and the estimated salvage value.

Problems 4.53 and 4.54 utilize the following estimates:

	Design A	Design B
First cost, $	−45,000	−90,000
Annual cost, $/year	−10,000	−4,000
Salvage value, $	13,000	15,000
Life, years	3	∞

4.53 The alternative designs are compared using capitalized costs at 10% per year compounded continuously. The relation that correctly determines the capitalized cost of Design A is:
   a. $-45{,}000 - 10{,}000(P/A,10.52\%,6) - 37{,}000(P/F,10.52\%,3) + 13{,}000(P/F,10.52\%,6)$

b. $-45{,}000 - 10{,}000(P/A,10.52\%,3) + 13{,}000(P/F,10.52\%,3)$
c. $[-45{,}000(A/P,10\%,3) - 10{,}000 + 13{,}000(A/F,10\%,3)] \div 0.10$
d. $[-45{,}000(A/P,10.52\%,3) - 10{,}000 + 13{,}000(A/F,10.52\%,3)] \div 0.1052$

4.54 The PW value for Design B at an interest rate of 8% per year is closest to
   a. $\$-128{,}020$.
   b. $\$-94{,}000$.
   c. $\$-57{,}200$.
   d. $\$-140{,}000$.

4.55 A rich graduate of the engineering program at your university wishes to start an endowment that will provide scholarship money of $40,000 per year beginning in year 5 and continuing indefinitely. If the university earns 10% per year on its investments, the single donation required now is closest to
   a. $225,470.
   b. $248,360.
   c. $273,200.
   d. $293,820.

4.56 A $50,000 bond with a coupon rate of 6% per year payable quarterly matures 10 years from now. At an interest rate of 8% per year compounded quarterly, the relation that correctly calculates the present worth of the remaining payments to the owner is
   a. $750(P/A,1.5\%,40) + 50{,}000(P/F,1.5\%,40)$
   b. $750(P/A,2\%,40) + 50{,}000(P/F,2\%,40)$
   c. $3000(P/A,8\%,10) + 50{,}000(P/F,8\%,10)$
   d. $1500(P/A,4\%,20) + 50{,}000(P/F,4\%,20)$

4.57 John purchased a $20,000 20-year bond at a 2% discount. It pays semiannually at a rate of 4% per year. If he wishes to make 6% per year on his investment, the semiannual dividend he will receive is
   a. $400.
   b. $390.
   c. $600.
   d. $590.

4.58 A 5% $5,000 5-year bond is for sale. The dividend is paid quarterly. If you keep the bond to maturity and want to realize a return of 6% per year compounded quarterly, the most you can pay now for the bond is closest to
   a. $5000.
   b. $4000.
   c. $4785.
   d. $5150.

Annual Worth Analysis

Andy Sotiriou/Getty Images

An AW analysis is commonly preferred over a PW analysis because the AW value is easy to calculate; the measure of worth—AW in dollars per year—is understood by most individuals; and its assumptions are essentially the same as those of the PW method.

Annual worth is known by other titles. Some are equivalent annual worth (EAW), equivalent annual cost (EAC), annual equivalent (AE), and EUAC (equivalent uniform annual cost). The alternative selected by the AW method will always be the same as that selected by the PW method, and all other alternative evaluation methods, provided they are performed correctly.

Objectives

1. Calculate capital recovery and AW over one life cycle.

    | AW calculation |

2. Select the best alternative on the basis of an AW analysis.

    | Alternative selection by AW |

3. Select the best permanent investment alternative using AW values.

    | Permanent investment AW |

4. Use a spreadsheet to perform an AW evaluation.

    | Spreadsheets |

5.1 AW VALUE CALCULATIONS

The annual worth (AW) method is commonly used for comparing alternatives. All cash flows are converted to an equivalent uniform annual amount over one life cycle of the alternative. The AW value is easily understood by all since it is stated in terms of dollars per year. The major advantage over all other methods is that the equal service requirement is met without using the least common multiple (LCM) of alternative lives. The AW value is calculated over one life cycle and is assumed to be exactly the same for any succeeding cycles, provided all cash flows change with the rate of inflation or deflation. If this cannot be reasonably assumed, a study period and specific cash flow estimates are needed for the analysis. The repeatability of the AW value over multiple cycles is demonstrated in Example 5.1.

New digital scanning graphics equipment is expected to cost $20,000, to be used for 3 years, and to have an annual operating cost (AOC) of $8000. Determine the AW values for one and two life cycles at $i = 22\%$ per year.

EXAMPLE 5.1

Solution

First use the cash flows for one life cycle (Figure 5.1) to determine AW.

$$AW = -20,000(A/P,22\%,3) - 8000 = \$-17,793$$

For two life cycles, calculate AW over 6 years. Note that the purchase for the second cycle occurs at the end of year 3, which is year zero for the second life cycle (Figure 5.1).

$$AW = -20,000(A/P,22\%,6) - 20,000(P/F,22\%,3)(A/P,22\%,6) - 8000$$
$$= \$-17,793$$

The same AW value can be obtained for any number of life cycles, thus demonstrating that the AW value for one cycle represents the equivalent annual worth of the alternative for every cycle.

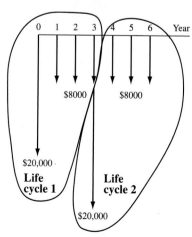

FIGURE 5.1

Cash flows over two life cycles of an alternative.

It is always possible to determine the AW, PW, and FW values from each other using the following relation. The n is the number of years for an equal-service comparison, which is the LCM of lives or a selected study period.

$$AW = PW(A/P,i,n) = FW(A/F,i,n) \qquad [5.1]$$

The AW value of an alternative is the addition of two distinct components: *capital recovery (CR)* of the initial investment and the *equivalent A value* of the annual operating costs (AOC).

$$AW = CR + A \text{ of AOC} \qquad [5.2]$$

The recovery of an amount of capital P committed to an asset, plus the time value of the capital at a particular interest rate, is a fundamental principle of economic analysis. *Capital recovery is the equivalent annual cost of owning the asset plus the return on the initial investment.* The A/P factor is used to convert P to an equivalent annual cost. If there is some anticipated positive salvage value S at the end of the asset's useful life, its equivalent annual value is removed using the A/F factor. This action reduces the equivalent annual cost of owning the asset. Accordingly, CR is

$$CR = -P(A/P,i,n) + S(A/F,i,n) \qquad [5.3]$$

The annual amount (A of AOC) is determined from uniform recurring costs (and possibly receipts) and nonrecurring amounts. The P/A and P/F factors may be necessary to first obtain a present worth amount, then the A/P factor converts this amount to the A value in Equation [5.2].

EXAMPLE 5.2   Lockheed Martin is increasing its booster thrust power in order to win more satellite launch contracts from European companies interested in new global communications markets. A piece of earth-based tracking equipment is expected to require an investment of $13 million. Annual operating costs for the system are expected to start the first year and continue at $0.9 million per year. The useful life of the tracker is 8 years with a salvage value of $0.5 million. Calculate the AW value for the system if the corporate MARR is currently 12% per year.

Solution

The cash flows (Figure 5.2a) for the tracker system must be converted to an equivalent AW cash flow sequence over 8 years (Figure 5.2b). (All amounts are expressed in $1 million units.) The AOC is $A = \$-0.9$ per year, and the capital recovery is calculated by using Equation [5.3].

$$
\begin{aligned}
CR &= -13(A/P,12\%,8) + 0.5(A/F,12\%,8) \\
&= -13(0.2013) + 0.5(0.0813) \\
&= \$-2.576
\end{aligned}
$$

The correct interpretation of this result is very important to Lockheed Martin. It means that each and every year for 8 years, the equivalent total revenue from

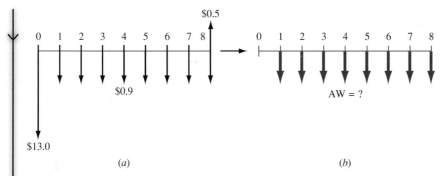

FIGURE 5.2   (a) Cash flow diagram for satellite tracker costs, and (b) conversion to an equivalent AW (in $1 million), Example 5.2.

the tracker must be at least $2,576,000 *just to recover the initial present worth investment plus the required return of 12% per year.* This does not include the AOC of $0.9 million each year. Total AW is found by Equation [5.2].

AW = −2.576 − 0.9 = $−3.476 million per year

This is the AW for all future life cycles of 8 years, provided the costs rise at the same rate as inflation, and the same costs and services apply for each succeeding life cycle.

There is a second, equally correct way to determine CR. Either method results in the same value.

$$CR = -(P - S)(A/P,i,n) - S(i) \qquad [5.4]$$

There is a basic logic to this formula. Subtracting S from the initial investment P before applying the A/P factor recognizes that the salvage value will be recovered. This reduces CR, the annual cost of asset ownership. However, the fact that S is not recovered until the last year of ownership is compensated for by charging the cost of the annual interest $S(i)$ against the CR. In Example 5.2, the use of this second way to calculate CR results in the same value.

$$CR = -(13.0 - 0.5)(A/P,12\%,8) - 0.5(0.12)$$
$$= -12.5(0.2013) - 0.06 = \$-2.576$$

Although either CR relation results in the same amount, it is better to consistently use the same method. The first method, Equation [5.3], is used in this text.

For solution by computer, use the PMT function to determine CR only in a single spreadsheet cell. The format is = PMT($i\%,n,P,-S$). As an illustration, the CR in Example 5.2 is displayed when = PMT(12%,8,13,−0.5) is entered.

The annual worth method is applicable in any situation where PW, FW, or Benefit/Cost analysis can be utilized. The AW method is especially useful in certain types of studies: asset replacement and retention studies to minimize overall annual costs, breakeven studies and make-or-buy decisions (all covered in later chapters), and all studies dealing with production or manufacturing where cost/unit is the focus.

5.2 EVALUATING ALTERNATIVES BASED ON ANNUAL WORTH

The annual worth method is typically the easiest of the evaluation techniques to perform, when the MARR is specified. The alternative selected has the lowest equivalent annual cost (cost alternatives), or highest equivalent income (revenue alternatives). The selection guidelines for the AW method are the same as for the PW method.

One alternative: AW ≥ 0, the alternative is financially viable.

Two or more alternatives: Choose the numerically largest AW value (lowest cost or highest income).

If a study period is used to compare two or more alternatives, the AW values are calculated using cash flow estimates over only the study period. For a study period shorter than the alternative's expected life, use an estimated market value for the salvage value.

EXAMPLE 5.3   PizzaRush, which is located in the general Los Angeles area, fares very well with its competition in offering fast delivery. Many students at the area universities and community colleges work part-time delivering orders made via the web at PizzaRush.com. The owner, a software engineering graduate of USC, plans to purchase and install five portable, in-car systems to increase delivery speed and accuracy. The systems provide a link between the web order-placement software and the in-car GPS system for satellite-generated directions to any address in the Los Angeles area. The expected result is faster, friendlier service to customers, and more income for PizzaRush.

Each system costs $4600, has a 5-year useful life, and may be salvaged for an estimated $300. Total operating cost for all systems is $650 for the first year, increasing by $50 per year thereafter. The MARR is 10% per year. Perform an annual worth evaluation that answers the following questions:

a. How much new annual revenue is necessary to recover only the initial investment at an MARR of 10% per year?

b. The owner conservatively estimates increased income of $5000 per year for all five systems. Is this project financially viable at the MARR? See cash flow diagram in Figure 5.3.

c. Based on the answer in part (b), determine how much new income PizzaRush must have to economically justify the project. Operating costs remain as estimated.

Solution

a. The CR value will answer this question. Use Equation [5.3] at 10%.

$$CR = -5(4600)(A/P,10\%,5) + 5(300)(A/F,10\%,5)$$
$$= \$-5822$$

b. The financial viability could be determined now without calculating the AW value, because the $5000 in new income is lower than the CR of $5822, which does not yet include the annual costs. So, the project is not economically justified. However, to complete the analysis, determine the total AW. The annual operating costs and incomes form an arithmetic gradient series with a base of $4350 in year 1, decreasing by $50 per year for 5 years. The AW relation is

$$AW = \text{capital recovery} + A \text{ of net income}$$
$$= -5822 + 4350 - 50(A/G,10\%,5) \qquad\qquad \textbf{[5.5]}$$
$$= \$-1562$$

This shows conclusively that the alternative is not financially viable at MARR = 10%.

c. An equivalent of the projected $5000 plus the AW amount are necessary to make the project economically justified at a 10% return. This is 5000 + 1562 = $6562 per year in new revenue. At this point AW will equal zero based on Equation [5.5].

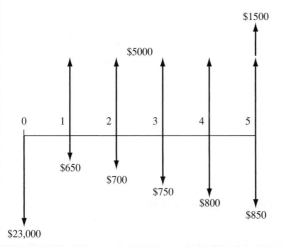

$1500   **FIGURE 5.3**
Cash flow diagram used to compute AW, Example 5.3.

EXAMPLE 5.4

A quarry outside of Austin, Texas wishes to evaluate two similar pieces of equipment by which the company can meet new state environmental requirements for dust emissions. The MARR is 12% per year. Determine which alternative is economically better using (*a*) the AW method, and (*b*) AW method with a 3-year study period.

Equipment	X	Y
First cost, $	40,000	75,000
AOC, $ per year	25,000	15,000
Life, years	4	6
Salvage value, $	10,000	7,000
Estimated value after 3 years, $	14,000	20,000

Solution

a. Calculating AW values over the respective lives indicates that *Y* is the better alternative.

$$AW_X = -40{,}000(A/P,12\%,4) - 25{,}000 + 10{,}000(A/F,12\%,4)$$
$$= \$-36{,}077$$
$$AW_Y = -75{,}000(A/P,12\%,6) - 15{,}000 + 7{,}000(A/F,12\%,6)$$
$$= \$-32{,}380$$

b. All *n* values are 3 years and the "salvage values" become the estimated market values after 3 years. Now *X* is economically better.

$$AW_X = -40{,}000(A/P,12\%,3) - 25{,}000 + 14{,}000(A/F,12\%,3)$$
$$= \$-37{,}505$$
$$AW_Y = -75{,}000(A/P,12\%,3) - 15{,}000 + 20{,}000(A/F,12\%,3)$$
$$= \$-40{,}299$$

5.3 AW OF A PERMANENT INVESTMENT

The annual worth equivalent of a very long-lived project is the AW value of its capitalized cost (CC), discussed in Section 4.4. The AW value of the first cost, *P*, or present worth, PW, of the alternative uses the same relation as Equation [4.2].

$$AW = CC(i) = Pi \qquad [5.6]$$

Cash flows that occur at regular intervals are converted to AW values over one life cycle of their occurrence. All other nonregular cash flows are first converted to a *P* value and then multiplied by *i* to obtain the AW value over infinity.

EXAMPLE 5.5

If you receive an inheritance of $10,000 today, how long do you have to invest it at 8% per year to be able to withdraw $2000 every year forever? Assume the 8% per year is a return that you can depend on forever.

Solution

Cash flow is detailed in Figure 5.4. Solving Equation [5.6] for P indicates that it is necessary to have $25,000 accumulated at the time that the $2000 annual withdrawals start.

$$P = 2000/0.08 = \$25{,}000$$

Find $n = 11.91$ years using the relation $\$25{,}000 = 10{,}000(F/P,8\%,n)$.

Comment: It is easy to use Excel to solve this problem. In any cell write the function $= \text{NPER}(8\%,,-10000,25000)$ to display the answer of 11.91 years.

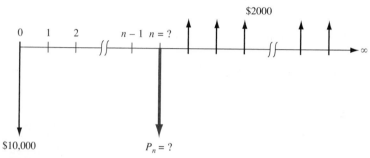

FIGURE 5.4   Diagram to determine n for a perpetual withdrawal, Example 5.5.

EXAMPLE 5.6

The state is considering three proposals for increasing the capacity of the main drainage canal in an agricultural region. Proposal A requires dredging the canal. The state is planning to purchase the dredging equipment and accessories for $650,000. The equipment is expected to have a 10-year life with a $17,000 salvage value. The annual operating costs are estimated to total $50,000. To control weeds in the canal itself and along the banks, environmentally safe herbicides will be sprayed during the irrigation season. The yearly cost of the weed control program is expected to be $120,000.

    Proposal B is to line the canal walls with concrete at an initial cost of $4 million. The lining is assumed to be permanent, but minor maintenance will be required every year at a cost of $5000. In addition, lining repairs will have to be made every 5 years at a cost of $30,000.

    Proposal C is to construct a new pipeline along a different route. Estimates are: an initial cost of $6 million, annual maintenance of $3000 for right-of-way, and a life of 50 years.

Compare the alternatives on the basis of annual worth, using an interest rate of 5% per year.

Solution

Since this is an investment for a permanent project, compute the AW for one cycle of all recurring costs. For proposals A and C, the CR values are found using Equation [5.3], with $n_A = 10$ and $n_C = 50$, respectively. For proposal B, the CR is simply $P(i)$.

Proposal A

CR of dredging equipment:

$-650,000(A/P,5\%,10) + 17,000(A/F,5\%,10)$	$ $-82,824$
Annual cost of dredging	$-50,000$
Annual cost of weed control	$-120,000$
	$\$-252,824$

Proposal B

CR of initial investment: $-4,000,000(0.05)$	$\$-200,000$
Annual maintenance cost	$-5,000$
Lining repair cost: $-30,000(A/F,5\%,5)$	$-5,429$
	$\$-210,429$

Proposal C

CR of pipeline: $-6,000,000(A/P,5\%,50)$	$\$-328,680$
Annual maintenance cost	$-3,000$
	$\$-331,680$

Proposal B is selected.

Comment: The A/F factor is used instead of A/P in B because the lining repair cost begins in year 5, not year 0, and continues indefinitely at 5-year intervals.

If the 50-year life of proposal C is considered infinite, CR $= P(i) = \$-300,000$, instead of $\$-328,680$ for $n = 50$. This is a small economic difference. How expected lives of 40 or more years are treated economically is a matter of local practice.

5.4 USING SPREADSHEETS FOR AW ANALYSIS

Annual worth evaluation of equal or unequal life, mutually exclusive alternatives is simplified using the PMT function. The general format for determining an alternative's AW is

$$= \text{PMT}(i\%,n,P,F) - A$$

PMT determines the required capital recovery (CR). Usually F is the estimated salvage value entered as $-S$, and $-A$ is the annual operating cost (AOC). Since AW evaluation does not require evaluation over the least common multiple of lives, the n values can be different for each alternative.

As with the PV function, the $+$ and $-$ signs on PMT values must be correctly entered to ensure the appropriate sense to the result. To obtain a negative AW value for a cost alternative, enter $+P$, $-S$, and $-A$. The next two examples illustrate spreadsheet-based AW evaluations for unequal life alternatives, including a permanent investment discussed in Section 5.3.

Herme, the quarry supervisor for Espinosa Stone, wants to select the more economical of two alternatives for quarry dust control. Help him perform a spreadsheet-based AW evaluation for the estimates with MARR $= 12\%$ per year.   **EXAMPLE 5.7**

Equipment	X	Y
First cost, $	40,000	75,000
AOC, $ per year	25,000	15,000
Life, years	4	6
Salvage value, $	10,000	7,000

Solution

Because these are the same estimates as Example 5.4a, you can see what is required to hand-calculate AW values by referring back. Spreadsheet evaluation is much faster and easier; it is possible to enter the two single-cell PMT functions for X and Y using the general format presented earlier. Figure 5.5 includes details. The decision is to select Y with the lower-cost AW value.

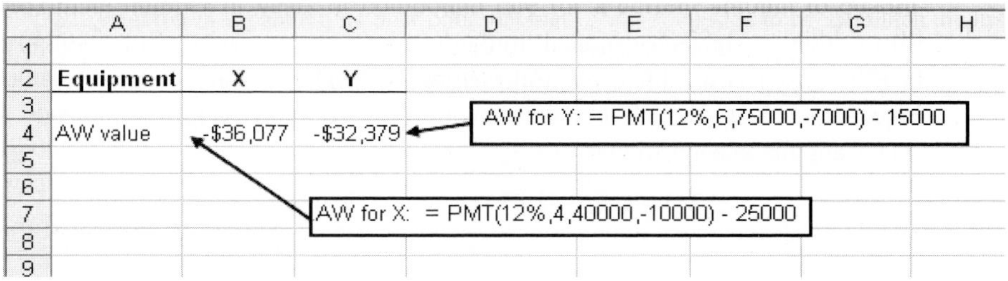

	A	B	C	D	E	F	G	H
1								
2	Equipment	X	Y					
3								
4	AW value	-$36,077	-$32,379	AW for Y: = PMT(12%,6,75000,-7000) - 15000				
5								
6								
7			AW for X: = PMT(12%,4,40000,-10000) - 25000					
8								
9								

FIGURE 5.5   Evaluation by AW method using PMT functions, Example 5.7.

Perform a spreadsheet-based AW evaluation of the three proposals in Example 5.6. Note that the lives vary from 10 years for A, to 50 for C, to permanent for B.   **EXAMPLE 5.8**

Solution

Figure 5.6 summarizes the estimates on the left. Obtaining AW values is easy using PMT and a few computations shown on the right. The PMT function determines capital recovery of first cost at 5% per year for A and C. For B, capital recovery is simply P times i, according to Equation [5.6], because proposal B (line walls with cement) is considered permanent. (Remember to enter $-4,000,000$ for P to ensure a minus sign on the CR amount.)

All annual costs are uniform series except proposal B's periodic lining repair cost of $30,000 every 5 years, which is annualized (in cell G7) using = PMT(5%,5,,30000). Adding all costs (row 8) indicates that proposal B has the lowest AW of costs.

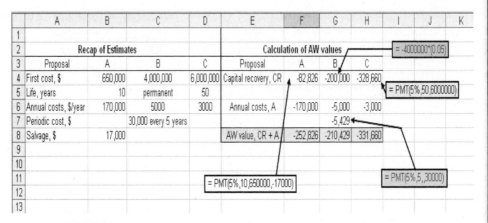

FIGURE 5.6   Use of PMT functions to perform an AW evaluation of three proposals with different lives, Example 5.8.

When estimated annual costs are not a uniform series A, the cash flows must be entered individually (using the negative sign for cash outflows) and the use of single-cell functions is somewhat limited. In these cases, first use the NPV function to find the PW value over one life cycle, followed by the PMT function to determine the AW value. Alternatively, though it is more involved, imbed the NPV function directly into PMT using the general format

$$= \text{PMT}(i\%,n,P+\text{NPV}(i\%,\text{year\_1\_cell:year\_}n\text{\_cell})).$$

Try it!

SUMMARY

The annual worth method of comparing alternatives is preferred to the present worth method, because the AW comparison is performed for only one life cycle. This is a distinct advantage when comparing different-life alternatives. When a study period is specified, the AW calculation is determined for that time period only, and the estimated value remaining in the alternative at the end of the study period becomes the salvage value.

For infinite-life alternatives, the initial cost is annualized by multiplying P by i. For finite-life alternatives, the AW through one life cycle is equal to the perpetual equivalent annual worth.

PROBLEMS

Capital Recovery and AW

5.1 Jim, the owner of Computers-on-Wheels, an on-site repair service, just purchased a $20,000 diagnostics system and software for PCs, laptops, and LANs. During the expected 5-year life, he determines that the annual recovery amount for this investment is $20,000/5 = $4000. (*a*) Explain why Jim has mistakenly underestimated the annual increase needed in revenue. (*b*) Determine the required annual revenue increase for a 15% per year return.

5.2 A 600-ton press used to produce composite-material fuel cell components for automobiles using proton exchange membrane (PEM) technology can reduce the weight of enclosure parts up to 75%. At MARR = 12% per year, calculate (*a*) capital recovery and (*b*) annual revenue required.

Installed cost = $-3.8 million   *n* = 12 years
Salvage value = $250,000

Annual operating costs = $350,000 to start
        increasing by $25,000 per year

5.3 Five years ago, Diamond Electronics, a division of De Beers, paid $3,150,000 for new diamond cutting tools. At that time, the company estimated an added revenue need of $500,000 to recover the investment at 10% per year. If there is an estimated 8 more years of service with a salvage value of $300,000, compare the revenue needed over the entire life with that estimated 5 years ago.

5.4 Brent owns Beck Trucking. Seven years ago, he purchased a large-capacity dump truck for $115,000 to provide short-haul earth moving services. He sold it today for $45,000. Operating and maintenance costs averaged $9500 per year. A complete overhaul at the end of year 4 cost an extra $3200. (*a*) Calculate the annual cost of the truck at 7% per year. (*b*) If Brent estimates that Beck cleared at most $20,000 per year added revenue from using the truck, was the purchase economically advantageous?

Evaluating Alternatives Using AW

5.5 You have the estimates for two alternatives, A and B, from which to select the cheaper one using an AW evaluation. Your boss selects a study period of 5 years, but, A has an expected life of 3 years and

B has a 10-year life. Concerning the estimates, what is necessary to correctly perform the AW calculations over the 5-year period?

5.6 BP Oil is in the process of replacing sections of its Prudhoe Bay, Alaska oil transit pipeline. This will reduce corrosion problems, while allowing higher line pressures and flow rates to downstream processing facilities. The installed cost is expected to be about $170 million. Alaska imposes a 22.5% tax on annual profits (net revenue over costs), which are estimated to average $85 million per year for a 20 year period. (*a*) At a corporate MARR of 10% per year, does the project AW indicate it will make at least the MARR? (*b*) Recalculate the AW at MARR values increasing by 10% per year, that is, 20%, 30%, etc. At what required return does the project become financially unacceptable?

5.7 Equipment needed at a Valero Corporation refinery for the conversion of corn stock to ethanol, a cleaner-burning gasoline additive, will cost $175,000 and have net cash flows of $35,000 the first year, increasing by $10,000 per year over the life of 5 years. (*a*) Determine the AW amounts at different MARR values to determine when the project switches from financially justified to unjustified. (Hint: Start at 12% per year and increase by 1% per year.) (*b*) If a spreadsheet is used, plot a graph of AW versus interest rate on the spreadsheet.

5.8 Your office mate performed an 8% per year PW evaluation of two alternatives before he left on vacation. Now, your supervisor wants you to provide the results in AW terms. The PW results are:

PW$_A$ = $-517,510 for a life of 4 years
PW$_B$ = $-812,100 for a life of 8 years

5.9 The TT Racing and Performance Motor Corporation wishes to evaluate two alternative machines for NASCAR motor tuneups. Use the AW method at 9% per year to select the better alternative.

	Machine R	Machine S
First cost, $	-250,000	-370,500
Annual operating cost, $ per year	-40,000	-50,000
Life, years	3	5
Salvage value, $	20,000	20,000

5.10 Rework Problem 5.9 if there is a major overhaul cost every 2 years that was not included in the original estimates. The cost is $60,000 for R and $25,000 for S.

5.11 The Haber Process, developed in the early 1900s, is commonly used to produce ammonia for industrial and consumer applications. Higher yields take place as temperature and pressure (in *atmospheres* or *atm*) are varied. New pressure chamber equipment is necessary at KGC Industries in India. Three alternatives are available, the third being contracting all services. Use the annual worth method for a study period of 4 years and $i = 6\%$ per year to select the lowest annual cost alternative. There are no positive salvage values expected, but the market values are expected to decrease by 20% of the first cost for each year of use.

	200 atm	500 atm	Contract
First cost, $	−310,000	−600,000	0
Annual cost, $ per year	−95,000	−60,000	−190,000
Life, years	5	7	4

5.12 Holly Farms is considering two environmental chambers to accomplish detailed laboratory confirmations of on-line bacteria tests in chicken meat for the presence of *E. coli* 0157:H7 and *Listeria monocytogenes*.

	Chamber D103	Chamber 490G
Installed cost, $	−400,000	−250,000
Annual operating cost, $ per year	−4,000	−3,000
Salvage value at 10% of *P*, $	40,000	25,000
Life, years	3	2

a. If this project will last for 6 years and $i = 10\%$ per year, perform an AW method evaluation to determine which chamber is more economical.

b. Chamber D103 can be purchased with different options and, therefore, at different installed costs. They range from $300,000 to $500,000. Will the selection change if one of these other models is installed?

5.13 Blue Whale Moving and Storage recently purchased a warehouse building in Santiago. The manager has two good options for moving pallets of stored goods in and around the facility. Alternative 1 includes a 4000-pound capacity, electric forklift ($P = \$-30,000$; $n = 12$ years; AOC = $\$-1000$ per year; $S = \$8000$) and 500 new pallets at $10 each. The forklift operator's annual salary and indirect benefits are estimated at $32,000.

Alternative 2 involves the use of two electric pallet movers ("walkies") each with a 3000-pound capacity (for each mover, $P = \$-2,000$; $n = 4$ years; AOC = $\$-150$ per year; no salvage) and 800 pallets at $10 each. The two operators' salaries and benefits will total $55,000 per year. For both options, new pallets are purchased now and every two years that the equipment is in use. If the MARR is 8% per year, which alternative is better?

5.14 Work Example 4.2 in Chapter 4 using the annual worth method and compare your conclusion with that reached using the PW method.

5.15 McLaughlin Services is on contract to Tuscany County, Florida. One of the McLaughlin engineers is evaluating alternatives that use a robotic, liquid-propelled "pig" to periodically inspect the interior of buried potable water pipes for leakage, corrosion, weld strength, movement over time, and a variety of other parameters. Two equivalent robot instruments are available.

Robot Joeboy: $P = \$-85,000$;
annual M&O costs = $\$-30,000$; $S = \$40,000$

Robot Watcheye: $P = \$-97,000$;
annual M&O costs = $\$-27,000$; $S = \$42,000$

a. Select one using an AW comparison, $i = 10\%$ per year, and a 3-year study period.

b. If the first cost of Watcheye is somewhat negotiable, by how many dollars must it increase or decrease to make the two instruments equally attractive financially? Use the 3-year study period.

Evaluating Long-Life Alternatives

5.16 Cheryl and Gunther wish to place into a retirement fund an equal amount each year for 20 consecutive years to accumulate just enough to withdraw $24,000 per year starting exactly one year after the last deposit is made. The fund has a reliable return of 8% per year. Determine the annual deposit for two withdrawal plans: (*a*) forever (years 21 to infinity); (*b*) 30 years (years 21 through 50).

(c) How much less per year is needed when the withdrawal horizon decreases from infinity to 30 years?

5.17 Baker|Trimline owned a specialized tools company for a total of 12 years when it was sold for $38 million cash. During the ownership, annual net cash flow (revenues over all costs) varied significantly.

Year	1	2	3	4	5	6	7	8	9	10	11	12
Net Cash Flow, $ million per year	4	0	−1	−3	−3	1	4	6	8	10	12	12

The company made 12% per year on its positive cash flows and paid the same rate on short-term loans to cover the lean years. The president wants to use the cash accumulated after 12 years to improve capital investments starting in year 13 and forward. If an 8% per year return is expected after the sale, what annual amount can Baker|Trimline invest forever?

5.18 A major repair on the suspension system of Jane's 3-year old car cost her $2,000 because the warranty expired after 2 years of ownership. Based on this experience, she will plan on additional $2000 expenses every 3 years henceforth. Also, she spends $800 every 2 years for maintenance now that the warranty is over. This is for years 2, 4, 6, 8, and 10 when she plans to donate the car to charity. Use these costs to determine Jane's equivalent perpetual annual cost (for years 1 through infinity) at $i = 5\%$ per year, if cars she owns in the future have the same cost pattern.

5.19 A West Virginia coal mining operation has installed an in-shaft monitoring system for oxygen tank and gear readiness for emergencies. Based on maintenance patterns for previous systems, costs are minimal for the first few years, increase for a time period, and then level off. Maintenance costs are expected to be $150,000 in year 3, $175,000 in year 4, and amounts increasing by $25,000 per year through year 6 and remain constant thereafter for the expected 10-year life of this system. If similar systems will replace the current one, determine the perpetual equivalent annual maintenance cost at $i = 10\%$ per year.

5.20 Harmony Auto Group sells and services imported and domestic cars. The owner is considering outsourcing all of its new auto warranty service work to Winslow, Inc., a private repair service that works on any make and year car. Either a 5-year contract basis or 10-year license agreement are available from Winslow. Revenue from the manufacturer will be shared with no added cost incurred by the car/warranty owner. Alternatively, Harmony can continue to do warranty work in-house. Use the estimates made by the Harmony owner to perform an annual worth evaluation at 12% per year to select the best option. All dollar values are in millions.

	Contract	**License**	**In-house**
First cost, $	0	−2	−30
Annual cost, $ per year	−2	−0.2	−5
Annual income, $ per year	2.5	1.5	9
Life, years	5	10	∞

5.21 ABC Drinks purchases its 355 ml cans in large bulk from Wald-China Can Corporation. The finish on the anodized aluminum surface is produced by mechanical finishing technology called brushing or bead blasting. Engineers at Wald are switching to more efficient, faster, and cheaper machines to supply ABC. Use the estimates and MARR = 8% per year to select between two alternatives.

Brush alternative: $P = \$-400{,}000$; $n = 10$ years; $S = \$50{,}000$; nonlabor AOC = $\$-50{,}000$ in year 1, decreasing by $5000 annually starting in year 2.

Bead blasting alternative: $P = \$-300{,}000$; n is large, assume permanent; no salvage; nonlabor AOC = $\$-50{,}000$ per year.

5.22 You are an engineer with Yorkshire Shipping in Singapore. Your boss, Zul, asks you to recommend one of two methods to reduce or eliminate rodent damage to silo-stored grain as it awaits shipment. Perform an AW analysis at 10% per year compounded quarterly. Dollar values are in millions.

	Alternative A Major Reduction	**Alternative B** Almost Eliminate
First cost, $	−10	−50
Annual operating cost, $ per year	−0.8	−0.6
Salvage value, $	0.7	0.2
Life, years	5	Almost permanent

PROBLEMS FOR TEST REVIEW AND FE EXAM PRACTICE

5.23 The least common multiple (LCM) of lives need not be used when evaluating two alternatives by the annual worth method because, if inflation and deflation effects are neglected,

    a. the assumptions for annual worth analysis are different from those for the present worth method.

    b. cost and revenue estimates never remain the same over more than one life cycle.

    c. the annual worth value used to evaluate the alternatives is determined over a large number of life cycles.

    d. the annual worth over one life cycle is assumed to be the same for all succeeding life cycles.

5.24 An automation asset with a high first cost of $10 million has a capital recovery (CR) of $1,985,000 per year. The correct interpretation of this CR value is that

    a. the owner must pay an additional $1,985,000 each year to retain the asset.

    b. each year of its expected life, a net revenue of $1,985,000 must be realized to recover the $10 million first cost and the required rate of return on this investment.

    c. each year of its expected life, a net revenue of $1,985,000 must be realized to recover the $10 million first cost.

    d. the services provided by the asset will stop if less than $1,985,000 in net revenue is reported in any year.

Problems 5.25 through 5.27 refer to the following estimates for three mutually exclusive alternatives. The MARR is 6% per year.

	Vendor 1	Vendor 2	Vendor 3
First cost, $	−200,000	−550,000	−1,000,000
Annual cost, $ per year	−50,000	−20,000	−10,000
Revenue, $ per year	120,000	120,000	110,000
Salvage value, $	25,000	0	500,000
Life, years	10	15	∞

5.25 The annual worth of vendor 2 cash flow estimates is closest to

    a. $67,000.

    b. $43,370.

    c. $ −43,370.

    d. $63,370.

5.26 Of the following three relations, the correct one(s) to calculate the annual worth of vendor 1 cash flow estimates is (note: All dollar values are in thousands)

    Relation 1: $AW_1 = -200(A/P,6\%,10) + 70 + 25(A/F,6\%,10)$

    Relation 2: $AW_1 = [-200 - 50(P/A,6\%,10) + 120(P/A,6\%,10) + 25(P/F,6\%,10)](A/P,6\%,10)$

    Relation 3: $AW_1 = -200(F/P,6\%,10) + 25 + (-50 + 120)(A/P,6\%,10)$.

    a. 1 and 3.

    b. 1.

    c. 1 and 2.

    d. 3.

5.27 The AW values for the alternatives are as listed. The vendor(s) that should be recommended are

$$AW_1 = \$44,723$$
$$AW_2 = \$43,370$$
$$AW_3 = \$40,000$$

    a. 1 and 2.

    b. 3.

    c. 2.

    d. 1.

5.28 If a revenue alternative has a negative AW value and it was correctly calculated, it means one or more of the following:

    1. The equivalent annual worth of revenues does not exceed that of the costs.

    2. The estimates are wrong somewhere.

    3. A minus or plus sign of a cash flow was entered incorrectly into the Excel PMT function.

    4. The alternative should have a longer life so revenues will exceed costs.

    a. 1

    b. 2

    c. 3

    d. 4

5.29 Estimates for one of two process upgrades are: first cost of $40,000; annual costs of $5000 per year; market value that decreases 5% per year to the salvage value of $20,000 after the expected life of 10 years. If a 4-year study period is used for AW analysis at 15% per year, the correct AW value is closest to

a. $-15,000.
b. $-11,985.
c. $-7,600.
d. $-12,600.

5.30 The perpetual annual worth of investing $50,000 now and $20,000 per year starting in year 16 and continuing forever at 12% per year is closest to

a. $4200.
b. $8650.
c. $9655.
d. $10,655.

5.31 You graduated with an MS degree in engineering and have a great job. Next month you start an automatic draft from your paycheck of $1000 each month, placing it into a mutual fund retirement account. You plan to continue at this rate for 20 years then stop making deposits. Assume a return of 6% per year compounded monthly is a good expectation for all years. If you retire

10 years after the last deposit and the first retirement benefit is withdrawn at the end of the first month of year 31, the equivalent monthly worth of each retirement payment to you and your heirs forever is closest to

a. $3290 per month.
b. $2820 per month.
c. $4200 per month.
d. $4180 per month.

5.32 If the present worth PW of a perpetual investment and the interest rate are known, the AW value is determined using the relation(s)

Relation 1: AW = PW divided by i

Relation 2: AW = PW times i

Relation 3: AW = PW times the $(A/P,i\%,\infty)$ factor

a. 1 only.
b. 2 and 3.
c. 2 only.
d. 1 and 3.

Rate of Return Analysis

Simon Fell/Getty Images

Although the most commonly quoted measure of economic worth for a project or alternative is the rate of return (ROR), its meaning is easily misinterpreted, and the methods to determine ROR are often applied incorrectly. This chapter presents the methods by which one, two or more alternatives are evaluated using the ROR procedure. The ROR is known by several other names: internal rate of return (IRR), return on investment (ROI), and profitability index (PI), to name three.

In some cases, more than one ROR value may satisfy the rate of return equation. This chapter describes how to recognize this possibility and presents an approach to find the multiple values. Alternatively, one unique ROR value can be obtained by using an external reinvestment rate established independently of the project cash flows.

Objectives

Purpose: Understand the meaning of rate of return (ROR) and perform ROR calculations when considering one or more alternatives.

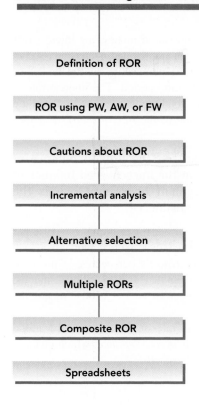

Definition of ROR	1. State the meaning of rate of return.
ROR using PW, AW, or FW	2. Calculate the rate of return using a present worth, annual worth, or future worth equation.
Cautions about ROR	3. Understand the difficulties of using the ROR method, relative to PW, AW, and FW methods.
Incremental analysis	4. Tabulate incremental cash flows and interpret ROR on the incremental investment.
Alternative selection	5. Select the best of multiple alternatives using an incremental ROR analysis.
Multiple RORs	6. Determine the maximum number of possible ROR values and their values for a cash flow series.
Composite ROR	7. Calculate the composite rate of return using a stated reinvestment rate.
Spreadsheets	8. Use a spreadsheet to perform ROR analysis of one or more alternatives.

6.1 INTERPRETATION OF ROR VALUES

As stated in Chapter 1, interest rate and rate of return refer to the same thing. We commonly use the term *interest rate* when discussing borrowed money and *rate of return* when dealing with investments.

From the perspective of someone who has borrowed money, the interest rate is applied to the *unpaid balance* so that the total loan amount and interest are paid in full exactly with the last loan payment. From the perspective of the lender or investor there is an *unrecovered balance* at each time period. The interest rate is the return on this unrecovered balance so that the total amount lent and interest are recovered exactly with the last receipt. Calculation of the *rate of return* describes both of these perspectives.

> **Rate of return (ROR) is the rate paid on the unpaid balance of borrowed money, or the rate earned on the unrecovered balance of an investment, so that the final payment or receipt brings the balance to exactly zero with interest considered.**

The rate of return is expressed as a percent per period, for example, $i = 10\%$ per year. It is stated as a positive percentage; the fact that interest paid on a loan is actually a negative rate of return from the borrower's perspective is not considered. The numerical value of i can range from -100% to infinity, that is, $-100\% < i < \infty$. In terms of an investment, a return of $i = -100\%$ means the entire amount is lost.

The definition above does *not* state that the rate of return is on the initial amount of the investment; rather it is on the *unrecovered balance,* which changes each time period. Example 6.1 illustrates this difference.

EXAMPLE 6.1   Wells Fargo Bank lent a newly graduated engineer $1000 at $i = 10\%$ per year for 4 years to buy home office equipment. From the bank's perspective (the lender), the investment in this young engineer is expected to produce an equivalent net cash flow of $315.47 for each of 4 years.

$$A = \$1000(A/P,10\%,4) = \$315.47$$

This represents a 10% per year rate of return on the bank's unrecovered balance. Compute the amount of the unrecovered investment for each of the 4 years using (*a*) the rate of return on the unrecovered balance (the correct basis) and (*b*) the return on the initial $1000 investment (the incorrect basis).

Solution

   a. Table 6.1 shows the unrecovered balance at the end of each year in column 6 using the 10% rate on the *unrecovered balance at the beginning of the year.*

TABLE 6.1 **Unrecovered Balances Using a Rate of Return of 10% on the Unrecovered Balance**

(1)	(2)	(3) = 0.10 × (2)	(4)	(5) = (4) − (3)	(6) = (2) + (5)
Year	Beginning Unrecovered Balance	Interest on Unrecovered Balance	Cash Flow	Recovered Amount	Ending Unrecovered Balance
0	—	—	$−1,000.00	—	—
1	$−1,000.00	$100.00	+315.47	$215.47	$−784.53
2	−784.53	78.45	+315.47	237.02	−547.51
3	−547.51	54.75	+315.47	260.72	−286.79
4	−286.79	28.68	+315.47	286.79	0
		$261.88		$1,000.00	

TABLE 6.2 **Unrecovered Balances Using a 10% Return on the Initial Amount**

(1)	(2)	(3) = 0.10 × (2)	(4)	(5) = (4) − (3)	(6) = (2) + (5)
Year	Beginning Unrecovered Balance	Interest on Initial Amount	Cash Flow	Recovered Amount	Ending Unrecovered Balance
0	—	—	$−1,000.00	—	—
1	$−1,000.00	$100	+315.47	$215.47	$−784.53
2	−784.53	100	+315.47	215.47	−569.06
3	−569.06	100	+315.47	215.47	−353.59
4	−353.59	100	+315.47	215.47	−138.12
		$400		$861.88	

After 4 years the total $1000 is recovered, and the balance in column 6 is exactly zero.

b. Table 6.2 shows the unrecovered balance if the 10% return is always figured on the *initial $1000*. Column 6 in year 4 shows a remaining unrecovered amount of $138.12, because only $861.88 is recovered in the 4 years (column 5).

Because rate of return is the interest rate on the unrecovered balance, the computations in *Table 6.1 present a correct interpretation of a 10% rate of return.* An interest rate applied to the original principal represents a higher rate than stated. From the standpoint of the borrower, it is better that interest is charged on the unpaid balance than on the initial amount borrowed.

6.2 ROR CALCULATION

The basis for calculating an unknown rate of return is an *equivalence* relation in PW, AW, or FW terms. The objective is to *find the interest rate,* represented as i^*, at which the cash flows are equivalent. The calculations are the reverse of those made in previous chapters, where the interest rate was known. For example, if you invest $1000 now and are promised payments of $500 three years from now and $1500 five years from now, the rate of return relation using PW factors is

$$1000 = 500(P/F,i^*,3) + 1500(P/F,i^*,5) \qquad \textbf{[6.1]}$$

The value of i^* is sought (see Figure 6.1). Move the $1000 to the right side in Equation [6.1].

$$0 = -1000 + 500(P/F,i^*,3) + 1500(P/F,i^*,5)$$

The equation is solved to obtain $i^* = 16.9\%$ per year. The rate of return will always be greater than zero if the total amount of receipts is greater than the total amount of disbursements.

It should be evident that rate of return relations are merely a rearrangement of a present worth equation. That is, if the above interest rate is known to be 16.9%, and it is used to find the present worth of $500 three years from now and $1500 five years from now, the PW relation is

$$PW = 500(P/F,16.9\%,3) + 1500(P/F,16.9\%,5) = \$1000$$

This illustrates that rate of return and present worth equations are set up in exactly the same fashion. The only differences are what is given and what is sought. The PW-based ROR equation can be generalized as

$$0 = -PW_D + PW_R \qquad \textbf{[6.2]}$$

where PW_D = present worth of costs or disbursements
    $PW_R \doteq$ present worth of incomes or receipts

Annual worth or future worth values can also be used in Equation [6.2].

There are two ways to determine i^* once the PW relation is established: solution via trial and error by hand and solution by spreadsheet function. The second is faster; the first helps in understanding how ROR computations work.

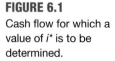

FIGURE 6.1
Cash flow for which a value of i^* is to be determined.

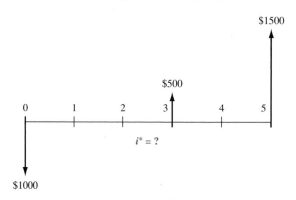

i* Using Trial and Error by Hand

The general procedure for using a PW-based equation is

1. Draw a cash flow diagram.
2. Set up the rate of return equation in the form of Equation [6.2].
3. Select values of i by trial and error until the equation is balanced.

The next two examples illustrate PW and AW equivalence relations to find i^*.

i* by Spreadsheet

The fastest way to determine an i^* value, when there is a series of equal cash flows (A series), is to apply the RATE function. This is a powerful one-cell function, where it is acceptable to have a separate P value in year 0 and an F value in year n that is separate from the series A amount. The format is = RATE(n,A,P,F). If cash flows vary over the years, the IRR function is used to determine i^*. These two functions are illustrated in the last section of this chapter.

EXAMPLE 6.2

The HVAC engineer for a company that constructed one of the world's tallest buildings (Burj Dubai in the United Arab Emirates) requested that $500,000 be spent on software and hardware to improve the efficiency of the environmental control systems. This is expected to save $10,000 per year for 10 years in energy costs and $700,000 at the end of 10 years in equipment refurbishment costs. Find the rate of return.

Solution

For trial-and-error use the procedure based on a PW equation.

1. Figure 6.2 shows the cash flow diagram.
2. Use Equation [6.2] format for the ROR equation.

$$0 = -500{,}000 + 10{,}000(P/A,i^*,10) + 700{,}000(P/F,i^*,10)$$

3. Try $i = 5\%$.

$$0 = -500{,}000 + 10{,}000(P/A,5\%,10) + 700{,}000(P/F,5\%,10)$$
$$0 < \$6946$$

The result is positive, indicating that the return is more than 5%. Try $i = 6\%$.

$$0 = -500{,}000 + 10{,}000(P/A,6\%,10) + 700{,}000(P/F,6\%,10)$$
$$0 > \$-35{,}519$$

Since 6% is too high, linearly interpolate between 5% and 6%.

$$i^* = 5.00 + \frac{6946 - 0}{6946 - (-35{,}519)}(1.0)$$
$$= 5.00 + 0.16 = 5.16\%$$

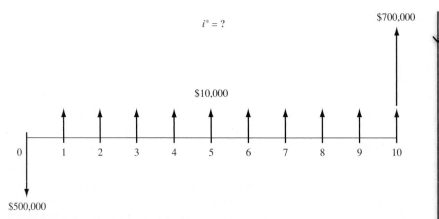

FIGURE 6.2   Cash flow diagram, Example 6.2.

Comment: For a spreadsheet solution, the single-cell entry = RATE(10,10000, −500000,700000) displays $i* = 5.16\%$.

EXAMPLE 6.3   Allied Materials needs $8 million in new capital for expanded composites manufacturing. It is offering small-denomination corporate bonds at a deep discount price of $800 for a 4% $1000 face value bond that matures in 20 years and pays the dividend semiannually. Find the nominal and effective annual rates, compounded semiannually, that Allied is paying an investor.

Solution

By Equaion [4.1], the semiannual income from the bond dividend is $I = 1000(0.04)/2 = \$20$. This will be received by the investor for a total of 40 6-month periods. The AW-based relation to calculate the effective semiannual rate is

$$0 = -800(A/P,i^*,40) + 20 + 1000(A/F,i^*,40)$$

By trial-and-error and linear interpolation, $i^* = 2.87\%$ semiannually. The nominal annual rate is i^* times 2.

Nominal $i = 2.87(2) = 5.74\%$ per year compounded semiannually

Using Equation [3.2], the effective annual rate is

Effective $i = (1.0287)^2 - 1 = 0.0582$ (5.82% per year)

Comment: The spreadsheet function = RATE(40,20,−800,1000) results in $i^* = 2.84\%$ semiannually. This can be compared with the trial-and-error result of 2.87%.

6.3 CAUTIONS WHEN USING THE ROR METHOD

The rate of return method is commonly used in engineering and business settings to evaluate one project, as discussed above, and to select one alternative from two or more, as explained later. When applied correctly, the ROR technique will always result in a good decision, indeed, the same one as with a PW, AW, or FW analysis. However, there are some assumptions and difficulties with ROR analysis that must be considered when calculating $i*$ and in interpreting its real-world meaning for a particular project. The summary that follows applies for solutions by hand and by spreadsheet.

- *Computational difficulty versus understanding.* Especially in obtaining a trial-and-error solution by hand, the computations rapidly become very involved. Spreadsheet solution is easier; however, there are no spreadsheet functions that offer the same level of understanding to the learner as that provided by hand solution of PW, AW, and FW relations.
- *Special procedure for multiple alternatives.* To correctly use the ROR method to choose from two or more mutually exclusive alternatives requires an analysis procedure significantly different from that used in other methods. Section 6.5 explains this procedure.
- *Multiple $i*$ values.* Depending upon the sequence of cash flow disbursements and receipts, there may be more than one real-number root to the ROR equation, resulting in more than one $i*$ value. There is a procedure to use the ROR method and obtain one unique $i*$ value. Sections 6.6 and 6.7 cover these aspects of ROR analysis.
- *Reinvestment at $i*$.* The PW, AW, and FW methods assume that any net positive investment (i.e., net positive cash flows once the time value of money is considered) are *reinvested at the MARR*. But the ROR method assumes reinvestment at the $i*$ *rate*. When $i*$ is not close to the MARR (e.g., if $i*$ is substantially larger than MARR), or if multiple $i*$ values exist, this is an unrealistic assumption. In such cases, the $i*$ value is not a good basis for decision making.

In general, it is good practice to use the MARR to determine PW, AW, or FW. If the ROR value is needed, find $i*$ while taking these cautions into consideration. As an illustration, if a project is evaluated at MARR = 15% and has PW < 0, there is no need to calculate $i*$, because $i* < 15\%$. However, if PW > 0, then calculate the exact $i*$ and report it along with the conclusion that the project is financially justified.

6.4 UNDERSTANDING INCREMENTAL ROR ANALYSIS

From previous chapters, we know that the PW (or AW or FW) value calculated at the MARR identifies the one mutually exclusive alternative that is best from the economic viewpoint. The best alternative is simply the one that has the numerically largest PW value. (This represents the equivalently lowest net cost or highest net revenue cash flow.) In this section, we learn that the ROR can also be used to identify the best alternative; however, it is *not* always as simple as selecting the highest rate of return alternative.

Let's assume that a company uses a MARR of 16% per year, that the company has $90,000 available for investment, and that two alternatives (A and B) are being evaluated. Alternative A requires an investment of $50,000 and has an internal rate of return i_A^* of 35% per year. Alternative B requires $85,000 and has an i_B^* of 29% per year. Intuitively we may conclude that the better alternative is the one that has the larger return, A in this case. However, this is not necessarily so. While A has the higher projected return, it requires an initial investment that is much less than the total money available ($90,000). What happens to the investment capital that is left over? It is generally assumed that excess funds will be invested at the company's MARR, as we learned earlier. Using this assumption, it is possible to determine the consequences of the alternative investments. If alternative A is selected, $50,000 will return 35% per year. The $40,000 left over will be invested at the MARR of 16% per year. The rate of return on the total capital available, then, will be the weighted average. Thus, if alternative A is selected,

$$\text{Overall ROR}_A = \frac{50,000(0.35) + 40,000(0.16)}{90,000} = 26.6\%$$

If alternative B is selected, $85,000 will be invested at 29% per year, and the remaining $5000 will earn 16% per year. Now the weighted average is

$$\text{Overall ROR}_B = \frac{85,000(0.29) + 5000(0.16)}{90,000} = 28.3\%$$

These calculations show that even though the i^* for alternative A is higher, alternative B presents the better overall ROR for the $90,000. If either a PW, AW, or FW comparison is conducted using the MARR of 16% per year as i, alternative B will be chosen.

This example illustrates a major dilemma of the rate of return method when comparing alternatives: Under some circumstances, alternative ROR (i^*) values do not provide the same ranking of alternatives as do the PW, AW, and FW analyses. To resolve the dilemma, conduct an *incremental analysis* between two alternatives at a time and base the alternative selection on the *ROR of the incremental cash flow series.*

A standardized format (Table 6.3) simplifies the incremental analysis. If the alternatives have *equal lives,* the year column will go from 0 to n. If the alternatives have *unequal lives,* the year column will go from 0 to the LCM (least common multiple) of the two lives. The use of the LCM is necessary because *incremental ROR analysis requires equal-service comparison* between alternatives. Therefore, all the assumptions and requirements developed earlier apply for any incremental ROR evaluation. When the LCM of lives is used, the salvage value and reinvestment in each alternative are shown at their respective times. If a study period is defined, the cash flow tabulation is for the specified period.

For the purpose of simplification, use the convention that between two alternatives, the one with the *larger initial investment* will be regarded as *alternative B.* Then, for each year in Table 6.3,

Incremental cash flow = cash flow$_B$ − cash flow$_A$ [6.3]

TABLE 6.3 **Format for Incremental Cash Flow Tabulation**

| Year | Cash Flow | | Incremental Cash Flow |
	Alternative A (1)	Alternative B (2)	(3) = (2) − (1)
0			
1			
.			
.			
.			

As discussed in Chapter 4, there are two types of alternatives.

Revenue alternative, where there are both negative and positive cash flows.

Cost alternative, where all cash flow estimates are negative.

In either case, Equation [6.3] is used to determine the incremental cash flow series with the sign of each cash flow carefully determined. The next two examples illustrate incremental cash flow tabulation of cost alternatives of equal and different lives. A later example treats revenue alternatives.

A tool and die company in Sydney is considering the purchase of a drill press with fuzzy-logic software to improve accuracy and reduce tool wear. The company has the opportunity to buy a slightly used machine for $15,000 or a new one for $21,000. Because the new machine is a more sophisticated model, its operating cost is expected to be $7000 per year, while the used machine is expected to require $8200 per year. Each machine is expected to have a 25-year life with a 5% salvage value. Tabulate the incremental cash flow.

EXAMPLE 6.4

Solution

Incremental cash flow is tabulated in Table 6.4 using Equation [6.3]. The subtraction performed is (new − used) since the new machine has a larger initial cost. The salvage values are separated from the year 25 cash flow for clarity.

TABLE 6.4 **Cash Flow Tabulation for Example 6.4**

| Year | Cash Flow | | Incremental Cash Flow (New − Used) |
	Used Press	New Press	
0	$ −15,000	$ −21,000	$ −6,000
1–25	−8,200	−7,000	+1,200
25	+750	+1,050	+300
Total	$−219,250	$−194,950	$+24,300

Comment: When the cash flow columns are subtracted, the difference between the totals of the two cash flow series should equal the total of the incremental cash flow column. This merely provides a check of the addition and subtraction in preparing the tabulation.

EXAMPLE 6.5

Sandersen Meat Processors has asked its lead process engineer to evaluate two different types of conveyors for the beef cutting line. Type A has an initial cost of $70,000 and a life of 3 years. Type B has an initial cost of $95,000 and a life expectancy of 6 years. The annual operating cost (AOC) for type A is expected to be $9000, while the AOC for type B is expected to be $7000. If the salvage values are $5000 and $10,000 for type A and type B, respectively, tabulate the incremental cash flow using their LCM.

Solution

The LCM of 3 and 6 is 6 years. In the incremental cash flow tabulation for 6 years (Table 6.5), note that the reinvestment and salvage value of A is shown in year 3.

TABLE 6.5   Incremental Cash Flow Tabulation, Example 6.5

Year	Cash Flow Type A	Cash Flow Type B	Incremental Cash Flow (B − A)
0	$ −70,000	$ −95,000	$−25,000
1	−9,000	−7,000	+2,000
2	−9,000	−7,000	+2,000
3	−70,000 / −9,000 / +5,000	−7,000	+67,000
4	−9,000	−7,000	+2,000
5	−9,000	−7,000	+2,000
6	−9,000 / +5,000	−7,000 / +10,000	+7,000
	$−184,000	$−127,000	$+57,000

Once the incremental cash flows are tabulated, determine the incremental rate of return on the extra amount required by the larger investment alternative. This rate, termed Δi^*, represents the return over n years expected on the optional extra investment in year 0. The general selection guideline is to make the extra investment if the incremental rate of return meets or exceeds the MARR. Briefly stated,

If $\Delta i^* \geq$ MARR, select the larger investment alternative (labeled B).

Otherwise, select the lower investment alternative (labeled A).

Use of this guideline is demonstrated in Section 6.5. The best rationale for understanding incremental ROR analysis is to think of only *one alternative* under consideration, that alternative being represented by the incremental cash flow series. Only if the return on the extra investment, which is the Δi^* value, meets or exceeds the MARR is it financially justified, in which case the larger investment alternative should be selected.

As a matter of efficiency, if the analysis is between *multiple revenue alternatives,* an acceptable procedure is to initially determine each alternative's i^* and remove those alternatives with $i^* <$ MARR, since their return is too low. Then complete the incremental analysis for the remaining alternatives. If no alternative i^* meets or exceeds the MARR, the do-nothing alternative is economically the best. This initial "weeding out" can't be done for cost alternatives since they have no positive cash flows.

When *independent projects* are compared, no incremental analysis is necessary. All projects with $i^* \geq$ MARR are acceptable. Limitations on the initial investment amount are considered separately, as discussed in Section 4.5.

6.5 ROR EVALUATION OF TWO OR MORE MUTUALLY EXCLUSIVE ALTERNATIVES

When selecting from two or more mutually exclusive alternatives on the basis of ROR, equal-service comparison is required, and an incremental ROR analysis must be used. The incremental ROR value between two alternatives (B and A) is correctly identified as Δi^*_{B-A}, but it is usually shortened to Δi^*. The selection guideline, as introduced in Section 6.4, is:

> **Select the alternative that:**
> 1. **requires the largest initial investment, and**
> 2. **has a $\Delta i^* \geq$ MARR, indicating that the extra initial investment is economically justified.**

If the higher initial investment is not justified, it should not be made as the extra funds could be invested elsewhere.

Before conducting the incremental evaluation, classify the alternatives as *cost* or *revenue* alternatives. The incremental comparison will differ slightly for each type.

Cost: Evaluate alternatives only against each other.

Revenue: First evaluate against do-nothing (DN), then against each other.

The following procedure for comparing multiple, mutually exclusive alternatives, using a PW-based equivalence relation, can now be applied.

1. Order the alternatives by increasing initial investment. *For revenue alternatives* add DN as the first alternative.
2. Determine the incremental cash flow between the first two ordered alternatives (B − A) over their least common multiple of lives. (For revenue alternatives, the first ordered alternative is DN.)

3. Set up a PW-based relation of this incremental cash flow series and determine Δi^*, the incremental rate of return.
4. If $\Delta i^* \geq$ MARR, eliminate A; B is the survivor. Otherwise, A is the survivor.
5. Compare the survivor to the next alternative. Continue to compare alternatives using steps (2) through (4) until only one alternative remains as the survivor.

The next two examples illustrate this procedure for cost and revenue alternatives, respectively, as well as for equal and different-life alternatives.

For completeness's sake, it is important to understand the procedural difference for comparing *independent projects*. If the projects are independent rather than mutually exclusive, the preceding procedure does not apply. As mentioned in Section 6.4, no incremental evaluation is necessary; all projects with $i^* \geq$ MARR are selected, thus comparing each project against the MARR, not each other.

EXAMPLE 6.6   As the film of an oil spill from an at-sea tanker moves ashore, great losses occur for aquatic life as well as shoreline feeders and dwellers, such as birds. Environmental engineers and lawyers from several international petroleum corporations and transport companies—Exxon-Mobil, BP, Shell, and some transporters for OPEC producers—have developed a plan to strategically locate throughout the world newly developed equipment that is substantially more effective than manual procedures in cleaning crude oil residue from bird feathers. The Sierra Club, Greenpeace, and other international environmental interest groups are in favor of the initiative. Alternative machines from manufacturers in Asia, America, Europe, and Africa are available with the cost estimates in Table 6.6. Annual cost estimates are expected to be high to ensure readiness at any time. The company representatives have agreed to use the average of the corporate MARR values, which results in MARR = 13.5%. Use incremental ROR analysis to determine which manufacturer offers the best economic choice.

Solution

Follow the procedure for incremental ROR analysis.

1. These are *cost alternatives* and are arranged by increasing first cost.
2. The lives are all the same at $n = 8$ years. The B − A incremental cash flows are indicated in Table 6.7. The estimated salvage values are shown separately in year 8.

TABLE 6.6   Costs for Four Alternative Machines, Example 6.6

	Machine A	Machine B	Machine C	Machine D
First cost, $	−5,000	−6,500	−10,000	−15,000
AOC, $/year	−3,500	−3,200	−3,000	−1,400
Salvage value, $	+500	+900	+700	+1,000
Life, years	8	8	8	8

TABLE 6.7 Incremental Cash Flow for Comparison of Machine B-to-A

Year	Cash Flow Machine A	Cash Flow Machine B	Incremental Cash Flow for (B − A)
0	$−5000	$−6500	$−1500
1–8	−3500	−3200	+300
8	+500	+900	+400

3. The following PW relation for (B − A) results in $\Delta i^* = 14.57\%$.

$$0 = -1500 + 300(P/A,\Delta i^*,8) + 400(P/F,\Delta i^*,8)$$

4. Since this return exceeds MARR $= 13.5\%$, A is eliminated and B is the survivor.
5. The comparison of C-to-B results in the elimination of C based on $\Delta i^* = -18.77\%$ from the incremental relation

$$0 = -3500 + 200(P/A,\Delta i^*,8) - 200(P/F,\Delta i^*,8)$$

The D-to-B incremental cash flow PW relation for the final evaluation is

$$0 = -8500 + 1800(P/A,\Delta i^*,8) + 100(P/F,\Delta i^*,8)$$

With $\Delta i^* = 13.60\%$, machine D is the overall, though marginal, survivor of the evaluation; it should be purchased and located in the event of oil spill accidents.

Harold owns a construction company that subcontracts to international power equipment corporations such as GE, ABB, Siemens, and LG. For the last 4 years he has leased crane and lifting equipment for $32,000 annually. He now wishes to purchase similar equipment. Use an MARR of 12% per year to determine if any of the options detailed in Table 6.8 are financially justified.

EXAMPLE 6.7

Solution

Apply the incremental ROR procedure with MARR $= 12\%$ per year.

1. Because these are revenue alternatives, add the do-nothing option as the first alternative and order the remaining. The comparison order is DN, 4, 2, 1, 3.

TABLE 6.8 Estimates for Alternative Equipment, Example 6.7

Alternative	1	2	3	4
First cost, $	−80,000	−50,000	−145,000	−20,000
Annual cost, $/year	−28,000	−26,000	−16,000	−21,000
Annual revenue, $/year	61,000	43,000	51,000	29,000
Life, years	4	4	8	4

2. Each annual cash flow for the DN alternative is $0. Therefore, the incremental cash flows for comparing 4-to-DN are the same as those for alternative 4.
3. The Δi^* for the comparison 4-to-DN is actually the project ROR. Since $n = 4$ years, the PW relation and return are

$$0 = -20{,}000 + (29{,}000 - 21{,}000)\,(P/A,\Delta i^*,4)$$
$$(P/A,\Delta i^*,4) = 2.5$$
$$\Delta i^* = 21.9\%$$

4. Since $21.9\% > 12\%$, eliminate DN and proceed with the 2-to-4 comparison.
5. Both alternatives 2 and 4 have $n = 4$. The incremental cash flows are $\$-30{,}000$ in year 0 and $(43{,}000-26{,}000) - (29{,}000-21{,}000) = \$+9000$ in years 1 to 4. Incremental analysis results in $\Delta i^* = 7.7\%$ from the PW relation

$$0 = -30{,}000 + 9000(P/A,\Delta i^*,4)$$
$$\Delta i^* = 7.7\%$$

Alternative 4 is, again, the survivor. Continue with the comparison 1-to-4 to obtain

$$0 = -60{,}000 + 25{,}000(P/A,\Delta i^*,4)$$
$$\Delta i^* = 24.1\%$$

Now alternative 1 is the survivor. The final comparison of 3 to 1 must be conducted over the LCM of 8 years for equal service. Table 6.9 details the incremental cash flows, including the alternative 1 repurchase in year 4. The PW relation is

$$0 = -65{,}000 + 2000(P/A,\Delta i^*,8) + 80{,}000(P/F,\Delta i^*,4)$$
$$\Delta i^* = 10.1\%$$

TABLE 6.9   Incremental Cash Flows for the Comparison of Alternatives 3-to-1, Example 6.7

| | Alternative Cash Flows | | Incremental |
| | Alternative | Alternative | Cash Flow |
Year	1	3	for (3–1)
0	$-80,000	$-145,000	$-65,000
1	+33,000	+35,000	+2,000
2	+33,000	+35,000	+2,000
3	+33,000	+35,000	+2,000
4	-47,000	+35,000	+82,000
5	+33,000	+35,000	+2,000
6	+33,000	+35,000	+2,000
7	+33,000	+35,000	+2,000
8	+33,000	+35,000	+2,000

Since $10.1\% < 12\%$, eliminate 3; declare alternative 1 the survivor and select it as the one that is economically justified.

The previous incremental analyses were performed using the PW relations. It is equally correct to us AW-based or FW-based analysis; however, the LCM of lives must be used since ROR analysis requires an equal-service comparison. Consequently, there is usually no advantage to developing AW relations to find Δi^* for different-life alternatives.

The use of the IRR spreadsheet function can greatly speed up incremental ROR comparison of multiple alternatives, especially for those with unequal lives. This is fully illustrated in the last section of this chapter.

6.6 MULTIPLE ROR VALUES

For some cash flow series (net for one project or incremental for two alternatives) it is possible that more than one unique rate of return i^* exists. This is referred to as *multiple i^* values*. It is difficult to complete the economic evaluation when multiple i^* values are present, since none of the values may be the *correct* rate of return. The discussion that follows explains how to predict the number of i^* values in the range -100% to infinity, how to determine their values, and how to resolve the difficulty of knowing the "true" ROR value (if this is important). If using ROR evaluation is not absolutely necessary, a simple way to avoid this dilemma is to use the PW, AW, or FW evaluation method at the MARR.

In actuality, finding the rate of return is solving for the root(s) of an nth order polynomial. *Conventional* or *simple* cash flows have only one sign change over the entire series, as shown in Table 6.10. Commonly, this is negative in year 0 to positive at some time during the series. There is a unique, real number i^* value for a conventional series. A *nonconventional* series (Table 6.10) has more than one sign change and multiple roots may exist. The *cash flow rule of signs* (based upon Descartes' rule) states:

> **The maximum number of i^* values is equal to the number of sign changes in the cash flow series.**

When applying this rule, zero cash flow values are disregarded.

Prior to determining the multiple i^* values, a second rule can be applied to indicate that a unique nonnegative i^* value exists for a nonconventional series. It is the *cumulative cash flow test* (also called Norstom's criterion). It states:

> **There is one nonnegative i^* value if the cumulative cash flow series $S_0, S_1, \ldots, S_n$ changes sign only once and $S_n \neq 0$.**

To perform the test, count the number of sign changes in the S_t series, where

$$S_t = \text{cumulative cash flows through period } t$$

TABLE 6.10 **Examples of Conventional and Nonconventional Net or Incremental Cash Flows for a 6-year Period**

Type of Series	Sign on Cash Flow							Number of Sign Changes
	0	1	2	3	4	5	6	
Conventional	−	+	+	+	+	+	+	1
Conventional	−	−	−	+	+	+	+	1
Conventional	+	+	+	+	+	−	−	1
Nonconventional	−	+	+	+	−	−	−	2
Nonconventional	+	+	−	−	−	+	+	2
Nonconventional	−	+	−	−	+	+	+	3

More than one sign change provides no information, and the rule of signs is applied to indicate the possible number of i^* values.

Now the unique or multiple i^* value(s) can be determined using trial-and-error hand solution or using the IRR spreadsheet function that incorporates the "guess" option to search for the multiple i^* values. The next example illustrates the two rules on sign changes and hand solution. The use of spreadsheets is shown in the last section of the chapter.

EXAMPLE 6.8 The engineering design and testing group for Honda Motor Corp. does contract-based work for automobile manufacturers throughout the world. During the last 3 years, the net cash flows for contract payments have varied widely, as shown below, primarily due to a large manufacturer's inability to pay its contract fee.

Year	0	1	2	3
Cash Flow ($1000)	+2000	−500	−8100	+6800

a. Determine the maximum number of i^* values that may satisfy the ROR relation.

b. Write the PW-based ROR relation and approximate the i^* value(s) by plotting PW versus i.

Solution

a. Table 6.11 shows the annual cash flows and cumulative cash flows. Since there are two sign changes in the cash flow sequence, the rule of signs indicates a maximum of two i^* values. The *cumulative* cash flow sequence has two sign changes, indicating there is not just one nonnegative root. The conclusion is that as many as two i^* values can be found.

TABLE 6.11 Cash Flow and Cumulative Cash Flow Sequences, Example 6.8

Year	Cash Flow ($1000)	Sequence Number	Cumulative Cash Flow ($1000)
0	+2000	S_0	+2000
1	−500	S_1	+1500
2	−8100	S_2	−6600
3	+6800	S_3	+200

b. The PW relation is

$$PW = 2000 - 500(P/F,i,1) - 8100(P/F,i,2) + 6800(P/F,i,3)$$

Select values of i to find the two i^* values, and plot PW versus i. The PW values are shown below and plotted in Figure 6.3 (using a smooth approximation) for i values of 0, 5, 10, 20, 30, 40, and 50%. The characteristic parabolic shape for a second-degree polynomial is obtained, with PW crossing the i axis at approximately $i_1^* = 8$ and $i_2^* = 41\%$.

$i\%$	0	5	10	20	30	40	50
PW ($1000)	+200	+51.44	−39.55	−106.13	−82.01	−11.83	+81.85

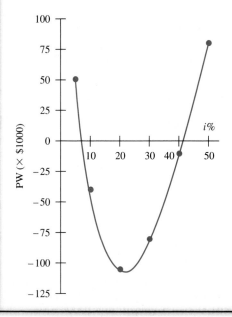

FIGURE 6.3 Present worth of cash flows at several interest rates, Example 6.8.

EXAMPLE 6.9   An American-Australian joint venture has been contracted to provide the train cars for a 25-mile subway system using new tunnel-boring and track-design technologies. Austin, Texas was selected as the proof-of-concept site based on its variety in landscape features (hilly terrain, lake and green space areas, and relatively low precipitation) and its public environmental mindedness. Selection of either a Swiss or Japanese contractor to provide the gears and power components for the electric transfer motor assemblies resulted in two cost alternatives. Table 6.12 gives the incremental cash flow estimates (in $1000) over the expected 10-year life of the motors. Determine the number of $i*$ values and estimate them graphically.

TABLE 6.12   **Incremental and Cumulative Cash Flow Series, Example 6.9**

	Cash Flow, $1000			Cash Flow, $1000	
Year	Incremental	Cumulative	Year	Incremental	Cumulative
0	−500	−500	6	+800	+200
1	−2000	−2500	7	+400	+600
2	−2000	−4500	8	+300	+900
3	+2500	−2000	9	+200	+1100
4	+1500	−500	10	+100	+1200
5	−100	−600			

Solution

The incremental cash flows form a nonconventional series with three sign changes in years 3, 5, and 6. The cumulative series has one sign change in year 6. This test indicates a single nonnegative root. The incremental ROR is determined from the PW relation (in $1000).

$$0 = -500 - 2000(P/F,\Delta i^*,1) - 2000(P/F,\Delta i^*,2) \cdots + 100(P/F,\Delta i^*,10)$$

Calculation of PW at various i values is plotted (Figure 6.4) to estimate the unique Δi^* of 8% per year.

FIGURE 6.4

Graphical estimation of Δi^* using PW values, Example 6.9.

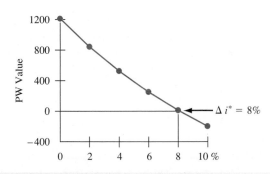

Commonly when multiple i^* or Δi^* values are indicated, there is only one realistic root. Others may be large negative or positive numbers that make no real-world sense and can be neglected. (One clear advantage of using a spreadsheet for ROR evaluation, as described in Section 6.8, is that realistic i^* values are commonly determined first by the functions.) As mentioned previously, the use of a PW, AW, or FW analysis eliminates the multiple ROR dilemma since the MARR is used in all equivalence relations, and excess funds are assumed to earn the MARR. (See Section 6.4 for a quick review.) It is because of the complexities of the ROR method, namely, incremental analysis, the use of LCM for equal service, reinvestment rate assumption, and possible multiple i^* values, that other methods are preferred to ROR. Yet, the ROR result is important in that some people wish to know the estimated return on proposed project(s).

6.7 REMOVING MULTIPLE ROR VALUES BY USING A REINVESTMENT RATE

All ROR values calculated thus far can be termed *internal rates of return (IRR)*. As discussed in Section 6.1, the IRR guarantees that the last receipt or payment is exactly zero with interest considered. No excess funds are generated in any year, so all funds are kept internal to the project. A project can generate excess funds prior to the end of the project when the net cash flow in any year is positive ($NCF_t > 0$). This can result in a nonconventional series, as we learned in Section 6.6. The ROR method assumes these excess funds can earn at any one of the multiple i^* values. This generates ambiguity when the ROR method is used to evaluate alternatives. For example, assume an alternative's nonconventional cash flows have two ROR roots at -2% and 40%. Further, assume that neither is a realistic reinvestment assumption. Instead, excess funds will likely earn at the MARR of 15%. The question "Is the project economically justified?" is not answered by the ROR analysis.

The correct way to find one useful ROR value is to set a reinvestment rate (also called an *external rate of return*) at which all excess cash earns interest, and use it to determine the unique *composite ROR* value, as described below.

> **Remember, the composite rate of return technique works when excess project funds (positive net cash flows in any year) earn at a rate that is stated and is determined externally to the project. This rate is commonly different from any of the multiple ROR values.**

In the preceding example, if excess funds can actually earn 40%, then the one correct, unique i^* is 40%. However, if excess funds can be expected to earn at, say, the MARR of 15%, this technique will find the correct and unique (composite) rate for the project. In either case, the dilemma of multiple ROR values is eliminated.

Consider the internal rate of return calculations for the following cash flows: $10,000 is invested at $t = 0$, $8000 is received in year 2, and $9000 is received in year 5. The PW equation to determine $i*$ is

$$0 = -10,000 + 8000(P/F,i,2) + 9000(P/F,i,5)$$
$$i* = 16.815\%$$

If this rate is used for the unrecovered balances, the investment will be recovered exactly at the end of year 5. The procedure to verify this is identical to that used in Table 6.1, which describes how the ROR works to exactly remove the unrecovered balance with the final cash flow.

In calculating $i*$, no consideration is given to the $8000 available after year 2. What happens if funds released from a project *are* considered in calculating the overall rate of return of a project? After all, something must be done with the released funds. One possibility is to assume the money is reinvested at some stated rate. The ROR method assumes excess funds are external to a project and earn at the $i*$ rate, but this may not be a realistic rate in everyday practice. The common approach is to assume that reinvestment occurs at the MARR. This approach has the advantage of converting a nonconventional cash flow series (with multiple $i*$ values) to a conventional series with one root, which can be considered *the* rate of return.

The rate of earnings used for the released funds is called the *reinvestment rate* (or *external rate of return*) and is symbolized by c. This rate depends upon the market rate available for investments. If a company is making, say, 8% on its daily investments, then $c = 8\%$. It is common practice to set c equal to the MARR. The one interest rate that now satisfies the rate of return equation is called the *composite rate of return (CRR)* and is symbolized by i'. By definition,

> The *composite rate of return i'* **is the unique rate of return for a project when net positive cash flows, which represent money not immediately needed by the project, are reinvested at the reinvestment rate c.**

The term *composite* is used for this unique rate of return because it is derived using another external rate, namely, the reinvestment rate c. If c happens to equal any one of the $i*$ values, then the composite rate i' will equal that $i*$ value. The CRR is also known by the term *return on invested capital (RIC)*. Once the unique i' is determined, it is compared to the MARR to decide on the project's financial viability.

The procedure to determine i' is called the *net-investment procedure*. The technique involves finding the future worth of the net investment amount 1 year in the future. Find the project's net-investment value F_t in year t from F_{t-1} by using the F/P factor for 1 year at the reinvestment rate c if the previous net investment is positive (extra money generated by project), or at the CRR rate i' if the net-investment value is negative (project used all available funds). Mathematically, for each year t set up the relation

$$F_t = F_{t-1}(1 + i) + C_t \qquad \text{[6.4]}$$

where $t = 1, 2, \ldots, n$

   $n =$ total years in project

   $C_t =$ net cash flow in year t

   $i = \begin{cases} c & \text{if } F_{t-1} > 0 \quad \text{(net positive investment)} \\ i' & \text{if } F_{t-1} < 0 \quad \text{(net negative investment)} \end{cases}$

Set the net-investment relation for year n equal to zero ($F_n = 0$) and solve for i'. The i' value obtained is unique for a stated reinvestment rate c.

The development of F_1 through F_3 for the cash flow series below, is illustrated for a reinvestment rate of $c =$ MARR $= 15\%$. Figure 6.5 tracks the development year-by-year.

Year	Cash Flow, $
0	50
1	−200
2	50
3	100

The net investment for year $t = 0$ is $F_0 = \$50$, which is positive, so it returns $c = 15\%$ during the first year. By Equation [6.4], F_1 is

$$F_1 = 50(1 + 0.15) - 200 = \${-142.50}$$

Since the project net investment is now negative (Figure 6.5b), the value F_1 earns interest at i' for year 2.

$$F_2 = F_1(1 + i') + C_2 = -142.50(1 + i') + 50$$

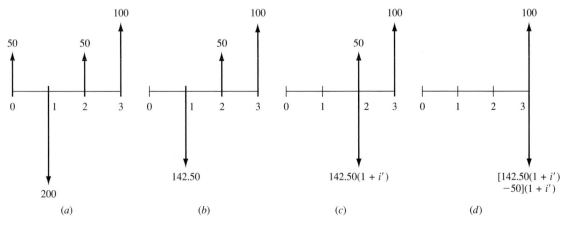

(a)   (b)   (c)   (d)

FIGURE 6.5 Cash flow series for which the composite rate of return i' is computed: (a) original form; equivalent form in (b) year 1, (c) year 2, and (d) year 3.

The i' value is to be determined (Figure 6.5c). Since F_2 will be negative for all $i' > 0$, use i' to set up F_3 as shown in Figure 6.5d.

$$F_3 = F_2(1 + i') + C_3 = [-142.50(1 + i') + 50](1 + i') + 100 \qquad \textbf{[6.5]}$$

Setting Equation [6.5] equal to zero and solving for i' will result in the unique composite rate of return i'. The resulting values are 3.13% and -168%, since Equation [6.5] is a quadratic relation (power 2 for i'). The value of $i' = 3.13\%$ is the correct $i*$ in the range -100% to ∞.

Several comments are in order. If the reinvestment rate c is equal to the internal rate of return $i*$ (or one of the $i*$ values when there are multiple ones), the i' that is calculated will be exactly the same as $i*$; that is, $c = i* = i'$. The closer the c value is to $i*$, the smaller the difference between the composite and internal rates. As mentioned earlier, it is correct to assume that $c =$ MARR, if all throw-off funds from the project can realistically earn at the MARR rate.

A summary of the relations between c, i', and $i*$ follows. These relations are demonstrated in Example 6.10.

Relation between Reinvestment Rate c and $i*$	Relation between i' and $i*$
$c = i*$	$i' = i*$
$c < i*$	$i' < i*$
$c > i*$	$i' > i*$

Remember: This entire net-investment procedure is used when multiple $i*$ values are indicated. Multiple $i*$ values are present when a nonconventional cash flow series does not have a single, unique root. Finally, it is important to remember that this procedure is unnecessary if the PW, AW or FW method is used to evaluate a project at the MARR.

EXAMPLE 6.10   Compute the composite rate of return for the Honda Motor Corp. engineering group in Example 6.8 if the reinvestment rate is (a) 7.47% and (b) the corporate MARR is 20%. The exact multiple $i*$ values are 7.47% and 41.35% per year.

Solution

a. Use the procedure to determine i' for $c = i* = 7.47\%$. All terms are in $1000. The first net-investment expression is $F_0 = \$+2000$. Since $F_0 > 0$, use $c = 7.47\%$ to write F_1 by Equation [6.4].

$$F_1 = 2000(1.0747) - 500 = \$1649.40$$

Since $F_1 > 0$, use $c = 7.47\%$ to determine F_2.

$$F_2 = 1649.40(1.0747) - 8100 = \$-6327.39$$

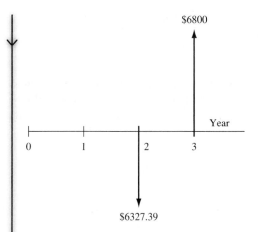

FIGURE 6.6 Equivalent cash flow (in thousands) in year 2 with reinvestment at $c = 7.47\%$, Example 6.10.

Figure 6.6 shows the equivalent cash flow at this time. Since $F_2 < 0$, use i' to express F_3.

$$F_3 = -6327.39(1 + i') + 6800$$

Set $F_3 = 0$ and solve for $i' = 7.47\%$ directly. The CRR is 7.47%, which is the same as c, the reinvestment rate, and one of the multiple i^* values.

b. For MARR $= c = 20\%$, the net-investment series is

$F_0 = +2000$	$(F_0 > 0,\ \text{use } c)$
$F_1 = 2000(1.20) - 500 = \$1900$	$(F_1 > 0,\ \text{use } c)$
$F_2 = 1900(1.20) - 8100 = \$-5820$	$(F_2 < 0,\ \text{use } i')$
$F_3 = -5820(1 + i') + 6800$	

Set $F_3 = 0$ and solve for $i' = 16.84\%$.

Note that since $i' < \text{MARR} = 20\%$, the project is not financially justified. This is verified by calculating PW $= \$-106$ at 20% for the original cash flows.

6.8 USING SPREADSHEETS FOR ROR ANALYSIS

Spreadsheets greatly reduce the time needed to perform a rate of return analysis through the use of the RATE or IRR functions. Coupled with the NPV function to develop a spreadsheet plot of PW versus i, the IRR function can perform virtually any analysis for one project, perform incremental analysis of multiple alternatives, and find multiple i^* values for nonconventional cash flows and incremental cash flows between two alternatives.

If the annual cash flows are all equal with separate P and/or F values, find i^* using

$$= \text{RATE}(n, A, P, F)$$

If cash flows vary throughout the n years, find one or multiple $i*$ values using

$$= \text{IRR(first\_cell:last\_cell, guess)}$$

For IRR, each cash flow must be entered in succession by spreadsheet row or column. A "zero" cash flow year must be entered as "0" so the year is accounted for. "Guess," an optional entry that starts the ROR analysis, is used most commonly to find multiple $i*$ values for nonconventional cash flows, or if the #NUM error is displayed when IRR is initiated without a guess entry. The next two examples illustrate RATE, IRR, and NPV use as follows:

> *One project*—RATE, IRR, and NPV for single and multiple $i*$ values (Example 6.11)
>
> *Multiple alternatives*—IRR for incremental evaluation (Example 6.12)

EXAMPLE 6.11   Two brothers, Gerald and Henry, own the Edwards Service Company in St. Johns, Newfoundland. It provides onshore services for spent lubricants from North Atlantic offshore platforms. The company needs immediate cash flow. Because the Edwards have an excellent reputation among major oil producers, they have been offered an 8-year contract that pays $200,000 total with 50% upfront and 50% at the end of the suggested 8-year contract. The estimated annual cost for Edwards to provide the services is $30,000. Assume you are the financial person for Edwards. Is the project justified if the brothers want to make at least 8% per year? Use spreadsheet functions and charts to perform a thorough ROR analysis.

Solution

The spreadsheet analysis can be accomplished in several ways, some more thorough than others. The four approaches illustrated here are of increasing thoroughness.

1. Refer to Figure 6.7 (left). The easiest and quickest approach is to develop the RATE function for $n = 8$, $A = \$-30,000$, $P = \$100,000$, and $F = \$100,000$

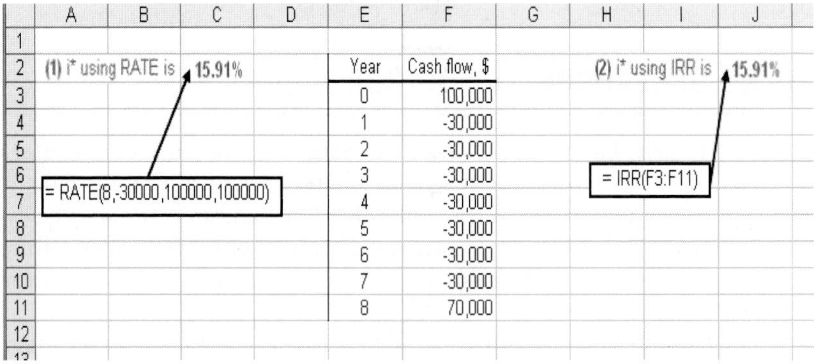

FIGURE 6.7   ROR analysis of a project using the (1) RATE and (2) IRR functions, Example 6.11

in a single cell. The value $i^* = 15.91\%$ is displayed. The project is definitely justified since MARR $= 8\%$.

2. Refer to Figure 6.7 (right). Alternatively, to obtain the same answer, enter the years and associated net cash flows in consecutive cells. Develop the IRR function for the 9 entries to display $i^* = 15.91\%$; the project is justified.

3. Refer to Figure 6.8 (top). The cash flow (upper left of the figure) is a non-conventional series with two sign changes. The cumulative series has one sign change, indicating that a unique nonnegative i^* exists. This is 15.91%, displayed when the function $=$ IRR(B3:B11) is input. However, as pre-dicted by the rule of signs, there is a second, though negative, i^* value at -23.98% found by using the guess option. A variety of guess percentages can be used; Figure 6.8 shows only four, but the results will always be $i_1^* = -23.98\%$ and $i_2^* = 15.91\%$.

   Is the project justified, knowing there are two roots? If reinvestment can be assumed to be at 15.91% rather than the MARR of 8%, the project is justified. If released funds are actually expected to make 8%, the pro-ject's return is between 8 and 15.91%, and it is still justified. However, if funds are actually expected to make less than the MARR, it is not justified

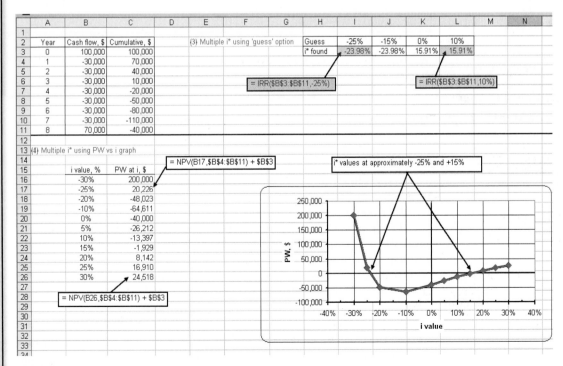

FIGURE 6.8 ROR analysis of a project with multiple i^* values using (3) the IRR function with "guess" option and (4) PW versus i graphical analysis, Example 6.11.

because the true rate is less than MARR, in fact between MARR and -23.98%. *The conclusion is that the ROR analysis does not provide a conclusive answer.* If reinvestment is not assumed at either i^* value, the procedure of Section 6.7 should be performed.

4. Refer to Figure 6.8 (bottom). An excellent graphical way to approximate i^* values is to generate an x-y scatter chart of PW versus i. Use the NPV function with different i values to determine where the PW curve crosses the PW = 0 line. Here i values from -30% to $+30\%$ were chosen. For details, refer to the cell tags for the NPV function formats using the cash flows entered at the top of the figure and different i values. The approximate i^* values are -25% and $+15\%$. Basically, this graphical analysis provides the same information as that of (3).

Comment: Don't make the mistake of thinking that the MIRR (modified IRR) spreadsheet function performs the same ROR analysis on nonconventional series as that which determines the unique composite rate i' as described in Section 6.7. The answer displayed using MIRR is not the i' value, except in the unique situation that both MIRR rates (loan finance rate and positive cash flow reinvestment rate) happen to be the same and this value is exactly one of the roots of the ROR relation. For Example 6.11, Figure 6.8, using the function = MIRR(B3:B11, 15.91%,15.91%), will display $i^* = 15.91\%$. So it is necessary to know the answer up-front *and* assume that reinvestment is at this rate to get the unique answer; not a likely situation!

EXAMPLE 6.12

Perform a spreadsheet analysis of the four alternatives for Harold's construction company in Example 6.7.

Solution

First, review the situation in Example 6.7 for the four revenue alternatives. Figure 6.9 performs the complete analysis in one spreadsheet starting with the addition of the do-nothing alternative and the ordered alternatives DN, 4, 2, 1, and 3.

The top portion provides the estimates, including net annual cash flows in row 6. The middle portion calculates the incremental investment and cash flows for each comparison of two alternatives. Equal lives of 4 years are present for the first three comparisons. Details for the final comparison of 3-to-1 over the LCM of 8 years are included in the cell tags. (Multiple i^* values are not indicated for any incremental series.)

The bottom portion shows conclusions after using the IRR function to find the incremental ROR, comparing it to MARR = 12%, and identifying the surviving alternative. The entire logic of the analysis is identical to that of the hand solution. Alternative 1 is selected

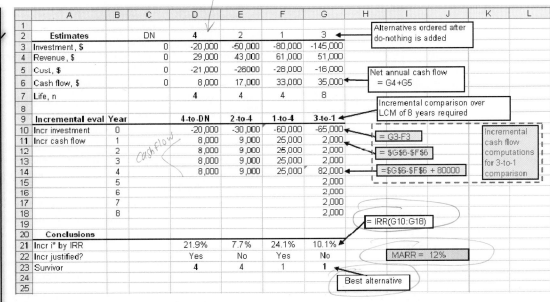

Handwritten annotation (top): ORDERED   -28 +16 = -12 -10 = -2

	A	B	C	D	E	F	G	H	I	J	K	L
1												
2	Estimates		DN	4	2	1	3		Alternatives ordered after do-nothing is added			
3	Investment, $		0	-20,000	-50,000	-80,000	-145,000					
4	Revenue, $		0	29,000	43,000	61,000	51,000					
5	Cust, $		0	-21,000	-26000	-28,000	-16,000		Net annual cash flow = G4+G5			
6	Cash flow, $		0	8,000	17,000	33,000	35,000					
7	Life, n			4	4	4	8					
8									Incremental comparison over LCM of 8 years required			
9	Incremental eval	Year		4-to-DN	2-to-4	1-to-4	3-to-1					
10	Incr investment	0		-20,000	-30,000	-60,000	-65,000		= G3-F3		Incremental cash flow computations for 3-to-1 comparison	
11	Incr cash flow	1		8,000	9,000	25,000	2,000					
12		2		8,000	9,000	25,000	2,000		= G6-F6			
13		3		8,000	9,000	25,000	2,000					
14		4		8,000	9,000	25,000	82,000		=G6-F6 + 80000			
15		5					2,000					
16		6					2,000					
17		7					2,000					
18		8					2,000					
19									= IRR(G10:G18)			
20	Conclusions											
21	Incr i* by IRR			21.9%	7.7%	24.1%	10.1%					
22	Incr justified?			Yes	No	Yes	No		MARR = 12%			
23	Survivor			4	4	1	1					
24									Best alternative			
25												

Handwritten: cash flow

FIGURE 6.9 Incremental ROR analysis of four revenue alternatives, Example 6.12.

SUMMARY

Just as present worth, annual worth, and future worth methods find the best alternative from among several, the ROR calculations serve the same purpose. With ROR analysis, however, the *incremental* cash flow series between two alternatives is evaluated. Alternatives are ordered initially by increasing initial investments, and the pairwise ROR evaluation proceeds from smallest to largest investment. The alternative with the larger incremental $i*$ value is selected as each comparison is conducted. Once eliminated, an alternative cannot be reconsidered.

Rate-of-return calculations performed by hand typically require trial-and-error solutions. Spread-sheets greatly speed up this process. The analysis may result in more than one ROR value depending upon the number of sign changes present in the cash flow series. The cash flow rule of signs and cumulative cash flow test assist in determining if a unique ROR value does exist. The dilemma of multiple rates can be effectively dealt with by calculating the composite rate of return, though this procedure can become cumbersome. In the end, if multiple rates are present, it is strongly recommended to rely on the PW, AW, or FW value determined at the MARR.

PROBLEMS

Understanding ROR

6.1 What is the lowest rate of return (in percent) that is possible?

6.2 What is the *nominal* rate of return per year on an investment that doubles in value every 6 months?

6.3 General Dynamics got a $100 million loan amortized over a 5-year period at 0.5% per month interest. What is the difference in the amount of interest in the second month's payment if interest is charged on the principal of the loan rather than on the unrecovered balance?

6.4 Assume a small engineering firm borrows $300,000 at 0.7% per month interest. If the firm makes a payment of $50,000 at the end of the first month, what is the unrecovered balance immediately after the payment is made?

ROR Calculation

6.5 For the equation $50,000 = 15,000(P/A,i,4) + 9000(P/F,i,4)$, determine the interest rate per period.

6.6 Determine the rate of return per year for the cash flow shown below.

Year	1	2	3	4
Cash Flow, $	−80,000	−9000	40,000	70,000

6.7 Val-lok Industries manufactures miniature fittings and valves. Over a 5-year period, the costs associated with one product line were as follows: initial investment cost of $24,000 and annual costs of $17,000. Annual revenue was $27,000. What rate of return did the company make on this product?

6.8 The Office of Naval Research sponsors a contest for college students to build underwater robots that can perform a series of tasks without human intervention. The University of Florida, with its SubjuGator robot, won the $7000 first prize (and serious bragging rights) over 21 other universities. If the team spent $2000 for parts (at time 0), what annual rate of return did the team make over the project life of two years?

6.9 U.S Census Bureau statistics show that the annual earnings for a person with a high-school diploma are $35,220 versus $57,925 for someone with a bachelor's degree. If the cost of attending college is assumed to be $30,000 per year for four years and the foregone earnings during those years is assumed to be $35,220 per year, what rate of return does earning a bachelor's degree represent, if a 35-year study period is considered? (Hint: The investment in years 1 through 4 is the cost of college plus the foregone earnings, and the income in years 5 through 35 is the difference in income between a high-school diploma and a bachelor's degree.)

6.10 A person who retires at age 62 can receive Social Security (SS) benefits of $1537 per month, but by waiting for the full retirement age of 66, the person will receive $2057 per month. What rate of return will the person make (*a*) per month, and (*b*) per year (nominal) by waiting for full retirement age, if the person collects SS benefits for a total of 20 years?

6.11 A contract between BF Goodrich and the Steelworkers Union of America called for the company to spend $100 million in capital investment to keep the facilities competitive. The contract also required the company to provide buyout packages for 400 workers. If the average buyout package is $100,000 and the company is able to reduce costs by $20 million per year, what rate of return will the company make over a 10-year study period? Assume all of the company's expenditures occur at time 0 and the savings begin one year later.

6.12 Betson Enterprises distributes and markets the Big Buck video game, which allows players to "hunt" for elk, antelope, moose, and bucks without shivering outside in the cold. E-sports entertainment in New York City purchased five machines for $6000 each and took in an average of $600 per week in sales. What rate of return does this represent (*a*) per week, and (*b*) per year (nominal), if a 3-year study period is used? Assume 52 weeks per year.

6.13 Rubber sidewalks made from ground-up tires are said to be environmentally friendly and easier on peoples' knees. Rubbersidewalks, Inc. of Gardena, CA, manufactures the small rubberized squares that are being installed where tree roots, freezing weather, and snow removal have required sidewalk replacement or major repairs every three years. The District of Columbia spent $60,000 for a rubber sidewalk to replace broken concrete in a residential neighborhood lined with towering willow oaks. If a concrete sidewalk costs $28,000 and lasts only 3 years versus a 9-year life for the rubber sidewalks, what rate of return does this represent?

6.14 Very light jets (VLJs) are smaller aircraft that may revolutionize the way people travel by plane. They cost between $1.5 and $3 million, seat 5 to 7 people, and can fly over 1000 miles at speeds approaching 460 mph. Eclipse Aviation was founded in 1998 and its sole business is making VLJs. The company invested $500 million (at time 0) and began taking orders 2 years later. If the company accepted orders for 2500 planes and received 10% down (in year 2) on planes having an average cost of $1.8 million, what rate of return will the company make over a 10-year planning period?

Assume 500 of the planes are delivered each year in years 6 through 10 and that the company's manufacturing and M&O costs average $10 million per year in years 1 through 10.

6.15 Steel cable barriers in highway medians are a low-cost way to improve traffic safety without busting state department of transportation budgets. Cable barriers cost $44,000 per mile, compared with $72,000 per mile for guardrail and $419,000 per mile for concrete barriers. Furthermore, cable barriers tend to snag tractor-trailer rigs, keeping them from ricocheting back into same-direction traffic. The state of Ohio spent $4.97 million installing 113 miles of cable barriers. If the cables prevent accidents totaling $1.3 million per year, (a) what rate of return does this represent if a 10-year study period is considered? (b) What is the rate of return for 113 miles of guardrail if accident prevention is $1.1 million per year over a 10-year study period?

6.16 A broadband service company borrowed $2 million for new equipment and repaid the loan in amounts of $200,000 in years 1 and 2 plus a lump sum amount of $2.2 million at the end of year 3. What was the interest rate on the loan?

6.17 A permanent endowment at the University of Alabama is to award scholarships to engineering students two times per year (end of June and end of December). The first awards are to be made beginning 5-1/2 years after the $20 million lump-sum donation is made. If the interest from the endowment is to fund 100 students each semester (i.e., twice a year) in the amount of $5000 each semester, what semiannual rate of return must the endowment fund earn?

6.18 Identify three possible difficulties with rate of return analyses compared to PW, AW, or FW analyses.

Incremental Analysis

6.19 In selecting between two mutually exclusive alternatives, under what circumstance will the one with the higher rate of return *not* necessarily represent the better investment?

6.20 If $200,000 is available for investment and $60,000 is invested at 20% per year with the remainder invested at 12%, what is the overall rate of return on the $200,000?

6.21 What is the purpose of a cash flow tabulation?

6.22 What is the difference between *revenue* and *cost* alternatives?

6.23 For the cash flows shown, what is the incremental cash flow between machines B and A (a) in year 0, (b) in year 3, and (c) in year 6?

	Machine A	Machine B
First cost, $	−15,000	−25,000
Annual operating cost, $ per year	−1,600	−400
Salvage value, $	3,000	6,000
Life, years	3	6

ROR Evaluation of Two or More Alternatives

6.24 For the alternatives shown, determine the sum of the incremental cash flows for Y − X.

	Alternative X	Alternative Y
First cost, $	−15,000	−25,000
Annual operating cost, $ per year	−1,600	−400
Salvage value, $	3,000	6,000
Life, years	5	5

6.25 For the alternatives shown, determine the sum of the incremental cash flows for Alternative B − Alternative A.

	Alternative A	Alternative B
First cost, $	−50,000	−85,000
Annual operating cost, $ per year	−8,600	−2000
Annual revenue, $ per year	22,000	45,000
Salvage value, $	3,000	8,000
Life, years	3	6

6.26 Standby power for pumps at water distribution booster stations can be provided by either gasoline- or diesel-powered engines. The estimates for the gasoline engines are as follows:

	Gasoline
First cost, $	−150,000
Annual M & O, $ per year	−41,000
Salvage value, $	23,000
Life, years	15

The incremental PW cash flow equation associated with (Diesel − Gasoline) is

$$0 = -40{,}000 + 11{,}000\,(P/A,\Delta i^*,15)$$
$$+ 16{,}000\,(P/F,\Delta i^*,15)$$

What is (a) the first cost of the diesel engines, (b) the annual M&O cost of the diesel engines, and (c) the salvage value of the diesel engines?

6.27 Certain parts for NASA's reusable space exploration vehicle can either be anodized or powder coated. Some of the costs for each process are shown.

	Anodize	Powder Coat
First cost, $	?	−65,000
Annual cost, $ per year	−21,000	?
Resale value, $	?	6,000
Life, years	3	3

The incremental AW cash flow equation associated with (Powder Coat − Anodize) is

$$0 = -14{,}000(A/P,\Delta i^*,3) + 5000$$
$$+ 2000(A/F,\Delta i^*,3).$$

What is (a) the first cost for anodizing, (b) the annual cost for powder coating, and (c) the resale value of the anodized parts?

6.28 Air Products & Chemicals manufactures nitrogen trifluoride, a highly specialized gas that is an industrial cleansing agent for flat panels used in laptop computers, televisions, and desktop monitors. The incremental cash flows (in $ thousands) associated with two alternatives for chemical storage and handling systems (identified as P3 and X3) are shown. Determine (a) the rate of return on the incremental cash flow, and (b) which alternative should be selected, if the company's MARR is 25% per year. Alternative X3 requires the larger initial investment.

Year	Incremental Cash Flow of X3 − P3, $
0	−2800
1–9	1100
10	2000

6.29 Konica Minolta plans to sell a copier that prints documents on both sides simultaneously, cutting in half the time it takes to complete big commercial jobs. The costs associated with producing chemically treated vinyl rollers and fiber-impregnated rubber rollers are estimated. Determine which of the two types should be selected by calculating the

rate of return on the incremental investment. Assume the company's MARR is 21% per year.

	Treated	Impregnated
First cost, $	−50,000	−95,000
Annual cost, $ per year	−100,000	−85,000
Salvage value, $	5,000	11,000
Life, years	3	6

6.30 The Texas Department of Transportation (TxDoT) is considering two designs for crash barriers along a reconstructed portion of I-10. Design 2B will cost $3 million to install and $120,000 per year to maintain. Design 4R will cost $3.8 million to install and $80,000 per year to maintain. Calculate the rate of return and determine which design is preferred, if TxDoT uses a MARR of 8% per year and a 20-year project period.

6.31 An industrial engineer working for American Manufactured Products was asked to make a recommendation about which of four mutually exclusive *cost* alternatives should be selected for improving a certain materials handling operation. The alternatives (identified as A, B, C, and D) were ranked in order of increasing initial investment and then compared by incremental rate of return analysis. The rate of return on each increment of investment was less than the company's MARR of 32% per year. Which alternative should be selected?

6.32 Two processes are under consideration for desalting irrigation return flows in the Rio Grande. Reverse osmosis (RO) will remove over 97% of the salt, allowing for a high percentage of blending with poor-quality well water. Nanofiltration (Nano) will not remove as much salt as RO (e.g., about 90%), but the operating pressure is lower and the flux is higher (allowing for fewer membranes). The costs associated with each system are shown. Calculate the rate of return on the incremental investment and determine which process should be selected, if the MARR is 6% per year. Assume one of the processes must be selected.

	Nano	RO
Capital cost, $	3,200,000	4,800,000
Operating cost, $ per year	250,000	490,000
Revenue, $ per year	1,900,000	2,600,000
Life, years	10	10

6.33 A WiMAX wireless network integrated with a satellite network can provide connectivity to any location within 10 km of the base station. The number of sectors per base station can be varied to increase the bandwidth. An independent cable operator is considering the three bandwith alternatives shown below (thousands $). Assume a life of 20 years and a MARR of 25% per year to determine which alternative is best using an incremental investment rate of return analysis.

Bandwidth, Mbps	First Cost, $	Operating Cost, $ per year	Annual Income, $ per year
44	−40,000	−2000	+4000
55	−46,000	−1000	+5000
88	−61,000	−500	+8000

6.34 Xerox's iGen3 high-speed commercial printers cost $1.5 billion to develop. The machines cost $500,000 to $750,000 depending on what options the client selects. Spectrum Imaging Systems is considering purchasing a new printer because of recent contracts it received for printing a weekly magazine and newspaper advertising materials. The operating costs and revenues generated are related to a large extent to the speed and other capabilities of the copier. The company is considering the four machines shown below. The company uses a 3-year planning period and a MARR of 15% per year. Determine which copier the company should acquire on the basis of an incremental rate of return analysis.

Copier	Initial Investment, $	Operating Cost, $ per year	Annual Income, $ per year	Salvage Value, $
iGen-1	−500,000	−350,000	+450,000	+70,000
iGen-2	−600,000	−300,000	+460,000	+85,000
iGen-3	−650,000	−275,000	+480,000	+95,000
iGen-4	−750,000	−200,000	+510,000	+120,000

6.35 Ashley Foods, Inc. has determined that any one of five machines can be used in one phase of its chili-canning operation. The costs of the machines are estimated, and all machines are expected to have a 4-year useful life. If the minimum attractive rate of return is 20% per year, determine which machine should be selected on the basis of a rate of return analysis.

Machine	First Cost, $	Annual Operating Cost, $ per year
1	−31,000	−16,000
2	−29,000	−19,300
3	−34,500	−17,000
4	−49,000	−12,200
5	−41,000	−15,500

6.36 Five *revenue* projects are under consideration by General Dynamics for improving material flow thorough an assembly line. The initial cost (in $ thousands) and life of each project are estimated. Revenue estimates are not known at this point.

	Project				
	A	B	C	D	E
Initial cost, $	−700	−2300	−900	−300	−1600
Life, yrs.	5	8	5	5	6

An engineer made the comparisons shown below. From the calculations, determine which project, if any, should be undertaken, if the company's MARR is (*a*) 11.5% per year, and (*b*) 13.5% per year. If other calculations must be made in order to make a decision, state which ones.

Comparison	Incremental Rate of Return, %
B vs. DN	13
A vs. B	19
D vs. DN	11
E vs. B	15
E vs. D	24
E vs. A	21
C vs. DN	7
C vs. A	19
E vs. DN	12
A vs. DN	10
E vs. C	33
D vs. C	33
D vs. B	29

6.37 Four different machines are under consideration for improving material flow in a certain production process. An engineer performed an economic analysis to select the best machine, but some of his calculations were erased from the report by a disgruntled employee. All machines are assumed to have a 10-year life.

a. Fill in the missing numbers in the report.

b. Which machine should the company select, if its MARR is 18% per year and one of the machines must be selected?

	Machine			
	1	**2**	**3**	**4**
Initial cost, $	?	−60,000	−72,000	−98,000
Annual cost, $ per year	−70,000	−64,000	−61,000	−58,000
Annual savings, $ per year	+80,000	+80,000	+80,000	+82,000
Overall ROR, %	18.6%	?	23.1%	20.8%
Machines compared		2 to 1	3 to 2	4 to 3
Incremental investment, $		−16,000	?	−26,000
Incremental cash flow, $ per year		+6,000	+3,000	?
ROR on increment, %		35.7%	?	?

6.38 Allstate Insurance Company is considering adopting one of the fraud detection systems shown below, all of which can be considered to last indefinitely. If the company's MARR is 14% per year, determine which one should be selected on the basis of a rate of return analysis.

	A	B	C	D	E
First cost, $	−10,000	−20,000	−15,000	−70,000	−50,000
Annual income, $ per year	2,000	4,000	2,900	10,000	6,000
Overall ROR, %	20	20	19.3	14.3	12

6.39 The four proposals described below are being evaluated.

a. If the proposals are *independent,* which one(s) should be selected when the MARR is 14.5% per year?

b. If the proposals are mutually exclusive, which one should be selected when the MARR is 15% per year?

c. If the proposals are mutually exclusive, which one should be selected when the MARR is 19% per year?

6.40 An engineer initiated the rate of return analysis for the infinite-life *revenue* alternatives shown on the next page.

a. Fill in the blanks in the rate of return column and incremental rate of return columns of the table.

	Problem 6.39		Incremental Rate of Return, %, when Compared with Proposal		
Proposal	**Initial Investment, $**	**Rate of Return, %**	**A**	**B**	**C**
A	−40,000	29			
B	−75,000	15	1		
C	−100,000	16	7	20	
D	−200,000	14	10	13	12

b. What alternative(s) should be selected if they are independent and the MARR is 21% per year?

c. What alternative should be selected if they are mutually exclusive and the MARR is 13% per year?

| | | Problem 6.40 | Incremental Rate of Return, %, on Incremental Cash Flow when Compared with Alternative | | | |
| | | Alternative's | | | | |
Alternative	Investment, $	Overall ROR, %	X1	X2	X3	X4
X1	−20,000	?	—	?	?	?
X2	−30,000	13.33	2	—	?	?
X3	−50,000	?	14	20	—	?
X4	−75,000	12	?	?	?	—

Multiple ROR Values

6.41 What is meant by a nonconventional cash flow series?

6.42 According to Descartes' rule of signs, what is the maximum number of real-number values that will balance a rate of return equation?

6.43 What cash flows are associated with Descartes' rule of signs and Norstrom's criterion?

6.44 According to the cumulative cash flow test (Norstrom's criterion), there are two requirements regarding the cumulative cash flows that must be satisfied to ensure that there is only one nonnegative root in a rate of return equation. What are they?

6.45 According to the cash flow (Descartes') rule of signs, how many possible i^* values are there for net cash flows that have the following signs:
  a. − − − + + + + +
  b. − − + − − − + + + + +
  c. + + − + − − − − − − + − + − − −

6.46 The cash flows associated with sales of handheld refractometers (instruments that determine the concentration of an aqueous solution by measuring its refractive index) are shown. Determine the cumulative net cash flow through year 4.

Year	1	2	3	4
Revenue, $	25,000	15,000	4,000	18,000
Cost, $	30,000	7,000	6,000	12,000

6.47 Boron nitride spray II (BNS II) from GE's Advanced Material Ceramics Division is a release agent and lubricant that prevents materials such as molten metal, rubber, plastics, and ceramic materials from sticking to or reacting with dies, molds,

or other surfaces. A European distributor of BNS II and other GE products had the net cash flows shown.

Year	Net Cash Flow, $
0	−17,000
1	−20,000
2	4,000
3	−11,000
4	32,000
5	47,000

a. Determine the number of possible rate of return values.

b. Find all rate of return values between −30 and 130%.

6.48 Faro laser trackers are portable contact measurement systems that use laser technology to measure large parts and machinery to accuracies of 0.0002 inches across a wide range of industrial applications. A customer that manufactures and installs cell phone relay dishes and satellite receiving stations reported the cash flows (in $ thousands) for one of its product lines.

a. Determine the number of possible rate of return values.

b. Find all rate of return values between 0 and 150%.

Year	Expense, $	Receipts, $
0	−3000	0
1	−1500	2900
2	−4000	5700
3	−2000	5500
4	−1300	1100

Removing Multiple ROR Values

6.49 What is the reinvestment interest rate used for?

6.50 A public-private initiative in Texas will significantly expand the wind-generated energy throughout the state. The cash flow for one phase of the project involving Central Point Energy, a transmission utility company, is shown. Calculate the composite rate of return using a reinvestment rate of 14% per year.

Year	Cash Flow, $ (thousands)
0	5000
1	−2000
2	−1500
3	−7000
4	4000

6.51 A commercial and residential builder who started using insulating concrete form (ICF) construction had the cash flows shown for the ICF portion of his business. Calculate (*a*) the internal rate of return, and (*b*) the composite rate of return, using a reinvestment rate of 8% per year.

Year	Cash Flow, $1000
0	−38
1	10
2	44
3	−5

6.52 The Food and Drug Administration granted the maker of artificial hearts the OK to sell up to 4000 per year to patients who are close to death and have no other treatment options. The implants are expected to cost $250,000. If the company's estimated cash flow is as shown, determine the composite rate of return, if the reinvestment rate is 20% per year. The cash flows are in $1000 units.

Year	Cash Flow, $
0	−5,000
1	4,000
2	0
3	0
4	20,000
5	−15,000

6.53 A new advertising campaign by a company that manufactures products that rely on biometrics, surveillance, and satellite technologies resulted in the cash flows (in $ thousands) shown. Calculate the composite rate of return using a reinvestment rate of 30% per year.

Year	Cash Flow, $
0	2000
1	1200
2	−4000
3	−3000
4	2000

PROBLEMS FOR TEST REVIEW AND FE EXAM PRACTICE

6.54 A chemical engineer working for a large chemical products company was asked to make a recommendation about which of three mutually exclusive *revenue* alternatives should be selected for improving the marketability of personal care products used for conditioning hair, cleansing skin, removing wrinkles, etc. The alternatives (X, Y, and Z) were ranked in order of increasing initial investment and then compared by incremental rate of return analysis. The rate of return on each increment of investment was less than the company's MARR of 17% per year. The alternative to select is
   a. DN.
   b. Alternative X.

   c. Alternative Y.
   d. Alternative Z.

6.55 When the net cash flow for an alternative changes signs only once, the cash flow is said to be
   a. conventional.
   b. simple.
   c. nonconventional.
   d. either (a) or (b).

6.56 Norstrom's criterion states that a unique nonnegative value exists if
   a. the net cash flow series starts negatively and changes sign only once.
   b. the cumulative cash flow series starts negatively and changes sign only once.

c. the net cash flow and cumulative cash flow series change sign only once.

d. none of the above.

6.57 In keeping with the increasingly popular merchandising theme of customization, an entrepreneurial engineer started a business making customized bobblehead dolls. He entered into a contract with Binkely Toys, Inc., the maker of the dolls, to purchase bobbleheads in lots of 250 for $2500. If the engineer's initial investment was $50,000 and his profit was $30,000 in year 1 and $38,000 in year 2, his rate of return was closest to

   a. <15%.

   b. 22%.

   c. 29%.

   d. 36%.

6.58 When the reinvestment rate is lower than the $i*$ rate of return, the composite rate of return will be

   a. lower than the composite rate.

   b. lower than the $i*$ rate.

   c. higher than the $i*$ rate.

   d. equal to the composite rate.

6.59 The cash flow for one of the product lines of Hardy Instruments, a manufacturer of electronic components, is shown. The number of possible rate of return values according to the cash flow rule of signs is

   a. one.

   b. two.   *expenses decrease 4 times*

   c. three.

   d. four.

Year	Expense, $	Receipts, $
0	−22,000	0
1	−41,000	14,000
2	−9,000	25,000
3	−34,000	29,500
4	−38,000	41,000
5	−42,000	33,000

6.60 Jewel-Osco evaluated three different pay-by-touch systems that identify customers by a finger scan and then deduct the amount of the bill directly from their checking accounts. The alternatives were ranked according to increasing initial investment cost and identified as Alternatives A, B, and

C. Based on the incremental rates of return shown and the company's MARR of 16% per year, the alternative that should be selected is

   a. Alternative A.

   b. Alternative B.

   c. Alternative C.

   d. Alternative DN.

Comparison	Rate of Return, %
A-to-DN	23.4
B-to-DN	8.1
C-to-DN	16.6
A-to-B	−5.1
A-to-C	12.0
B-to-C	83.9

6.61 For the cash flows shown, the correct equation to determine the composite rate of return for a reinvestment rate of 12% per year is

   a. $0 = \{[5000 − 10,000(1 + i')](1 + i')\} + 8000$

   b. $0 = \{[5000 − 10,000(1 + i')](1 + i')\} + 8000$

   c. $0 = \{[5000(1 + 0.12) − 10,000](1 + i')\} + 8000$

   d. $0 = \{[5000(1 + i') − 10,000](1 + i')\} + 8000$

Year	Cash Flow, $
0	5,000
1	−10,000
2	8,000

6.62 For the cash flows shown, the correct equation for F_2 at the reinvestment rate of 20% per year is

Year	Cash Flow, $
0	10,000
1	6,000
2	−8,000
3	−19,000
4	4,000

   a. $[10,000(1 + i') + 6000](1.20) − 8000$

   b. $[10,000(1.20) + 6000(1 + i')](1.20) − 8000$

   c. $[10,000(1.20) + 6000](1.20) − 8000$

   d. $[10,000(1.20) + 6000](1 + i') − 8000$

Benefit/Cost Analysis and Public Sector Projects

PhotoLink/Getty Images

The evaluation methods of previous chapters are usually applied to alternatives in the private sector, that is, for-profit and not-for-profit corporations and businesses. This chapter introduces *public sector alternatives* and their economic consideration. Here the owners and users (beneficiaries) are the citizens of the government unit—city, county, state, province, or nation. There are substantial differences in the characteristics of public and private sector alternatives and their economic evaluation. Partnerships of the public and private sector have become increasingly common, especially for large infrastructure construction projects such as major highways, power generation plants, water resource developments, and the like.

The benefit/cost (B/C) ratio was developed, in part, to introduce objectivity into the economic analysis of public sector evaluation, thus reducing the effects of politics and special interests. The different formats of B/C analysis are discussed. The B/C analysis can use equivalency computations based on PW, AW, or FW values.

Objectives

Purpose: Understand public sector economics; evaluate alternatives using the benefit/cost ratio method.

Public sector	1. Identify fundamental differences between public and private sector alternatives.
B/C for single project	2. Use the benefit/cost ratio to evaluate a single project.
Alternative selection	3. Select the best of two or more alternatives using the incremental B/C ratio method.
Spreadsheets	4. Use a spreadsheet to perform B/C analysis of one or more alternatives.

7.1 PUBLIC VERSUS PRIVATE SECTOR PROJECTS

Public sector projects are owned, used, and financed by the citizenry of any government level, whereas projects in the private sector are owned by corporations, partnerships, and individuals. Virtually all the examples in previous chapters have been from the private sector. Notable exceptions occur in Chapters 4 and 5 where capitalized cost was introduced for long-life alternatives and perpetual investments.

Public sector projects have a primary purpose to provide services to the citizenry for the public good at no profit. Areas such as health, safety, economic welfare, and utilities comprise a majority of alternatives that require engineering economic analysis. Some public sector examples are

Hospitals and clinics	Transportation: highways, bridges,
Parks and recreation	waterways
Utilities: water, electricity, gas,	Police and fire protection
sewer, sanitation	Courts and prisons
Schools: primary, secondary,	Food stamp and rent relief programs
community colleges, universities	Job training
Economic development	Public housing
Convention centers	Emergency relief
Sports arenas	Codes and standards

There are significant differences in the characteristics of private and public sector alternatives.

Characteristic	Public Sector	Private Sector
Size of investment	Larger	Some large; more medium to small

Often alternatives developed to serve public needs require large initial investments, possibly distributed over several years. Modern highways, public transportation systems, airports, and flood control systems are examples.

Life estimates	Longer (30–50+ years)	Shorter (2–25 years)

The long lives of public projects often prompt the use of the capitalized cost method, where infinity is used for n and annual costs are calculated as $A = P(i)$.

Annual cash flow estimates	No profit; costs, benefits, and disbenefits are estimated	Revenues contribute to profits; costs are estimated

Public sector projects (also called publicly owned) do not have profits; they do have costs that are paid by the appropriate government unit; and they benefit the citizenry. Public sector projects often have undesirable consequences (*disbenefits*). It is these consequences that can cause public controversy about the projects, because benefits to one group of taxpayers might be disbenefits to other taxpayers, as discussed more fully below. To perform an economic analysis of public alternatives, the costs (initial and annual), the benefits, and the disbenefits, if considered, must be estimated as accurately as possible in *monetary units*.

> **Costs**—estimated expenditures *to the government entity* for construction, operation, and maintenance of the project, less any expected salvage value.
>
> **Benefits**—advantages to be experienced *by the owners, the public*. Benefits can include income and savings.
>
> **Disbenefits**—expected undesirable or negative consequences *to the owners* if the alternative is implemented. Disbenefits may be indirect economic disadvantages of the alternative.

In many cases, it is difficult to estimate and agree upon the economic impact of benefits and disbenefits for a public sector alternative. For example, assume a short bypass around a congested traffic area is recommended. How much will it benefit a driver in *dollars per driving minute* to be able to bypass five traffic lights, as compared to stopping at an average of two lights for 45 seconds each? The bases and standards for benefits estimation are always difficult to establish and verify. Relative to revenue cash flow estimates in the private sector, benefit estimates are much harder to make, and they vary more widely around uncertain averages. And the disbenefits that accrue from an alternative are harder to estimate.

The examples in this chapter include straightforward identification of benefits, disbenefits, and costs. However, in actual situations, judgments are subject to interpretation, particularly in determining which elements of cash flow should be included in the economic evaluation. For example, improvements to the condition of pavement on city streets might result in fewer accidents, an obvious benefit to the taxpaying public. But fewer damaged cars and personal injuries mean less work and money for auto repair shops, towing companies, car dealerships, doctors and hospitals—also part of the taxpaying public. It may be necessary to take a limited viewpoint, because in the broadest viewpoint benefits are usually offset by approximately equal disbenefits.

Funding	Taxes, fees, bonds, private funds	Stocks, bonds, loans, individual owners

The capital used to finance public sector projects is commonly acquired from taxes, bonds, fees, and gifts from private donors. Taxes are collected from those who are the owners—the citizens (e.g., gasoline taxes for highways are paid by all gasoline users). This is also the case for fees, such as toll road fees for drivers. Bonds are often issued: municipal bonds and special-purpose bonds, such as utility district bonds.

| Interest rate | Lower | Higher, based on market cost of capital |

The interest rate for public sector projects, also called the discount rate, is virtually always lower than for private sector alternatives. Government agencies are exempt from taxes levied by higher-level government units. For example, municipal projects do not have to pay state taxes. Also, many loans are government subsidized and carry low interest rates.

| Selection criteria | Multiple criteria | Primarily based on MARR |

Multiple categories of users, economic as well as noneconomic interests, and special-interest political and citizen groups make the selection of one alternative over another much more difficult in public sector economics. Seldom is it possible to select an alternative on the sole basis of a criterion such as PW or ROR. Multiple attribute evaluation is discussed further in Appendix C.

| Environment of the evaluation | Politically inclined | Primarily economic |

There are often public meetings and debates associated with public sector projects. Elected officials commonly assist with the selection, especially when pressure is brought to bear by voters, developers, environmentalists, and others. The selection process is not as "clean" as in private sector evaluation.

The *viewpoint* of the public sector analysis must be determined before cost, benefit, and disbenefit estimates are made. There are always several viewpoints that may alter how a cash flow estimate is classified. Some example viewpoints are the citizen; the tax base; number of students in the school district; creation and retention of jobs; economic development potential; or a particular industry interest. Once established, the viewpoint assists in categorizing the costs, benefits, and disbenefits of each alternative.

EXAMPLE 7.1   The citizen-based Capital Improvement Projects (CIP) Committee for the city of Dundee has recommended a $25 million bond issue for the purchase of greenbelt/floodplain land to preserve low-lying green areas and wildlife habitat. Developers oppose the proposal due to the reduction of available land for commercial development. The city engineer and economic development director have made preliminary estimates for some obvious areas over a projected 15-year planning horizon. The inaccuracy of these estimates is made very clear in a report to the Dundee City Council. The estimates are not yet classified as costs, benefits, or disbenefits.

Economic Dimension	Estimate
1. Annual cost of $5 million in bonds over 15 years at 6% bond interest rate	$ 300,000 (years 1–14) $5,300,000 (year 15)
2. Annual maintenance, upkeep, and program management	$ 75,000 + 10% per year
3. Annual parks development budget	$ 500,000 (years 5–10)
4. Annual loss in commercial development	$2,000,000 (years 8–10)
5. State sales tax rebates not realized	$ 275,000 + 5% per year (years 8 on)
6. Annual municipal income from park use and regional sports events	$ 100,000 + 12% per year (years 6 on)
7. Savings in flood control projects	$ 300,000 (years 3–10) $1,400,000 (years 11–15)
8. Property damage (personal and city) not incurred due to flooding	$ 500,000 (years 10 and 15)

Identify different viewpoints for an economic analysis of the proposal, and classify the estimates accordingly.

Solution

There are many perspectives to take; three are addressed here. The viewpoints and goals are identified and each estimate is classified as a cost, benefit, or disbenefit. (How the classification is made will vary depending upon who does the analysis. *This solution offers only one logical answer.*)

Viewpoint 1: *Citizen of the city.* Goal: Maximize the quality and wellness of citizens with family and neighborhood as prime concerns.

Costs: 1, 2, 3   Benefits: 6, 7, 8   Disbenefits: 4, 5

Viewpoint 2: *City budget.* Goal: Ensure the budget is balanced and of sufficient size to fund rapidly growing city services.

Costs: 1, 2, 3, 5   Benefits: 6, 7, 8   Disbenefits: 4

Viewpoint 3: *Economic development.* Goal: Promote new commercial and industrial economic development for creation and retention of jobs.

Costs: 1, 2, 3, 4, 5   Benefits: 6, 7, 8   Disbenefits: none

If the analyst favors the economic development goals of the city, commercial development losses (4) are considered real costs, whereas they are undesirable consequences (disbenefits) from the citizen and budget viewpoints. Also, the loss of sales tax rebates from the state (5) is interpreted as a real cost from the budget and economic development perspectives, but as a disbenefit from the citizen viewpoint.

Increasingly, large public sector projects are developed through public-private partnerships. This is the trend in part because of the greater efficiency of the private sector and in part because full funding is not possible using traditional means of government financing—fees, taxes, and bonds. Examples of these projects are major highways, tunnels, airports, water resources, and public transportation. In these joint ventures, the government cannot make a profit, but the corporate partner can realize a reasonable profit.

Historically, public projects are designed for and financed by a government unit with a contractor doing the construction under either a *fixed price* (lump sum) or *cost plus* (cost reimbursement) contract with a profit margin specified. Here the contractor does not share the risk of success with the government "owner." When a partnership of public and private interests is developed, the project is commonly contracted under an arrangement called *build-operate-transfer (BOT),* which may also be referred to as BOOT, where the first O is for *own.* The BOT-administered project may require that the contractor be responsible partially or completely for design and financing, and completely responsible for the construction (the build element), operation (operate), and maintenance activities for a specified number of years. After this time period, the owner becomes the government unit when the title of ownership is transferred (transfer) at no or very low cost. Many of the projects in international settings and in developing countries utilize the BOT form of partnership.

A variation of the BOT/BOOT method is BOO (build-own-operate), where the transfer of ownership never takes place. This form of partnership is used when the project has a relatively short life or the technology deployed is changing quickly.

7.2 BENEFIT/COST ANALYSIS OF A SINGLE PROJECT

The benefit/cost ratio, a fundamental analysis method for public sector projects, was developed to introduce more objectivity into public sector economics. It was developed in response to the U.S. Flood Control Act of 1936. There are several variations of the B/C ratio; however, the basic approach is the same. All cost and benefit estimates must be converted to a common equivalent monetary unit (PW, AW, or FW) at the discount rate (interest rate). The B/C ratio is then calculated using one of these relations:

$$\text{B/C} = \frac{\text{PW of benefits}}{\text{PW of costs}} = \frac{\text{AW of benefits}}{\text{AW of costs}} = \frac{\text{FW of benefits}}{\text{FW of costs}} \qquad \textbf{[7.1]}$$

The sign convention for B/C analysis is positive signs, so *costs are preceded by a + sign.* Salvage values, when they are estimated, are subtracted from costs. Disbenefits are considered in different ways depending upon the model used. Most commonly, disbenefits are subtracted from benefits and placed in the numerator. The different formats are discussed below. The decision guideline for a single project is simple:

If B/C ≥ 1.0, accept the project as economically acceptable for the estimates and discount rate applied.

If B/C < 1.0, the project is not economically acceptable.

The *conventional B/C ratio* is the most widely used. It subtracts disbenefits from benefits.

$$B/C = \frac{\text{benefits} - \text{disbenefits}}{\text{costs}} = \frac{B - D}{C} \qquad [7.2]$$

The B/C value would change considerably were disbenefits added to costs. For example, if the numbers 10, 8, and 5 are used to represent the PW of benefits, disbenefits, and costs, respectively, Equation [7.2] results in B/C = $(10 - 8)/5$ = 0.40. The incorrect placement of disbenefits in the denominator results in B/C = $10/(8 + 5)$ = 0.77, which is approximately twice the correct B/C value. Clearly, then, the method by which disbenefits are handled affects the magnitude of the B/C ratio. However, no matter whether disbenefits are (correctly) subtracted from the numerator or (incorrectly) added to costs in the denominator, *a B/C ratio of less than 1.0 by the first method will always yield a B/C ratio less than 1.0 by the second method, and vice versa.*

The *modified B/C ratio* places benefits (including income and savings), disbenefits, and maintenance and operation (M&O) costs in the numerator. The denominator includes only the equivalent PW, AW, or FW of the initial investment.

$$\text{Modified B/C} = \frac{\text{benefits} - \text{disbenefits} - \text{M\&O costs}}{\text{initial investment}} \qquad [7.3]$$

Salvage value is included in the denominator with a negative sign. The modified B/C ratio will obviously yield a different value than the conventional B/C method. However, as discussed above with disbenefits, *the modified procedure can change the magnitude of the ratio but not the decision to accept or reject the project.*

The *benefit and cost difference* measure of worth, which does not involve a ratio, is based on the difference between the PW, AW, or FW of benefits (including income and savings) and costs, that is, $B - C$. If $(B - C) \geq 0$, the project is acceptable. This method has the advantage of eliminating the discrepancies noted above when disbenefits are regarded as costs, because B represents *net benefits.* Thus, for the numbers 10, 8, and 5 the same result is obtained regardless of how disbenefits are treated.

Subtracting disbenefits from benefits: $B - C = (10 - 8) - 5 = -3$

Adding disbenefits to costs: $B - C = 10 - (8 + 5) = -3$

EXAMPLE 7.2

The Ford Foundation expects to award $15 million in grants to public high schools to develop new ways to teach the fundamentals of engineering that prepare students for university-level material. The grants will extend over a 10-year period and will create an estimated savings of $1.5 million per year in faculty salaries and student-related expenses. The Foundation uses a discount rate of 6% per year.

This grants program will share Foundation funding with ongoing activities, so an estimated $200,000 per year will be removed from other program funding.

To make this program successful, a $500,000 per year operating cost will be incurred from the regular M&O budget. Use the B/C method to determine if the grants program is economically justified.

Solution

Use annual worth as the common monetary equivalent. For illustration only, all three B/C models are applied.

AW of investment cost.	$15,000,000(A/P,6%,10) = $2,038,050 per year
AW of M&O cost.	$500,000 per year
AW of benefit.	$1,500,000 per year
AW of disbenefit.	$200,000 per year

Use Equation [7.2] for conventional B/C analysis, where M&O is placed in the denominator as an annual cost. The project is not justified, since B/C < 1.0.

$$B/C = \frac{\overset{B}{1,500,000} - \overset{D}{200,000}}{\underset{Thus}{2,038,050} + \underset{M\&O}{500,000}} = \frac{1,300,000}{2,538,050} = 0.51$$

By Equation [7.3] the modified B/C ratio treats the M&O cost as a reduction to benefits.

$$\text{Modified B/C} = \frac{1,500,000 - 200,000 - 500,000}{2,038,050} = 0.39$$

For the $(B - C)$ model, B is the net benefit, and the annual M&O cost is included with costs.

$$
\begin{aligned}
B - C &= (1,500,000 - 200,000) - (2,038,050 - 500,000) \\
&= \$-1.24 \text{ million}
\end{aligned}
$$

7.3 INCREMENTAL B/C EVALUATION OF TWO OR MORE ALTERNATIVES

Incremental B/C analysis of two or more alternatives is very similar to that for incremental ROR analysis in Chapter 6. The incremental B/C ratio, $\Delta B/C$, between two alternatives is based upon the PW, AW, or FW equivalency of costs and benefits. Selection of the survivor of pairwise comparison is made using the following guideline:

If $\Delta B/C \geq 1$, select the larger-cost alternative.

Otherwise, select the lower-cost alternative.

Note that the decision is based upon incrementally justified *total* costs, not incrementally justified *initial* cost.

There are several special considerations for B/C analysis of multiple alternatives that make it slightly different from ROR analysis. As mentioned earlier, all costs have a positive sign in the B/C ratio. Also, the *ordering of alternatives is done on the basis of total costs* in the denominator of the ratio. Thus, if two alternatives have equal initial investments and lives, but 2 has a larger equivalent annual cost, then 2 must be incrementally justified against 1. If this convention is not correctly followed, it is possible to get a negative cost value in the denominator, which can incorrectly make $\Delta B/C < 1$ and reject a higher-cost alternative that is actually justified. In the unusual circumstance that the two alternatives have equal costs (yielding a $\Delta B/C$ of infinity), the alternative with the larger benefits is selected by inspection.

Like the ROR method, B/C analysis requires *equal-service comparison* of alternatives. Usually, the expected useful life of a public project is long (25 or 30 or more years), so alternatives generally have equal lives. However, when alternatives do have unequal lives, the use of PW to determine the equivalent costs and benefits requires that the LCM of lives be used.

There are two types of benefits. Before conducting the incremental evaluation, classify the alternatives as *usage cost* estimates or *direct benefit* estimates. Usage cost estimates have implied benefits based on the difference in costs between alternatives. Direct benefit alternatives have benefit amounts estimated. The incremental comparison differs slightly for each type; direct benefit alternatives are initially compared with the DN alternative. (This is the same treatment made for revenue alternatives in an ROR evaluation, but the term *revenue* is not used for public projects.)

Usage cost: Evaluate alternatives only against each other.

Direct benefit: First evaluate against do-nothing, then against each other.

Apply the following procedure for comparing multiple, mutually exclusive alternatives using the conventional B/C ratio:

1. For each alternative, determine the equivalent PW, AW, or FW values for costs C and benefits B [or $(B - D)$ if disbenefits are considered].
2. Order the alternatives by increasing total equivalent cost. For *direct benefit* alternatives, add DN as the first alternative.
3. Determine the incremental costs and benefits between the first two ordered alternatives (that is, $2 - 1$) over their least common multiple of lives. For *usage cost* alternatives, incremental benefits are determined as the difference in usage costs.

$$\Delta B = \text{usage cost of 2} - \text{usage cost of 1} \qquad \text{[7.4]}$$

4. Calculate the incremental conventional B/C ratio using Equation [7.2]. If disbenefits are considered, this is

$$\Delta B/C = \Delta(B - D)/\Delta C \qquad \text{[7.5]}$$

5. If $\Delta B/C \geq 1$, eliminate 1; 2 is the survivor; otherwise 1 is the survivor.
6. Continue to compare alternatives using steps 2 through 5 until only one alternative remains as the survivor.

In step 3, prior to calculating the $\Delta B/C$ ratio, visually check the PW, AW, or FW values to ensure that the larger cost alternative also yields larger benefits. If benefits are not larger, the comparison is unnecessary.

The next two examples illustrate this procedure; the first for two direct benefit estimate alternatives, and the second for four usage fee estimate alternatives.

EXAMPLE 7.3   The city of Garden Ridge, Florida has received two designs for a new wing to the municipal hospital. The costs and benefits are the same in most categories, but the city financial manager decided that the following estimates should be considered to determine which design to recommend at the city council meeting next week.

	Design 1	Design 2
Construction cost, $	10,000,000	15,000,000
Building maintenance cost, $/year	35,000	55,000
Patient benefits, $/year	800,000	1,050,000

The patient benefit is an estimate of the amount paid by an insurance carrier, not the patient, to occupy a hospital room with the features included in the design of each room. The discount rate is 5% per year, and the life of the addition is estimated at 30 years.

  a. Use conventional B/C ratio analysis to select design 1 or 2.

  b. Once the two designs were publicized, the privately owned hospital in the adjacent city of Forest Glen lodged a complaint that design 1 will reduce its own municipal hospital's income by an estimated $600,000 per year because some of the day-surgery features of design 1 duplicate its services. Subsequently, the Garden Ridge merchants' association argued that design 2 could reduce its annual revenue by an estimated $400,000 because it will eliminate an entire parking lot used for short-term parking. The city financial manager stated that these concerns would be entered into the evaluation. Redo the B/C analysis to determine if the economic decision is still the same.

Solution

  a. Apply the incremental B/C procedure with no disbenefits included and direct benefits estimated.

    1. Since most of the cash flows are already annualized, ΔB/C is based on AW values. The AW of costs is the sum of construction and maintenance costs.

$$AW_1 = 10,000,000(A/P,5\%,30) + 35,000 = \$685,500$$
$$AW_2 = 15,000,000(A/P,5\%,30) + 55,000 = \$1,030,750$$

    2. Since the alternatives have direct benefits estimated, the DN option is added as the first alternative with AW of costs and benefits of $0. The comparison order is DN, 1, 2.

    3. The comparison 1–to–DN has incremental costs and benefits exactly equal to those of alternative 1.

    4. Calculate the incremental B/C ratio.

$$\Delta B/C = 800,000/685,500 = 1.17$$

    5. Since $1.17 > 1.0$, design 1 is the survivor over DN.

The comparison continues for 2-to-1 using incremental AW values.

$$\Delta B = 1,050,000 - 800,000 = \$250,000$$
$$\Delta C = 1,030,750 - 685,500 = \$345,250$$
$$\Delta B/C = 250,000/345,250 = 0.72$$

Since $0.72 < 1.0$, design 2 is eliminated, and design 1 is the selection for the construction bid.

b. The revenue loss estimates are considered disbenefits. Since the disbenefits of design 2 are $200,000 less than those of 1, this positive difference is added to the $250,000 benefits of 2 to give it a total benefit of $450,000.

$$\Delta B/C = \frac{\$450,000}{\$345,250} = 1.30$$

Design 2 is now favored. The inclusion of disbenefits reversed the decision.

EXAMPLE 7.4

The Economic Development Corporation (EDC) for the city of Bahia, California and Moderna County is operated as a not-for-profit corporation. It is seeking a developer that will place a major water park in the city or county area. Financial incentives will be awarded. In response to a request for proposal (RFP) to the major water park developers in the country, four proposals have been received. Larger and more intricate water rides and increased size of the park will attract more customers, thus different levels of initial incentives are requested in the proposals.

Approved and in-place economic incentive guidelines allow entertainment industry prospects to receive up to $1 million cash as a first-year incentive award and 10% of this amount each year for 8 years in property tax reduction. Each proposal includes a provision that residents of the city or county will benefit from reduced entrance (usage) fees when using the park. This fee reduction will be in effect as long as the property tax reduction incentive continues. The EDC has estimated the annual total entrance fees with the reduction included for local residents. Also, EDC estimated the benefits of extra sales tax revenue. These estimates and the costs for the initial incentive and annual 10% tax reduction are summarized in the top section of Table 7.1.

Perform an incremental B/C study to determine which park proposal is the best economically. The discount rate is 7% per year.

Solution

The viewpoint taken for the economic analysis is that of a resident of the city or county. The first-year cash incentives and annual tax reduction incentives are real costs to the residents. Benefits are derived from two components: the decreased entrance fee estimates and the increased sales tax receipts. These will

TABLE 7.1   Estimates of Costs and Benefits, and the Incremental B/C Analysis for Four Water Park Proposals, Example 7.4

	Proposal 1	Proposal 2	Proposal 3	Proposal 4
Initial incentive, $	250,000	350,000	500,000	800,000
Tax incentive cost, $/year	25,000	35,000	50,000	80,000
Resident entrance fees, $/year *usage costs*	500,000	450,000	425,000	250,000
Extra sales taxes, $/year	310,000	320,000	320,000	340,000
Study period, years	8	8	8	8
AW of total costs, $/year	66,867	93,614	133,735	213,976
Alternatives compared		2-to-1	3-to-2	4-to-2
Incremental costs ΔC, $/year		26,747	40,120	120,360
Entrance fee reduction, $/year		50,000	25,000	200,000
Extra sales tax, $/year		10,000	0	20,000
Incremental benefits ΔB, $/year		60,000	25,000	220,000
Incremental B/C ratio		2.24	0.62	1.83
Increment justified?		Yes	No	Yes
Alternative selected		2	2	4

benefit each citizen indirectly through the increase in money available to those who use the park and through the city and county budgets where sales tax receipts are deposited. Since these benefits must be calculated indirectly from these two components, the alternatives are classified as usage cost estimates.

Table 7.1 includes the results of applying an AW-based incremental B/C procedure.

1. For each alternative, the capital recovery amount over 8 years is determined and added to the annual property tax incentive cost. For proposal #1,

$$\text{AW of total costs} = \text{initial incentive } (A/P,7\%,8) + \text{tax cost}$$
$$= \$250,000 \, (A/P,7\%,8) + 25,000 = \$66,867$$

2. The usage-cost alternatives are ordered by the AW of total costs in Table 7.1.
3. Table 7.1 shows incremental costs. For the 2-to-1 comparison,

$$\Delta C = \$93,614 - 66,867 = \$26,747$$

Incremental benefits for an alternative are the sum of the resident entrance fees compared to those of the next-lower-cost alternative, plus the increase in sales tax receipts over those of the next-lower-cost alternative. Thus, the benefits are determined incrementally for each pair of alternatives. For the 2-to-1 comparison, resident entrance fees decrease by $50,000 per year and the sales tax receipts increase by $10,000. The total benefit is the sum, $\Delta B = \$60,000$ per year.

4. For the 2-to-1 comparison, Equation [7.5] results in

$$\Delta B/C = \$60,000/\$26,747 = 2.24$$

5. Alternative #2 is clearly justified. Alternative #1 is eliminated.
6. This process is repeated for the 3-to-2 comparison, which has $\Delta B/C < 1.0$ because the incremental benefits are substantially less than the increase in costs. Proposal #3 is eliminated, and the 4-to-2 comparison results in

$$\Delta B = 200,000 + 20,000 = \$220,000$$
$$\Delta C = 213,976 - 93,614 = \$120,362$$
$$\Delta B/C = \$220,000/\$120,362 = 1.83$$

Since $\Delta B/C > 1.0$, proposal #4 survives as the one remaining alternative.

When the lives of alternatives are so long that they can be considered infinite, the capitalized cost is used to calculate the equivalent PW or AW values. Equation [5.6], $A = P(i)$, determines the equivalent AW values in the incremental B/C analysis.

If two or more *independent projects* are evaluated using B/C analysis and there is no budget limitation, no incremental comparison is necessary. The only comparison is between each project separately with the do-nothing alternative. The project B/C values are calculated, and those with $B/C \geq 1.0$ are accepted.

The Army Corps of Engineers wants to construct a dam on a flood-prone river. **EXAMPLE 7.5**
The estimated construction cost and average annual dollar benefits are listed below. A 6% per year rate applies and dam life is infinite for analysis purposes. (*a*) Select the one best location using the B/C method. (*b*) If the sites are now considered independent projects, which sites are acceptable?

Site	Construction Cost, $ millions	Annual Benefits, $
A	6	350,000
B	8	420,000
C	3	125,000
D	10	400,000
E	5	350,000
F	11	700,000

Solution

a. The capitalized cost $A = Pi$ is used to obtain AW values of the construction cost, as shown in the first row of Table 7.2. Since benefits are estimated

TABLE 7.2   Use of Incremental B/C Ratio Analysis for Example 7.5 (Values in $1000)

	DN	C	E	A	B	D	F
AW of cost, $/year	0	180	300	360	480	600	660
Annual benefits, $/year	0	125	350	350	420	400	700
Site B/C	—	0.69	1.17	0.97	0.88	0.67	1.06
Comparison		C-to-DN	E-to-DN	A-to-E	B-to-E	D-to-E	F-to-E
Δ Annual cost, $/year		180	300	60	180	300	360
Δ Annual benefits, $/year		125	350	0	70	50	350
Δ B/C ratio		0.69	1.17	—	0.39	0.17	0.97
Increment justified?		No	Yes	No	No	No	No
Site selected		DN	E	E	E	E	E

directly, initial comparison with DN is necessary. For the analysis, sites are ordered by increasing AW-of-cost values. This is DN, C, E, A, B, D, and F. The analysis between the ordered mutually exclusive alternatives is detailed in the lower portion of Table 7.2. Since only site E is incrementally justified, it is selected.

   b. The dam site proposals are now independent projects. The site B/C ratio is used to select from none to all six sites. In Table 7.2, third row, B/C > 1.0 for sites E and F only; they are acceptable.

7.4 USING SPREADSHEETS FOR B/C ANALYSIS

Formatting a spreadsheet to apply the incremental B/C procedure for mutually exclusive alternatives is basically the same as that for an incremental ROR analysis (Section 6.8). Once all estimates are expressed in terms of either PW, AW, or FW equivalents using Excel functions, the alternatives are *ordered by increasing total equivalent cost*. Then, the incremental B/C ratios from Equation [7.5] assist in selecting the one best alternative.

EXAMPLE 7.6   A significant new application of nanotechnology is the use of thin-film solar panels applied to houses to reduce the dependency on fossil-fuel generated electrical energy. A community of 400 new all-electric public housing units will utilize the technology as anticipated proof that significant reductions in overall utility costs can be attained over the expected 15-year life of the housing. Table 7.3 details the three bids received. Also included are estimated annual electricity usage costs for the community with the panels in use, and the PW

TABLE 7.3 Estimates for Alternatives Using Nanocrystal-Layered Thin-Film Solar Panel Technology for Energy, Example 7.6

Bidder	Geyser, Inc.	Harris Corp.	Crumbley, LLP
Bid identification	G	H	C
PW of initial cost, $	2,400,000	1,850,000	6,150,000
Annual maintenance, $/year	500,000	650,000	450,000
Annual utility bill, $/year	960,000	1,000,000	550,000
PW of backup systems, $	650,000	750,000	950,000

of backup systems required in case of panel failures. Use a spreadsheet, the conventional B/C ratio, and $i = 5\%$ per year to select the best bid.

Solution

Benefits will be estimated from utility bill differences, so the alternatives are classified as usage cost. Initial comparison to DN is unnecessary. The step-by-step procedure (Section 7.3) is included in the Figure 7.1 solution.

1. Base the analysis on present worth. With the use of PV functions to determine PW of construction costs and utility bills (rows 9 and 10), all benefit, disbenefit, and cost terms are prepared. (Note the inclusion of the minus sign on the PV function to ensure that cost terms have a positive sense.)

	A	B	C	D	E
1					
2	**Estimates**		Geyser	Harris	Crumbley
3		B/C category	G	H	C
4	PW of initial cost, $	Cost	2,400,000	1,850,000	6,150,000
5	Annual maintenance, $/year	Cost	500,000	650,000	450,000
6	Annual utility bill, $/year	Usage cost benefit	960,000	1,000,000	550,000
7	**PW of back-up systems, $**	Disbenefit	650,000	750,000	950,000
8					
9	**PW of costs, $**	Cost	7,589,829	8,596,778	10,820,846
10	**PW of utility bills, $**	Benefit	9,964,472	10,379,658	5,708,812
11					
12	**Comparison**			**H-to-G**	**C-to-G**
13	PW of ΔB (incr utility bills), $			-415,186	4,255,660
14	PW of ΔD (back-up), $			100,000	300,000
15	PW of ΔC, $			1,006,949	3,231,017
16					
17	Δ(B-D)/C			-0.51	1.22
18	Increment justified?			No	Yes
19	Selected contractor			G	C
20					

Original estimates

= -PV(5%,15,500000) + 2400000

= -PV(5%,15,960000)

= -(D10-C10)

= -(E10-C10)

= D7-C7

= D9-C9

= (E13-E14)/E15

FIGURE 7.1 Spreadsheet evaluation of multiple alternatives using incremental B/C analysis, Example 7.6.

2. Based on PW of costs (row 9), the order of evaluation is G, H, and finally C. It is important to realize that even though H has a smaller initial cost, G has a smaller equivalent total cost.

3. The first comparison is H-to-G. The form of Equation [7.4] helps determine ΔB with the minus sign on utility bill costs changed to a plus for the ΔB/C computation.

$$\Delta B = \text{utility bills for H} - \text{utility bills for G} = -(10{,}379{,}658 - 9{,}964{,}472) = \$-415{,}186$$

In this case $\Delta B < 0$ due to the larger PW of utility bills for H. There is no need to complete the comparison, since $\Delta(B - D)/C$ will be negative.

4. The value $\Delta(B - D)/C = -0.51$ is determined for illustration only.

5. G is the survivor; now compare C-to-G.

The comparison in column E results in $\Delta(B - D)/C = 1.22$, indicating that the Crumbley bid is incrementally justified and is the selected bid. This is the most expensive bid, especially in initial cost, but the savings in utility bills help make it the most economic over the 15-year planning horizon.

SUMMARY

The benefit/cost method is used primarily to evaluate alternatives in the public sector. When one is comparing mutually exclusive alternatives, the incremental B/C ratio must be greater than or equal to 1.0 for the incremental equivalent total cost to be economically justified. The PW, AW, or FW of the initial costs and estimated benefits can be used to perform an incremental B/C analysis.

Public sector economics are substantially different from those of the private sector. For public sector projects, the initial costs are usually large, the expected life is long (25, 35, or more years), and the sources for capital are usually a combination of taxes levied on the citizenry, user fees, bond issues, and private lenders. *It is very difficult to make accurate estimates of benefits and disbenefits for a public sector project.* The discount rates in the public sector are lower than those for corporate projects.

PROBLEMS

B/C Considerations

7.1 What is the primary purpose of public sector projects?

7.2 Identify the following as primarily public or private sector undertakings: eBay, farmer's market, state police department, car racing facility, social security, EMS, ATM, travel agency, amusement park, gambling casino, and swap meet.

7.3 State whether the following characteristics are primarily associated with public or private sector projects: Large initial investment, short life projects, profit, disbenefits, tax-free bonds, subsidized loans, low interest rate, income tax, and water quality regulations.

7.4 What is the difference between disbenefits and costs?

7.5 Identify the following cash flows as a benefit, disbenefit, or cost.

    a. $400,000 annual income to local businesses because of tourism created by a national park.

    b. Cost of fish from a hatchery to stock the lake at a state park.

    c. Less tire wear because of smoother road surfaces.

    d. Decrease in property values due to closure of a government research lab.

    e. School overcrowding because of military base expansion.

    f. Revenue to local motels because of an extended national park season.

7.6 What is meant by a BOT project?

7.7 In taking the broadest viewpoint for benefits and disbenefits, state why they might exactly offset each other.

7.8 Where is the salvage value placed in a conventional B/C ratio? Why?

Single Project B/C

7.9 What is the difference between a conventional and a modified B/C analysis?

7.10 A public water utility spent $12 million to build a treatment plant to remove certain heavy metals from well water. The plant's operating cost is $1.1 million per year. Public health benefits are expected to be $285,000 per year. What is the B/C ratio, if the plant is amortized over a 40-year period at 6% per year interest?

7.11 Calculate the conventional B/C ratio for the cash flow estimates shown at a discount rate of 8% per year.

Item	Cash Flow
PW of benefits, $	3,800,000
AW of disbenefits, $ per year	65,000
First cost, $	1,200,000
M&O costs, $ per year	300,000
Life of project, years	20

7.12 On-site granular ferric hydroxide (GFH) systems can be used to remove arsenic from water when

daily flow rates are relatively low. If the operating cost is $600,000 per year, what would the initial investment cost of the GFH system have to be in order to get a B/C ratio of 1.0? Public health benefits are estimated at $800,000 per year. Assume the equipment life is 10 years and the interest rate is 6% per year.

7.13 The cash flows associated with a Bexar County Arroyo improvement project are as follows: initial cost $650,000; life 20 years; maintenance cost $150,000 per year; benefits $600,000 per year; disbenefits $190,000 per year. Determine the conventional B/C ratio at an interest rate of 6% per year.

7.14 The Parks and Recreation Department of Burkett County estimates that the initial cost of a "barebones" permanent river park will be $2.3 million. Annual upkeep costs are estimated at $120,000. Benefits of $340,000 per year and disbenefits of $40,000 per year have also been identified. Using a discount rate of 6% per year, calculate (*a*) the conventional B/C ratio and (*b*) the modified B/C ratio.

7.15 From the following data, calculate the (*a*) conventional and (*b*) modified benefit/cost ratios using an interest rate of 6% per year and a study period of 40 years.

	To the People	To the Government	
Benefits:	$200,000 now and $100,000 per year for forty years	Costs:	$1.2 million now and $200,000 three years from now
		Savings:	$90,000 five years from now
Disbenefits:	$18,000 per year		

7.16 From the following data, calculate the (*a*) conventional and (*b*) modified benefit/cost ratios using an interest rate of 6% per year and an infinite project period.

	To the People	To the Government	
Benefits:	$300,000 now and $100,000 per year	Costs:	$1.5 million now and $200,000 three years from now.
Disbenefits:	$40,000 per year	Savings:	$70,000 per year

7.17 When red light cameras are installed at high-risk intersections, rear-end collisions go up, but all other types of accidents go down, including those involving pedestrians. Analysis of traffic accidents in a northwestern city revealed that the total number of collisions at photo-enhanced intersections decreased from 33 per month to 18. At the same time, the number of traffic tickets issued for red light violations averaged 1100 per month, at a cost to violators of $85 per citation. The cost to install the basic camera system at selected intersections was $750,000. If the cost of a collision is estimated at $41,000 and traffic ticket costs are considered as disbenefits, calculate the B/C ratio for the camera system. Use an interest rate of 0.5% per month and a 3-year study period.

7.18 The solid waste department of a certain southwestern city has embarked on a program to recycle solid waste materials that previously ended up in a landfill. The program involves distributing blue plastic containers that cost $45 each to 172,000 households. The city gets 10% of the revenue from the sale of recyclables, and the company that manages the recovery operation receives 90%. The smaller volume of disposable solid waste materials will extend the life of the landfill, resulting in equivalent savings of $80,000 per year. If the containers are amortized over a 10-year period at 6% per year, calculate the B/C ratio. The city's share of the revenue is $194,000 per year.

7.19 The B/C ratio for a mosquito control program proposed by the Harris County Department of Health is reported as 2.3. The person who prepared the report stated that the health benefits were estimated at $500,000, and that disbenefits of $45,000 per year were used in the calculation. He also stated that the costs for chemicals, machinery, maintenance, and labor were estimated at $150,000 per year, but he forgot to list the cost for initiating the program (trucks, pumps, tanks, etc.). If the initial cost was amortized over a 10-year period at 7% per year, what is the estimated initial cost?

7.20 A water utility is trying to decide between installing equipment for conducting analyses for endocrine disrupting substances, pharmaceuticals, and personal-care products or sending the samples to a private contract lab. The equipment to conduct the tests will cost $595,000 and will have a 5-year life. In addition, a full-time chemist will have to be hired at a cost of $56,000 per year. Expendable supplies (gases, chemicals, glassware, etc.) will cost $19,000 per year. The utility anticipates 140 samples per year. If the analyses are sent to an outside lab, the cost per sample will average $1250. If the utility uses an interest rate of 6% per year, calculate the B/C ratio associated with purchasing the equipment.

Alternative Comparison

7.21 Two relatively inexpensive alternatives are available to reduce potential earthquake damage at a top secret government research site. The cash flow estimates for each alternative are shown. At an interest rate of 8% per year, use the B/C ratio method to select one. Use a 20-year study period, and assume the damage costs would occur in the middle of the period, that is, in year 10.

	Alternative 1	Alternative 2
Initial cost, $	600,000	1,100,000
Annual maintenance, $ per year	50,000	70,000
Potential damage costs, $	950,000	250,000

7.22 The two alternatives shown below are under consideration for improving security at a county jail. Select one based on a B/C analysis at an interest rate of 7% per year and a 10-year study period.

	Extra Cameras (EC)	New Sensors (NS)
First cost, $	38,000	87,000
Annual M&O, $ per year	49,000	64,000
Benefits, $ per year	110,000	160,000
Disbenefits, $ per year	26,000	—

7.23 Two methods are being evaluated for constructing a second-story floor onto an existing building at the Pentagon. Method A will use lightweight expanded shale on a metal deck with open web joists and steel beams. For this method, the costs will be $14,100 for concrete, $6000 for metal decking, $4300 for joists, and $2600 for beams. Method B will be a reinforced concrete slab that will cost $5200 for concrete, $1400 for rebar, $2600 for equipment rental, and $1200 for expendable supplies. Special additives will be included in the lightweight concrete that will improve the heat transfer properties of the floor. If the energy costs for method A will be $600 per year lower than for method B, which one is more attractive at an interest rate of 7% per year over a 20-year study period? Use the B/C method.

7.24 A public utility in a medium-sized city is considering two cash rebate programs to achieve water conservation. Program 1, which is expected to cost an average of $60 per household, would involve a rebate of 75% of the purchase and installation costs of an ultra low-flush toilet ($100 maximum). This program is projected to achieve a 5% reduction in overall household water use over a 5-year evaluation period. This will benefit the citizenry to the extent of $1.25 per household per month. Program 2 would involve turf replacement with desert landscaping. This is expected to cost $500 per household, but it will result in reduced water cost estimated at $8 per household per month (on the average). At a discount rate of 0.5% per month, which programs should the utility undertake if the programs are (a) mutually exclusive, and (b) independent? Use the B/C method.

7.25 Solar and conventional alternatives are available for providing energy at a remote U.S. Army training site. The costs associated with each alternative are shown. Use the B/C method to determine which should be selected at an interest rate of 7% per year over a 5-year study period.

	Conventional	Solar
Initial cost, $	200,000	1,300,000
Annual power cost, $ per year	80,000	9,000
Salvage value, $	10,000	150,000

7.26 Four methods are available to recover lubricant from an automated milling system. The investment costs and income associated with each are shown. Assuming that all methods have a 10-year life with zero salvage value, determine which one should be selected using a MARR of 10% per year and the B/C method. Consider operating cost as a disbenefit.

	Method			
	1	**2**	**3**	**4**
First cost, $	15,000	19,000	25,000	33,000
Annual operating cost, $/year	10,000	12,000	9,000	11,000
Annual income, $/year	16,000	20,000	19,000	22,000

7.27 Engineers working in the Transportation Planning Department for the State of West Virginia have identified three viable sites for a new bridge across the Ohio River. All three sites are within 15 miles of each other, but the land and construction costs vary. The bridge that would directly link the cities on opposite sides of the river would have the lowest travel costs (shortest distance), but it would also have the highest construction cost because of property values. The bridge location just north of the two cities would involve less travel time than the one to the south. The costs associated with each bridge are shown. At an interest rate of 6% per year, determine which location is preferred, assuming the bridges will be permanent.

	Location		
	S	**D**	**N**
Construction cost, $	50,000,000	75,000,000	60,000,000
Maintenance cost, $ per year	150,000	130,000	140,000
User travel costs, $ per year	7,600,000	4,100,000	5,900,000

7.28 A consulting engineer is currently evaluating 4 different projects for the U. S. government. The present worths of the costs, benefits, disbenefits, and cost savings are shown. Assuming the interest rate is 10% per year compounded continuously, determine which of the projects, if any, should be selected if the projects are (a) independent, and (b) mutually exclusive.

	Good	Better	Best	Best of All
PW of costs, $	10,000	8,000	20,000	14,000
PW of benefits, $	15,000	11,000	25,000	42,000
PW of disbenefits, $	6,000	1,000	20,000	31,000
PW of cost savings, $	1,500	2,000	16,000	3,000

7.29 From the AW data shown for projects regarding solid waste handling at a military training facility, determine which project, if any, should be selected from the 6 mutually exclusive projects. If the proper ΔB/C comparisons have not been made, state which ones should be done.

Project Identification					
A	**B**	**C**	**D**	**E**	**F**
AW of cost, $ 20,000	60,000	36,000	48,000	32,000	26,000
Life, years 20	20	20	20	20	20
AW of benefits, $?	?	?	?	?	?
B/C Ratio 0.8	1.17	1.44	1.38	1.13	1.08

Selected incremental B/C ratios:

A vs. E = 2.0
B vs. C = 0.75
B vs. D = 0.33
E vs. D = 1.88
F vs. E = 1.33
E vs. C = 4.00
A vs. E = 1.67
A vs. B = 1.35
C vs. F = 2.40
C vs. D = 1.17
B vs. F = 1.24

7.30 From the AW data shown below for projects regarding campgrounds and lodging at a national park, determine which project, if any, should be selected from the 6 mutually exclusive projects. If the proper comparisons have not been made, state which one(s) should be done.

Project Identification					
G	**H**	**I**	**J**	**K**	**L**
Cost, $ 20,000	45,000	50,000	35,000	85,000	70,000
B/C Ratio 1.15	0.89	1.10	1.11	0.94	1.06

Selected incremental B/C ratios:

G vs. H = 0.68
G vs. I = 0.73
H vs. J = 0.10
J vs. I = 1.07
G vs. J = 1.07
H vs. K = 1.00
H vs. L = 1.36
J vs. K = 0.82
J vs. L = 1.00
K vs. L = 0.40
G vs. L = 1.02

7.31 The 4 mutually exclusive alternatives shown are being compared by the B/C method. Which alternative, if any, should be selected?

Alter-native	Alternative Cost, $ Millions	Direct B/C Ratio	Incremental B/C when Compared with Alternative			
			X	**Y**	**Z**	**ZZ**
X	20	0.75	—			
Y	30	1.07	1.70	—		
Z	50	1.20	1.50	1.40	—	
ZZ	90	1.11	1.21	1.13	1.00	—

PROBLEMS FOR TEST REVIEW AND FE EXAM PRACTICE

7.32 All of the following are examples of public sector projects except
  a. bridges.
  b. emergency relief.
  c. prisons.
  d. oil wells.

7.33 All of the following are usually associated with public sector projects except
  a. funding from taxes.
  b. profit.
  c. disbenefits.
  d. infinite life.

7.34 The *conventional* B/C ratio is written as
  a. B/C = (Benefits − Disbenefits)/− Costs
  b. B/C = (Benefits − Disbenefits)/Costs
  c. B/C = (Benefits + Disbenefits)/Costs
  d. B/C = Benefits/(Costs + Disbenefits)

7.35 From the values shown, the conventional B/C ratio is closest to
  a. 1.28.
  b. 1.33.
  c. 1.54.
  d. 2.76.

	PW, $	AW, $/year	FW, $
First cost	$100,000	$16,275	$259,370
M&O cost	$61,446	$10,000	$159,374
Benefits	$245,784	$40,000	$637,496
Disbenefits	$30,723	$5,000	$79,687

7.36 The four *mutually exclusive* alternatives shown are compared using the B/C method. The alternative, if any, to select is
  a. J.
  b. K.
  c. L.
  d. M.

Alter-native	Cost, $ millions	B/C Ratio vs. DN	Incremental B/C when Compared with Alternative			
			J	**K**	**L**	**M**
J	20	1.1	—			
K	25	0.96	0.40	—		
L	33	1.22	1.42	2.14	—	
M	45	0.89	0.72	0.80	0.08	—

Problems 7.37 and 7.38 are based on the following information

The Corps of Engineers compiled the following data to determine which of two flood control dams it should build in a certain residential area.

	Ice	T
Flood damage, $/year	220,000	140,000
Disbenefits, $/year	30,000	30,000
Costs, $/year	300,000	450,000

7.37 In conducting a B/C analysis of this data,
  a. the DN alternative *is not* an option.
  b. the DN alternative *is* an option.
  c. there is not enough information given to know if DN is an option or not.
  d. DN is an option only if the alternatives Ice and T are mutually exclusive.

7.38 The *conventional* B/C ratio between alternatives Ice and T is closest to
  a. 0.24.
  b. 0.33.
  c. 0.53.
  d. 0.73.

7.39 The following 4 *independent* alternatives are being compared using the B/C method. The alternative(s), if any, to select are

Alter-native	Cost, $ millions	B/C Ratio vs. DN	Incremental B/C when Compared with Alternative			
			J	**K**	**L**	**M**
J	20	1.1	—			
K	25	0.96	0.40	—		
L	33	1.22	1.42	2.14	—	
M	45	0.89	0.72	0.80	0.08	

  a. J.
  b. J and K.
  c. L.
  d. J and L.

7.40 For the data shown, the *conventional* benefit-to-cost ratio at $i = 10\%$ per year is closest to

Benefits of $20,000 in year 0 and $30,000 in year 5
Disbenefits of $7000 at year 3
Savings (government) of $25,000 in years 1 through 4
Cost of $100,000 in year 0
Project life is 5 years

  a. less than 1.5.
  b. 1.61.
  c. 1.86.
  d. 1.98.

7.41 If benefits are $10,000 per year forever, starting in year 1, and costs are $50,000 at time zero and $50,000 at the end of year 2, the B/C ratio at $i = 10\%$ per year is closest to
  a. 1.1.
  b. 1.8.
  c. 0.90.
  d. less than 0.75.

8 Chapter

Breakeven, Sensitivity, and Payback Analysis

Randy Allbritton/Getty Images

This chapter covers several related topics that assist in evaluating the effects of varying estimated values for one or more parameters present in an economic study. Since all estimates are for the future, it is important to understand which parameter(s) may make a significant impact on the economic justification of a project.

Breakeven analysis is performed for one project or two alternatives. For a single project, it determines a parameter value that makes revenue equal cost. Two alternatives break even when they are equally acceptable based upon a calculated value of one parameter common to both alternatives. *Make-or-buy decisions* for most subcontractor services, manufactured components, or international contracts, are routinely based upon the outcome of a breakeven analysis.

Sensitivity analysis is a technique that determines the impact on a measure of worth—most commonly PW, AW, ROR, or B/C—caused by varying estimated values. In short, it answers the question "What if?" It can be applied to one project, to two or more alternatives, and to one or more parameters. Since every study depends upon good estimation of costs (and revenue), sensitivity analysis is a vital tool to learn and utilize.

Payback period is a good technique to initially determine if a project may be financially acceptable. The technique determines the amount of time necessary for a proposed project to do two things—recover its initial investment (first cost) and generate enough annual net cash flow to meet or exceed the required MARR. As discussed in the chapter, there are a couple of drawbacks to payback analysis that must be remembered when the technique is used.

Objectives

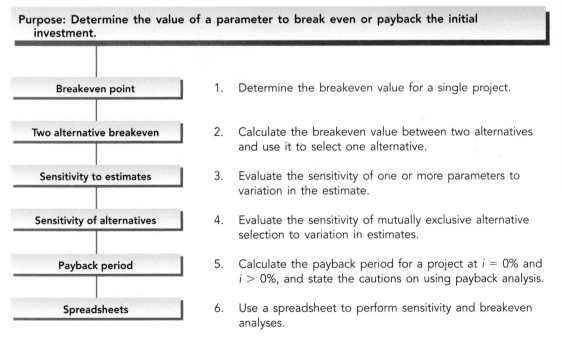

Purpose: Determine the value of a parameter to break even or payback the initial investment.

Breakeven point	1. Determine the breakeven value for a single project.
Two alternative breakeven	2. Calculate the breakeven value between two alternatives and use it to select one alternative.
Sensitivity to estimates	3. Evaluate the sensitivity of one or more parameters to variation in the estimate.
Sensitivity of alternatives	4. Evaluate the sensitivity of mutually exclusive alternative selection to variation in estimates.
Payback period	5. Calculate the payback period for a project at $i = 0\%$ and $i > 0\%$, and state the cautions on using payback analysis.
Spreadsheets	6. Use a spreadsheet to perform sensitivity and breakeven analyses.

8.1 BREAKEVEN ANALYSIS FOR A SINGLE PROJECT

Breakeven analysis determines the value of a parameter or decision variable that makes two relations equal. For example, breakeven analysis can determine the required years of use to recover the initial investment and annual operating costs. There are many forms of breakeven analysis; some equate PW or AW equivalence relations, some involve equating revenue and cost relations, others may equate demand and supply relations. However, they all have a common approach, that is, to equate two relations, or to set their difference equal to zero, and solve for the breakeven value of one variable that makes the equation true.

The need to determine the *breakeven value of a decision variable* without including the time value of money is common. For example, the variable may be a design capacity that will minimize cost, or the sales volume necessary to cover costs, or the cost of fuel to maximize revenue from electricity generation.

Figure 8.1a presents different shapes of a revenue relation identified as *R*. A linear revenue relation is commonly assumed, but a nonlinear relation is often more realistic with increasing per unit revenue for larger volumes (curve 1), or decreasing per unit revenue (curve 2).

FIGURE 8.1

Linear and nonlinear revenue and cost relations.

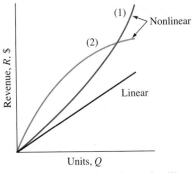

(*a*) Revenue relations—linear, increasing (1) and decreasing (2) per unit

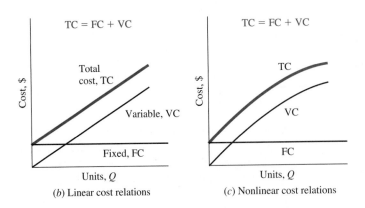

(*b*) Linear cost relations   (*c*) Nonlinear cost relations

Costs, which may be linear or nonlinear, usually include two components—fixed and variable—as indicated in Figures 8–1*b* and *c*.

Fixed costs (FC). Includes costs such as buildings, insurance, fixed overhead, some minimum level of labor, equipment capital recovery, and information systems.

Variable costs (VC). Includes costs such as direct labor, subcontractors, materials, indirect costs, marketing, advertisement, legal, and warranty.

The fixed cost component is essentially constant for all values of the variable, so it does not vary significantly over a wide range of operating parameters. Even if no output is produced, fixed costs are incurred at some threshold level. (Of course, this situation cannot last long before the operation must shut down.) Fixed costs are reduced through improved equipment, information systems and workforce utilization, less costly fringe benefit packages, subcontracting specific functions, and so on.

A simple VC relation is vQ, where v is the variable cost per unit and Q is the quantity. Variable costs change with output level, workforce size, and many other parameters. It is usually possible to decrease variable costs through improvements in design, efficiency, automation, materials, quality, safety, and sales volume.

When FC and VC are added, they form the total cost relation TC. Figure 8.1*b* illustrates linear fixed and variable costs. Figure 8.1*c* shows TC for a nonlinear VC in which unit variable costs decrease as Q rises.

At some value of Q, the revenue and total cost relations will intersect to identify the breakeven point Q_{BE} (Figure 8.2*a*). If $Q > Q_{BE}$, there is a profit; but if $Q < Q_{BE}$, there is a loss. For linear R and TC, the greater the quantity, the larger the profit. Profit is calculated as

$$\textbf{Profit} = \textbf{revenue} - \textbf{total cost} = R - \text{TC} \qquad [8.1]$$

A closed-form solution for Q_{BE} may be derived when revenue and total cost are linear functions of Q by equating the relations, indicating a profit of zero.

$$R = \text{TC}$$
$$rQ = \text{FC} + \text{VC} = \text{FC} + vQ$$

where r = revenue per unit
       v = variable cost per unit

Solve for Q to obtain the breakeven quantity.

$$Q_{BE} = \frac{\text{FC}}{r - v} \qquad [8.2]$$

The breakeven graph is an important management tool because it is easy to understand. For example, if the variable cost per unit is reduced, the linear TC line has a smaller slope (Figure 8.2*b*), and the breakeven point decreases. This is an advantage because the smaller the value of Q_{BE}, the greater the profit for a given amount of revenue.

FIGURE 8.2

(*a*) Breakeven point and (*b*) effect on breakeven point when the variable cost per unit is reduced

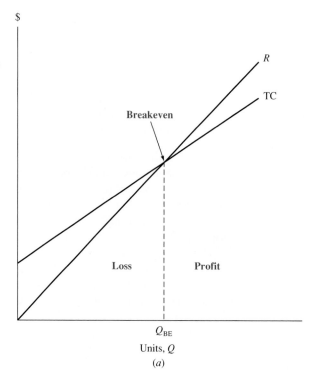

(*a*)

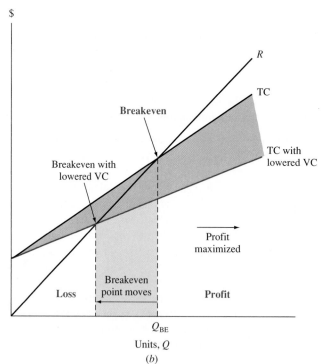

(*b*)

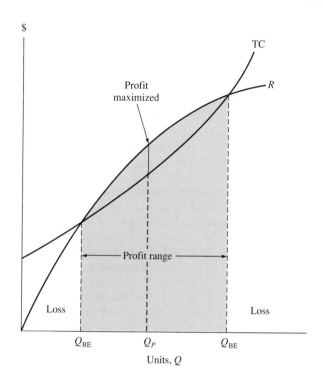

$

Profit
maximized

TC

R

←——— Profit range ———→

Loss   Loss

Q_{BE} Q_P Q_{BE}

Units, Q

FIGURE 8.3
Breakeven points and
maximum profit point
for a nonlinear
analysis.

If nonlinear R or TC models are used, there may be more than one breakeven point. Figure 8.3 presents this situation for two breakeven points. The maximum profit occurs at Q_P where the distance between the R and TC curves is greatest.

EXAMPLE 8.1

Nicholea Water LLC dispenses its product Nature's Pure Water via vending machines with most current locations at food markets and pharmacy or chemist stores. The average monthly fixed cost per site is $900, while each gallon costs 18¢ to purify and sells for 30¢. (*a*) Determine the monthly sales volume needed to break even. (*b*) Nicholea's president is negotiating for a sole-source contract with a municipal government where several sites will dispense larger amounts. The fixed cost and purification costs will be the same, but the sales price per gallon will be 30¢ for the first 5000 gallons per month and 20¢ for all above this threshold level. Determine the monthly breakeven volume at each site.

Solution

a. Use Equation [8.2] to determine the monthly breakeven quantity of 7500 gallons.

$$Q_{BE} = \frac{900}{0.30 - 0.18} = 7500$$

b. At 5000 gallons, the profit is negative at -300, as determined by Equation [8.1]. The revenue curve has a lower slope above this threshold gallonage level. Since Q_{BE} can't be determined directly from Equation [8.2], it is found by equating revenue and total cost relations with the threshold level of 5000 included. If Q_U is termed the breakeven quantity above threshold, the equated R and TC relations are

$$0.30(5000) + 0.20(Q_U) = 900 + 0.18(5000 + Q_U)$$

$$Q_U = \frac{900 + 900 - 1500}{0.20 - 0.18} = 15,000$$

Therefore, the required volume per site is 20,000 gallons per month, the point at which revenue and total cost break even at $4500. Figure 8.4 details the relations and points.

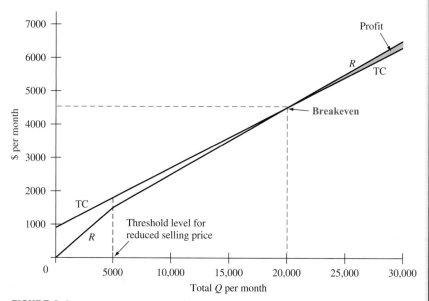

FIGURE 8.4   Breakeven graph with a volume discount for sales, Example 8.1*b.*

In some circumstances, breakeven analysis of revenue and cost is better if performed on a per unit basis. The revenue per unit is $R/Q = r$, and the TC relation is divided by Q to obtain cost per unit, also termed *average cost per unit* C_u.

$$C_u = \frac{TC}{Q} = \frac{FC + vQ}{Q} = \frac{FC}{Q} + v \qquad [8.3]$$

The relation $R/Q = TC/Q$ is solved for Q. The result for Q_{BE} is the same as Equation [8.2].

When an engineering economy study for a single project includes a P, F, A, i, or n value that cannot be reliably estimated, a breakeven quantity for one of the parameters can be determined by setting an equivalence relation for PW, FW, or AW equal to zero and solving for the unknown variable. This is the approach utilized in Chapter 7 to determine the breakeven rate of return i^*.

8.2 BREAKEVEN ANALYSIS BETWEEN TWO ALTERNATIVES

Breakeven analysis is an excellent technique with which to determine the value of a parameter that is common to two alternatives. The parameter can be the interest rate, capacity per year, first cost, annual operating cost, or any parameter. We have already performed breakeven analysis between alternatives in Chapter 7 for the incremental ROR value (Δi^*).

Breakeven analysis usually involves revenue or cost variables common to both alternatives. Figure 8.5 shows two alternatives with linear total cost (TC) relations. The fixed cost of alternative 2 is greater than that of alternative 1. However, alternative 2 has a smaller variable cost, as indicated by its lower slope. If the number of units of the common variable is greater than the breakeven amount, alternative 2 is selected, since the total cost will be lower. Conversely, an anticipated level of operation below the breakeven point favors alternative 1.

It is common to find the breakeven value by *equating PW or AW equivalence relations*. The AW is preferred when the variable units are expressed on a yearly basis or when alternatives have unequal lives. The following steps determine the breakeven point of the common variable.

1. Define the common variable and its dimensional units.
2. Use AW or PW analysis to express the total cost of each alternative as a function of the common variable.
3. Equate the two relations and solve for the breakeven value.
4. If the anticipated level is below the breakeven value, select the alternative with the higher variable cost (larger slope). If the level is above breakeven, select the alternative with the lower variable cost. Refer to Figure 8.5.

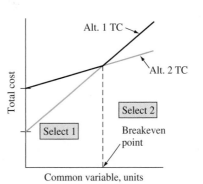

FIGURE 8.5
Breakeven between two alternatives with linear cost relations.

EXAMPLE 8.2   A small aerospace company is evaluating two alternatives: the purchase of an automatic feed machine and a manual feed machine for a finishing process. The auto-feed machine has an initial cost of $23,000, an estimated salvage value of $4000, and a predicted life of 10 years. One person will operate the machine at a rate of $24 per hour. The expected output is 8 tons per hour. Annual maintenance and operating cost is expected to be $3500.

The alternative manual feed machine has a first cost of $8000, no expected salvage value, a 5-year life, and an output of 6 tons per hour. However, three workers will be required at $12 per hour each. The machine will have an annual maintenance and operation cost of $1500. All projects are expected to generate a return of 10% per year. How many tons per year must be finished in order to justify the higher purchase cost of the auto-feed machine?

Solution

Use the steps above to calculate the breakeven point between the two alternatives.

1. Let x represent the number of tons per year.
2. For the auto-feed machine the annual variable cost is

$$\text{Annual VC} = \frac{\$24}{\text{hour}} \frac{1 \text{ hour}}{8 \text{ tons}} \frac{x \text{ tons}}{\text{year}} = 3x$$

The AW expression for the auto-feed machine is

$$\begin{aligned} \text{AW}_{\text{auto}} &= -23{,}000(A/P,10\%,10) + 4000(A/F,10\%,10) - 3500 - 3x \\ &= \$-6992 - 3x \end{aligned}$$

Similarly, the annual variable cost and AW for the manual feed machine are

$$\text{Annual VC} = \frac{\$12}{\text{hour}}(3 \text{ operators})\frac{1 \text{ hour}}{6 \text{ tons}} \frac{x \text{ tons}}{\text{year}} = 6x$$

$$\begin{aligned} \text{AW}_{\text{manual}} &= -8000(A/P,10\%,5) - 1500 - 6x \\ &= \$-3610 - 6x \end{aligned}$$

3. Equate the two cost relations and solve for x.

$$\text{AW}_{\text{auto}} = \text{AW}_{\text{manual}}$$
$$-6992 - 3x = -3610 - 6x$$
$$x = 1127 \text{ tons per year}$$

4. If the output is expected to exceed 1127 purchase the auto-feed machine, since its VC slope of 3 is smaller.

The breakeven approach is commonly used for make-or-buy decisions. The alternative to buy (or subcontract) often has no fixed cost, but has a larger variable cost. Where the two cost relations cross is the make-buy decision quantity. Amounts above this indicate that the item should be made, not purchased outside.

Guardian is a national manufacturing company of home health care appliances. It **EXAMPLE 8.3**
is faced with a make-or-buy decision. A newly engineered lift can be installed in a
car trunk to raise and lower a wheelchair. The steel arm of the lift can be purchased
for $0.60 per unit or made inhouse. If manufactured on site, two machines will be
required. Machine A is estimated to cost $18,000, have a life of 6 years, and a
$2000 salvage value; machine B will cost $12,000, have a life of 4 years, and a
$−500 salvage value (carry-away cost). Machine A will require an overhaul after
3 years costing $3000. The AOC for A is expected to be $6000 per year and for B
$5000 per year. A total of four operators will be required for the two machines at
a rate of $12.50 per hour per operator. One thousand units will be manufactured in
a normal 8-hour period. Use an MARR of 15% per year to determine the following:

a. Number of units to manufacture each year to justify the inhouse (make) option.
b. The maximum capital expense justifiable to purchase machine A, assuming
all other estimates for machines A and B are as stated. The company expects
to produce 125,000 units per year.

Solution

a. Use steps 1 to 3 stated previously to determine the breakeven point.
  1. Define x as the number of lifts produced per year.
  2. There are variable costs for the operators and fixed costs for the two
  machines for the make option.

$$\text{Annual VC} = (\text{cost per unit})(\text{units per year})$$
$$= \frac{4 \text{ operators}}{1000 \text{ units}} \frac{\$12.50}{\text{hour}}(8 \text{ hours})x$$
$$= 0.4x$$

The annual fixed costs for machines A and B are the AW amounts.

$$\text{AW}_A = -18,000(A/P,15\%,6) + 2000(A/F,15\%,6)$$
$$-6000 - 3000(P/F,15\%,3)(A/P,15\%,6)$$
$$\text{AW}_B = -12,000(A/P,15\%,4) - 500(A/F,15\%,4) - 5000$$

Total cost is the sum of AW_A, AW_B, and VC.
  3. Equating the annual costs of the buy option $(0.60x)$ and the make option
  yields

$$-0.60x = \text{AW}_A + \text{AW}_B - \text{VC}$$
$$= -18,000(A/P,15\%,6) + 2000(A/F,15\%,6) - 6000$$
$$-3000(P/F,15\%,3)(A/P,15\%,6) - 12,000(A/P,15\%,4)$$
$$-500(A/F,15\%,4) - 5000 - 0.4x$$
$$-0.2x = -20,352.43$$
$$x = 101,762 \text{ units per year}$$

A minimum of 101,762 lifts must be produced each year to justify the
make option, which has the lower variable cost of $0.40x$.

b. The production level is above breakeven. To find the maximum justifiable P_A, substitute 125,000 for x and P_A for the first cost of machine A. Solution yields $P_A = \$35,588$. This means that approximately twice the estimated first cost of \$18,000 could be spent on A.

8.3 SENSITIVITY ANALYSIS FOR VARIATION IN ESTIMATES

There is always some degree of risk in undertaking any project, often due to uncertainty and variation in parameter estimates. The effect of variation may be determined by using *sensitivity analysis*. Usually, one factor at a time is varied, and independence with other factors is assumed. This assumption is not completely correct in real-world situations, but it is practical since it is difficult to accurately account for dependencies among parameters.

Sensitivity analysis determines how a measure of worth—PW, AW, ROR, or B/C—and the alternative may be altered if a particular parameter varies over a stated range of values. For example, some variation in MARR will likely not alter the decision when all alternatives have $i^* > \text{MARR}$; thus, the decision is relatively insensitive to MARR. However, variation in the estimated P value may make selection from the same alternatives sensitive.

There are several types of sensitivity analyses. It is possible to examine the sensitivity to variation for one, two, or more parameters for one alternative, as well as evaluating the impact on selection between mutually exclusive alternatives. *Breakeven analysis is actually a form of sensitivity analysis* in that the accept/reject decision changes depending upon where the parameter's most likely estimate lies relative to the breakeven point. There are three types of sensitivity analyses covered in this and the next section:

- Variation of one parameter at a time for a single project (Example 8.4) or for selecting between mutually exclusive alternatives
- Variation of more than one parameter for a single project (Example 8.5)
- Sensitivity of mutually exclusive alternative selection to variation of more than one parameter (Section 8.4).

In all cases, the targeted parameter(s) and measure of worth must be selected prior to initiating the analysis. A general procedure to conduct a sensitivity analysis follows these steps:

1. Determine which parameter(s) of interest might vary from the most likely estimated value.
2. Select the probable range (numerical or percentage) and an increment of variation for each parameter.
3. Select the measure of worth.
4. Compute the results for each parameter using the measure of worth.

5. To better interpret the sensitivity, graphically display the parameter versus the measure of worth.

It is best to routinely use the AW or PW measure of worth. The ROR should not be used for multiple alternatives because of the incremental analysis required (discussed in Section 6.4). Spreadsheets are very useful in performing sensitivity analysis. One or more values can be easily changed to determine the effect. See Section 8.6 for a complete example.

EXAMPLE 8.4

Hammond Watches, Inc. is about to purchase equipment that will significantly improve the light reflected by the watch dial, thus making it easier to read the time in very low light environments. Most likely estimates are $P = \$80,000$, $n = 10$ years, $S = 0$, and increased net revenue after expenses of $25,000 the first year decreasing by $2000 per year thereafter. The VP of Manufacturing is concerned about the project's economic viability if the equipment life varies from the 10-year estimate, and the VP of Marketing is concerned about sensitivity to the revenue estimate if the annual decrease is larger or smaller than the $2000. Use MARR $= 15\%$ per year and PW equivalence to determine sensitivity to (*a*) variation in life for 8, 10, and 12 years, and (*b*) variation in revenue decreases from 0 to $3000 per year. Plot the resulting PW values for each parameter.

Solution

a. Follow the procedure outlined above for sensitivity to life estimates.
  1. Asset life is the parameter of interest.
  2. Range for n is 8 to 12 in 2-year increments.
  3. PW is the measure of worth.
  4. Use the most likely estimate of $G = \$-2000$ in revenue, and set up the PW relation for varying n values. Then insert values of $n = 8$, 10, and 12 to obtain the PW values.

$$PW = -80,000 + 25,000(P/A,15\%,n) - 2000(P/G,15\%,n)$$

n	PW
8	$ 7,221
10	11,511
12	13,145

  5. Figure 8.6*a* plots PW versus n. The nonlinear result indicates minimal sensitivity to this range of life estimates. All PW values are clearly positive, indicating that the MARR of 15% is well exceeded.

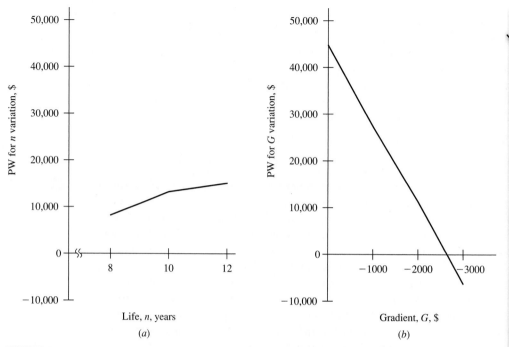

FIGURE 8.6 Sensitivity of PW to variation in *n* and *G* estimates, Example 8.4.

b. Now the revenue gradient is examined using the same PW relation with $n = 10$. Set the increment at $1000 to obtain sensitivity to variation in G.

$$PW = -80{,}000 + 25{,}000(P/A,15\%,10) - G(P/G,15\%,10)$$

G	PW
$ 0	$45,470
−1000	28,491
−2000	11,511
−3000	−5,469

Figure 8.6*b* graphs PW versus *G*. The resulting linear plot indicates high sensitivity over the range. In fact, if revenue does drop by $3000 per year, and other estimates prove to be accurate, the project will not make the 15% return.

Comments: It is important to recognize a significant limitation of one-at-a-time analysis. All other parameters are fixed at their most likely estimate when the PW value is calculated. This is a good approach when one parameter contributes most of the sensitivity, but not so good when several parameters are expected to contribute significantly to the sensitivity.

Sensitivity plots involving *n* and *i* will not be linear because of the mathematical form of the factor. Those involving *P*, *G*, and *S* will normally be linear since n and i are fixed. In Figure 8.6b, the decrease in PW is $16,980 per $1000 change in G. Also, the breakeven point can be estimated from the plot at about $-2700. Solving for G in the PW relation results in the exact value of $-2678. If the annual decrease in revenue is expected to exceed this amount, the project will not be economically justifiable at MARR $= 15\%$.

As a parameter estimate varies, selection between alternatives can change. To determine if the possible parameter range can alter the selection, calculate the PW, AW, or other measure of worth at different parameter values. This determines the sensitivity of the selection to the variation. Alternatively, the two PW or AW relations can be set equal and the parameter's breakeven value determined. If breakeven is in the range of the parameter's expected variation, the decision is considered sensitive. If this sensitivity/breakeven analysis of two alternatives (A and B) is performed using a spreadsheet, Excel's SOLVER tool can be used to equate the two equivalency relations by setting up the constraint $AW_A = AW_B$. Reference the Excel help function for details on using SOLVER.

When the sensitivity of *several parameters* is considered for *one project* using a *single measure of worth,* it is helpful to graph *percentage change* for each parameter versus the measure of worth. The variation in each parameter is indicated as a percentage deviation from the most likely estimate on the horizontal axis. If the response curve is flat and approaches horizontal over the expected range of variation, there is little sensitivity. As the curve increases in slope (positive or negative), the sensitivity increases. Those with large slopes indicate estimates that may require further study.

EXAMPLE 8.5

Janice is a process engineer with Upland Chemicals currently working in the research and planning department. She has performed an economic analysis for a process involving alternative fuels that is to be undertaken within the next 12 months. An ROR of 10% balanced the PW equivalence relation for the project. The required MARR for similar projects has been 5%, since some government subsidy is available for new fuels research. In an effort to evaluate the sensitivity of ROR to several variables, she constructed the graph in Figure 8.7 for variation ranges of selected parameters from the single-point estimates provided to her.

Sales price, $ per gallon	$\pm 20\%$
Materials cost, $ per ton	-10% to $+30\%$
Labor cost, $ per hour	$\pm 30\%$
Equipment maintenance cost, $ per year	$\pm 10\%$

What should Janice observe and do about the sensitivity of each parameter studied?

Solution

First of all, the range studied for each parameter varies from as little as 10% to as much as 30%. Therefore, sensitivity outside these ranges cannot be determined by this analysis. For the ranges studied, *labor and annual maintenance costs do*

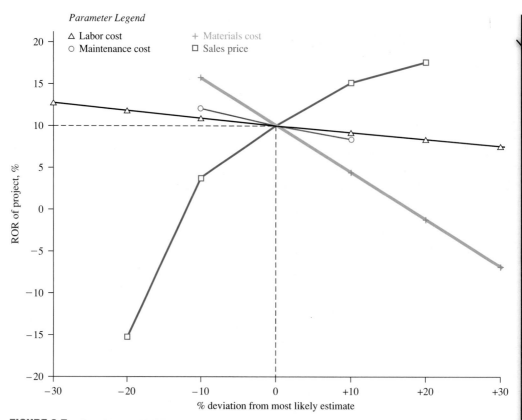

FIGURE 8.7   Sensitivity of ROR to four parameters, Example 8.5.

not cause much change in the ROR of the project. In fact, a 30% increase in labor cost will decrease the ROR by only an estimated 2 to 3%.

Materials cost can cause a significant decrease in ROR. If raw materials do increase by 10%, the 5% MARR will not be met, and a 20% increase will likely drive the ROR negative. Knowledge about and control of these costs are essential, especially in a cost inflationary environment.

Sales price can also have a large effect on the project's profitability. A relatively small change of ±10% from the estimated price can make ROR swing from about 3% to as much as 15%. Decreases beyond −10% can force the project into a negative return situation, and price increases above 10% should have a diminishing effect on the overall ROR. Sales price is a very powerful parameter; efforts by Upland's management to reduce predictable variation are clearly needed.

Janice needs to ensure that the results of her analysis are brought to the attention of appropriate management personnel. She should be sure that the sample data she used to perform the analysis are included in her report so the results can be verified and/or expanded. If no response is received in a couple of weeks, she should do a personal followup regarding reactions to her analysis.

8.4 SENSITIVITY ANALYSIS OF MULTIPLE PARAMETERS FOR MULTIPLE ALTERNATIVES

The economic advantages and disadvantages among two or more mutually exclusive alternatives can be studied by making three estimates for each parameter expected to vary enough to affect the selection: *a pessimistic, a most likely, and an optimistic estimate*. Depending upon the nature of a parameter, the pessimistic estimate may be the lowest value (alternative life is an example) or the largest value (such as asset first cost). This approach analyzes the measure of worth and alternative selection sensitivity within a predicted range of variation for each parameter. Usually the most likely estimate is used for all other parameters when the measure of worth is calculated for each alternative.

EXAMPLE 8.6

An engineer is evaluating three alternatives for which he has made three estimates for salvage value, annual operating cost, and life (Table 8.1). For example, alternative B has pessimistic estimates of $S = \$500$, $AOC = \$-4000$, and $n = 2$ years. The first costs are known, so they have the same value. Perform a sensitivity analysis and determine the most economic alternative using AW analysis at an MARR of 12% per year.

TABLE 8.1 **Competing Alternatives with Three Estimates for Selected Parameters**

Strategy		First Cost, $	Salvage Value, $	AOC, $	Life n, Years
Alternative A					
	P	−20,000	0	−11,000	3
Estimates	ML	−20,000	0	−9,000	5
	O	−20,000	0	−5,000	8
Alternative B					
	P	−15,000	500	−4,000	2
Estimates	ML	−15,000	1,000	−3,500	4
	O	−15,000	2,000	−2,000	7
Alternative C					
	P	−30,000	3,000	−8,000	3
Estimates	ML	−30,000	3,000	−7,000	7
	O	−30,000	3,000	−3,500	9

P = pessimistic; ML = most likely; O = optimistic.

Solution

For each alternative in Table 8.1, calculate the AW value. For example, the AW relation for A, pessimistic estimates, is

$$AW_A = -20,000(A/P,12\%,3) - 11,000 = \$-19,327$$

There are a total of nine AW relations for the three alternatives and three estimates. Table 8.2 presents all AW values. Figure 8.8 is a plot of AW versus the three estimates of life for each alternative (plots of AW for any parameter with variation, that is n, AOC or S, give the same conclusion.) Since the AW calculated using the ML estimates for B ($\$-8229$) is economically better than even the optimistic AW values for A and C, alternative B is clearly favored.

Comment: While the alternative that should be selected here is quite obvious, this is not always the case. Then it is necessary to select one set of estimates (P, ML, or O) upon which to base the selection.

TABLE 8.2   Annual Worth Values for Varying Parameters, Example 8.6

Estimates	Alternative AW Values		
	A	B	C
P	$\$-19,327$	$\$-12,640$	$\$-19,601$
ML	$-14,548$	$-8,229$	$-13,276$
O	$-9,026$	$-5,089$	$-8,927$

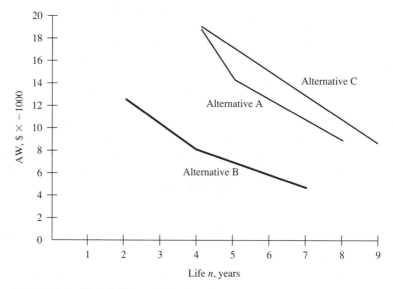

FIGURE 8.8   Plot of AW sensitivity for different life estimates, Example 8.6.

8.5 PAYBACK PERIOD ANALYSIS

Payback analysis (also called payout analysis) is another form of sensitivity analysis that uses a PW equivalence relation. Payback can take two forms: one for $i > 0\%$ (also called *discounted payback*) and another for $i = 0\%$ (also called *no-return payback*). *The payback period n_p is the time, usually in years, it* will take for estimated revenues and other economic benefits *to recover the initial investment P and a specific rate of return i%*. The n_p value is generally not an integer.

The payback period should be calculated using a required return that is greater than 0%. In practice, however, the payback period is often determined with a no-return requirement ($i = 0\%$) to initially screen a project and determine whether it warrants further consideration.

To find the discounted payback period at a stated rate $i > 0\%$, calculate the years n_p that make the following expression correct.

$$0 = -P + \sum_{t=1}^{t=n_p} \text{NCF}_t(P/F,i,t) \qquad [8.4]$$

As discussed in Chapter 1, NCF is the estimated net cash flow for each year t, where NCF = receipts $-$ disbursements. If the NCF values are equal each year, the P/A factor may be used to find n_p.

$$0 = -P + \text{NCF}(P/A,i,n_p) \qquad [8.5]$$

After n_p years, the cash flows will recover the investment and a return of $i\%$. If, in reality, the asset or alternative is used for more than n_p years, a larger return may result; but if the useful life is less than n_p years, there is not enough time to recover the initial investment and the $i\%$ return. It is very important to realize that in payback analysis *all net cash flows occurring after n_p years are neglected*. This is significantly different from the approach of all other evaluation methods (PW, AW, ROR, B/C) where all cash flows for the entire useful life are included. As a result, payback analysis can unfairly bias alternative selection. Therefore, the payback period n_p *should not be used as the primary measure of worth to select an alternative*. It provides initial screening or supplemental information in conjunction with an analysis performed using the PW or AW method.

No-return payback analysis determines n_p at $i = 0\%$. This n_p value serves merely as an initial indicator that a proposal is viable and worthy of a full economic evaluation. To determine the payback period, substitute $i = 0\%$ in Equation [8.4] and find n_p.

$$0 = -P + \sum_{t=1}^{t=n_p} \text{NCF}_t \qquad [8.6]$$

For a uniform net cash flow series, Equation [8.6] is solved for n_p directly.

$$n_p = \frac{P}{\text{NCF}} \qquad [8.7]$$

An example use of *no-return payback* as an initial screening of proposed projects is a corporation president who absolutely insists that every project must recover the investment in 3 years or less. Therefore, no proposed project with $n_p > 3$ at $i = 0\%$ should be considered further.

As with n_p for $i > 0\%$, it is incorrect to use the no-return payback period to make final alternative selections. It *neglects any required return*, since the time value of money is omitted, and it *neglects all net cash flows after time n_p*, including positive cash flows that may contribute to the return on the investment.

EXAMPLE 8.7   This year the owner/founder of J&J Health allocated a total of $18 million to develop new treatment techniques for sickle cell anemia, a blood disorder that primarily affects people of African ancestry and other ethnic groups, including people who are of Mediterranean and Middle Eastern descent. The results are estimated to positively impact net cash flow starting 6 years from now and for the foreseeable future at an average level of $6 million per year.

a. As an initial screening for economic viability, determine both the no-return and $i = 10\%$ payback periods.

b. Assume that any patents on the process will be awarded during the sixth year of the project. Determine the project ROR if the $6 million net cash flow were to continue for a total of 17 years (through year 22), when the patents legally expire.

Solution

a. The NCF for years 1 through 5 is $0 and $6 million thereafter. Let $x =$ number of years beyond 5 when NCF > 0. For no-return payback, apply Equation [8.6], and for $i = 10\%$, apply Equation [8.4]. In $ million units,

$$i = 0\%: \quad 0 = -18 + 5(0) + x(6)$$
$$n_p = 5 + x = 5 + 3 = 8 \text{ years}$$
$$i = 10\%: \quad 0 = -18 + 5(0) + 6(P/A,10\%,x)(P/F,10\%,5)$$
$$(P/A,10\%,x) = \frac{18}{6(0.6209)} = 4.8317$$
$$x = 6.9$$
$$n_p = 5 + x = 5 + 7 = 12 \text{ years (rounded up)}$$

b. The PW relation to determine i^* over the 22 years is satisfied at $i^* = 15.02\%$. In $ million units, PW is

$$PW = -18 + 6(P/A,i\%,17)(P/F,i\%,5)$$

The conclusions are: a 10% return requirement increases payback from 8 to 12 years; and when cash flows expected to occur after the payback period are considered, project return increases to 15% per year.

If two or more alternatives are evaluated using payback periods to indicate initially that one may be better than the other(s), the primary shortcoming of payback analysis (neglect of cash flows after n_p) may lead to an economically incorrect decision. When cash flows that occur after n_p are neglected, it is possible to favor short-lived assets when longer-lived assets produce a higher return. In these cases, PW or AW analysis should always be the primary selection method.

8.6 USING SPREADSHEETS FOR SENSITIVITY OR BREAKEVEN ANALYSIS

Spreadsheets are excellent tools to perform sensitivity, breakeven, and payback analyses. Estimates can be varied repeatedly one at a time to determine how PW, AW, or ROR changes. The following examples illustrate spreadsheet development for both sensitivity and breakeven analyses.

To include an Excel chart, care must be taken to arrange the data of interest in rows or columns. The *xy* scatter chart is a commonly used tool for engineering economic analyses. See Appendix A and Excel Help for details on constructing charts.

Halcrow, Inc., Division of Road and Highway Consultancy, is anticipating the purchase of concrete strength test equipment for use in highway construction. Estimates are:

EXAMPLE 8.8

First cost, $P = \$-100,000$ Annual operating costs, $AOC = \$-20,000$
Life, $n = 5$ years   Annual revenue, $R = \$50,000$

The Halcrow MARR for such projects is 10% per year. With an annual net cash flow (NCF) of \$30,000, the IRR function of Excel indicates the project is economically justified with $i^* = 15.2\%$ for these most likely estimates. Use spreadsheets to perform sensitivity analysis in ROR to variation in first cost and revenue.

a. Depending upon the model purchased, P may vary as much as 25% from $\$-75,000$ to $\$-125,000$.
b. Revenue variations up to 20% are not unexpected, but R may go as low as \$25,000 per year in the worst case.

Solution

a. Figure 8.9*a* details the ROR values using the IRR function for first cost values from $\$-75,000$ to $\$-125,000$ with the most likely estimate of $\$-100,000$ and $i^* = 15.2\%$ in the center column. (*P* and ROR values placed in rows facilitate plotting via Excel's *xy* scatter chart.) The plot of rows 7

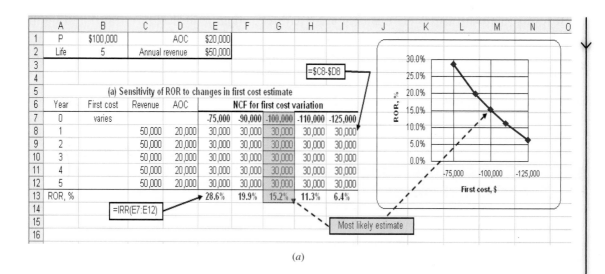

(a)

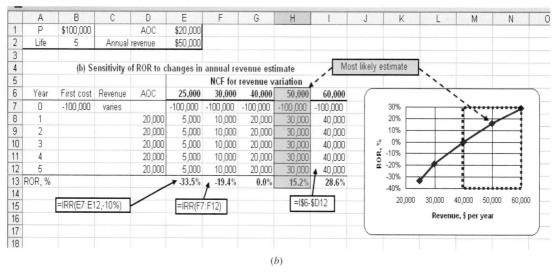

(b)

FIGURE 8.9 Sensitivity analysis of ROR to changes in (a) first cost estimate and (b) annual revenue estimate, Example 8.8

and 13 shows the significant reduction in ROR from about 30% to 6% over the possible range for P. If P increases by 25%, a project return of less than 10% is predicted.

b. Figure 8.9b uses the same spreadsheet format to evaluate ROR sensitivity to $\pm 20\%$ variation in revenue (framed area) and down to $25,000 per year. (The IRR function optional [guess] field of -10% avoids a #NUM error when $i^* = -33.5\%$ is determined at $R = \$25,000$.) Variation of -20% to

$40,000, as indicated on the chart, reduces ROR to 0%, which is much less than MARR = 10%. A drop in R to $25,000 per year indicates a large negative return of -33.5%. The ROR is very sensitive to revenue variation.

The Helstrom Corp. is a subcontractor in field operations to Boeing Company, manufacturer of the F18 Super Hornet multiple strike fighter aircraft. Helstrom is about to purchase diagnostics equipment to test on-board electronics systems. Estimates are:

EXAMPLE 8.9

First cost	$8,000,000
Annual operating cost	$100,000
Life	5 years
Variable cost per test	$800 (most likely)
Fixed contract income per test	$1600

Use Excel's GOAL SEEK tool to determine the breakeven number of tests per year if the variable cost per test (*a*) equals $800, the most likely estimate, and (*b*) varies from $800 to $1400. For simplicity purposes only, utilize a 0% interest rate.

Solution

a. There are a couple of ways to approach breakeven problems using spreadsheets. Equation [8.2] finds Q_{BE} directly. The annual fixed cost and breakeven relations are

$$FC = 8,000,000\,(A/P,0\%,5) + 100,000$$
$$= 8,000,000\,(0.2) + 100,000$$
$$= \$1,700,000$$
$$Q_{BE} = \frac{1,700,000}{1600 - 800} = 2125 \text{ tests}$$

A second approach, used in the spreadsheets here, applies Equation [8.1] with Q = number of tests per year.

$$\text{Profit} = \text{revenue} - \text{total cost}$$
$$= (1600 - 800)Q - (8,000,000(0.2) + 100,000)$$

Figure 8.10 (left side) shows all estimated parameters and the profit at 0 tests. The GOAL SEEK tool is set up to change the negative profit (cell B9) to $0 by increasing Q (cell C7). The right side presents the results with breakeven at 2125 tests per year. (When applying GOAL SEEK, the breakeven value 2125 will be displayed in cell B9; the results are shown separately here only for illustration.)

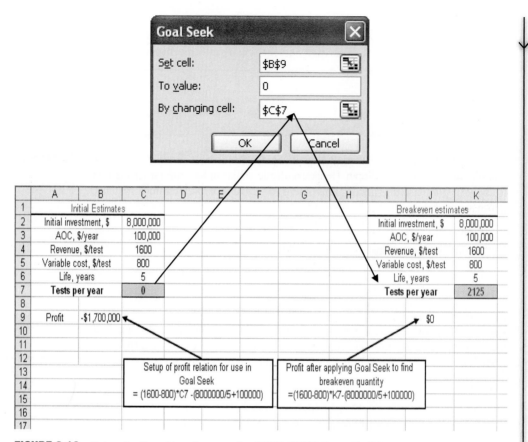

FIGURE 8.10 Determination of breakeven using GOAL SEEK, Example 8.9*a*.

b. To evaluate breakeven's sensitivity to a range of variable costs, the spreadsheet is reconfigured to Figure 8.11. It now includes values from $800 to $1400 per test in the profit relations in columns C and E. Column D displays breakeven values as GOAL SEEK is applied repeatedly for each variable cost. The tool setup for $800 (row 8) is detailed. (As before, in routine applications the results from GOAL SEEK would be in columns B and C; we have added columns D and E only to indicate before and after values.)

Breakeven is very sensitive to test cost. Breakeven tests increase fourfold (2125 to 8500) while the cost per test increases less than twofold ($800 to $1400).

Comment: GOAL SEEK is discussed further in Appendix A. The more powerful Excel tool SOLVER is better when multiple values (cells) are to be examined and when equality and inequality constraints are placed on the cell values. Consult the Excel Help system on your computer for details.

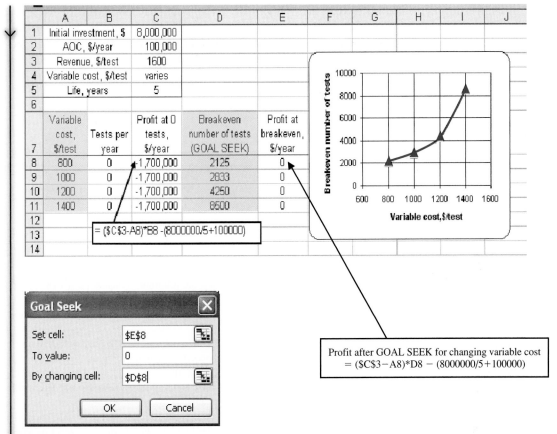

	A	B	C	D	E	F	G	H	I	J
1	Initial investment, $		8,000,000							
2	AOC, $/year		100,000							
3	Revenue, $/test		1600							
4	Variable cost, $/test		varies							
5	Life, years		5							
6										
7	Variable cost, $/test	Tests per year	Profit at 0 tests, $/year	Breakeven number of tests (GOAL SEEK)	Profit at breakeven, $/year					
8	800	0	-1,700,000	2125	0					
9	1000	0	-1,700,000	2833	0					
10	1200	0	-1,700,000	4250	0					
11	1400	0	-1,700,000	8500	0					
12										
13		= (C3-A8)*B8 -(8000000/5+100000)								
14										

Profit after GOAL SEEK for changing variable cost
$= (\$C\$3 - A8)*D8 - (8000000/5 + 100000)$

Goal Seek

Set cell: E8

To value: 0

By changing cell: D8

OK   Cancel

FIGURE 8.11 Determination of breakeven using GOAL SEEK repeatedly for changing variable cost estimates, Example 8.9*b*.

SUMMARY

This chapter treated sensitivity analysis and the related topics of breakeven analysis and payback period. Sensitivity to changing estimates for a parameter or for competing alternatives is determined via changes in a measure of worth. The variation in PW, AW, ROR, or B/C value for one or more parameters over a range of values is calculated (and may be plotted). Alternative selection sensitivity can be determined using three estimates of a parameter—optimistic, most likely, and pessimistic. In each sensitivity study, one parameter at a time is varied and independence between parameters is assumed.

The breakeven value Q_{BE} identifies a point of economic indifference. The acceptance criterion for one project is:

Accept the project if the estimated quantity exceeds Q_{BE}

For two alternatives determine the breakeven amount for a common parameter. The selection guideline is:

Select the alternative with the lower variable cost if the estimated quantity exceeds the breakeven amount.

Payback analysis estimates the number of years necessary to recover the initial investment plus a stated MARR. This is a supplemental analysis tool used for initial screening of a project prior to full economic justification. The technique has several drawbacks, especially for no-return payback where MARR is set to 0%. The main drawback is that payback analysis does not consider cash flows that may occur after the payback period has expired.

PROBLEMS

Breakeven Analysis—One Project

8.1 Explain in words why the breakeven quantity goes down when the variable cost is reduced. Relate your logic to Equation [8.2].

8.2 White Appliances has the following cost and revenue estimates for its new refrigerator model:

> Fixed cost = $2.58 million per year
> Cost per unit = $395
> Revenue per unit = $550

    a. Write the total cost relation.
    b. Determine the annual quantity needed to break even.
    c. Estimate the profit at 20% above breakeven.

8.3 Claudia has sold her car and received approval from the garage owner to re-lease her downtown reserved parking spot for the next four months so she can make some extra money. The rental fee is $200 per month, and she expects to charge $18 per day. Transportation in a car pool will cost her $6 per day. If there are a maximum of 20 work days per month for re-leasing the spot, determine the following:

    a. Total cost and revenue relations
    b. Breakeven quantity per month
    c. Amount of money she will make (or lose) if the number of re-leased days per month over the four-month period are 18, 12, 17, and 20.

The following information is used for Problems 8.4 through 8.8.

Hambry Enterprises produces a component for recycling uranium used as a nuclear fuel in power plant generators in France and the United States. Use the following cost and revenue figures, quoted in U.S. dollars per hundredweight (hwt), recorded for this year to calculate the answers for each plant.

Location	Fixed Cost, $ million	Revenue, $ per hwt	Cost, $ per hwt
France	3.50	8,500	3,900
United States	2.65	12,500	9,900

8.4 Determine the breakeven point for each plant.

8.5 Estimate the minimum revenue per hwt required for next year if breakeven values and variable costs remain constant, but fixed costs increase by 10%.

8.6 During this year, the French plant sold 950 units in Europe (which is more committed to nuclear power than the U.S.) and the U.S. plant sold 850 units. Determine the year's profit (loss) for each plant.

8.7 Hambry's president has a goal of $1 million profit next year at each plant with no revenue or fixed cost increases. Determine the decreases (dollar amounts and percentages) in variable cost necessary to meet this goal if the number of units sold is the same as this year.

8.8 Use Excel and the original cost data for the U.S. plant to plot the sensitivity of the breakeven point to revenue changes that range from 20% below to 30% above the $12,500 value.

8.9 At Ryerson Aggregates the main product is Rock-Glow, a natural looking pebble-matrix mix applied to swimming pool decks. It sells installed for $14.50 per square foot (sq ft). Ryerson can contract with several vendors for raw materials, which can cause the cost to vary from a high of $9.25 to a low of $7.50 per sq ft. Fixed costs average $2 million per year for the product.

    a. Determine the range of breakeven values as the cost varies from high to low. Does the breakeven point decrease as the variable cost is reduced as shown in Figure 8.2b?

b. Ryerson's owner wants a $500,000 profit per year. If sales average 350,000 sq ft annually, find the minimum allowed difference between revenue and cost, and determine if the stated amounts will generate this level of profit.

8.10 Benjamin used regression analysis to fit quadratic relations to monthly revenue and cost data with the following results:

$$R = -0.007Q^2 + 32Q$$
$$TC = 0.004Q^2 + 2.2Q + 8$$

a. Plot R and TC. Estimate the quantity Q at which the maximum profit should occur. Estimate the amount of profit at this quantity.

b. The profit relation $P = R - TC$ and calculus can be used to determine the quantity Q_p at which the maximum profit will occur, and the amount of this profit. The equations are:

$$\text{Profit} = aQ^2 + bQ + c$$
$$Q_p = -b/2a$$
$$\text{Maximum profit} = -b^2/4a + c$$

Use these relations to confirm the graphical estimates you made in (a). (Your instructor may ask you to derive the relations above.)

8.11 Brittany is co-oping this semester at Regency Aircraft, which customizes the interiors of private and corporate jets. Her first assignment is to develop the specifications for a new machine to cut, shape, and sew leather or vinyl covers and trims. The first cost is not easy to estimate due to many options, but the annual revenue and M&O costs should net out at $+15,000 per year over a 10-year life. Salvage is expected to be 20% of the first cost. Determine what can be paid for the machine now and recover the cost and an MARR of 8% per year under two scenarios:

I: No outside revenue will be developed.

II: Outside contracting will occur with estimated revenue of $10,000 the first year, increasing by $5000 per year thereafter.

8.12 The National Potato Cooperative purchased a de-skinning machine last year for $150,000. Revenue for the first year was $50,000. Over the total estimated life of 8 years, what must the remaining annual revenues (years 2 through 8) equal to recover the investment, if costs are constant at $42,000 and a return of 10% per year is expected? A salvage value of $20,000 is anticipated.

8.13 ABB purchased fieldbus communication equipment for a project in South Africa for $3.15 million. If net cash flow is estimated at $500,000 per year, and a salvage value of $400,000 is anticipated, determine how many years the equipment must be used to just break even at interest rates ranging from 8% to 15% per year.

Breakeven Analysis—Two Alternatives

The following information is used for Problems 8.14 through 8.16

Wilson Partners manufactures thermocouples for electronics applications. The current system has a fixed cost of $300,000 per year, has a variable cost of $10 per unit, and sells for $14 per unit. A newly proposed process will add on-board features that allow the revenue to increase to $16 per unit, but the fixed cost will now be $500,000 per year. The variable cost will be based on a $48 per hour rate with 0.2 hour dedicated to produce each unit.

8.14 Determine the annual breakeven quantity for the (a) current system and (b) the new system.

8.15 Plot the two profit relations and estimate graphically the breakeven quantity between the two alternatives.

8.16 Mathematically determine the breakeven quantity between the two alternatives and compare it with the graphical estimate.

8.17 A rural subdivision has several miles of access road that needs a new surface treatment. Alternative 1 is a gravel base and pavement with an initial cost of $500,000 that will last for 15 years and has an annual upkeep cost of $100 per mile. Alternative 2 is to enhance the gravel base now at a cost of $50,000 and immediately coat the surface with a durable hot oil mix, which costs $130 per barrel applied. Annual reapplication of the mix is required. A barrel covers 0.05 mile. (a) If the discount rate is 6% per year, determine the number of miles at which the two alternatives break even. (b) A drive in a pickup indicates a total of 12.5 miles of road. Which is the more economical alternative?

8.18 An engineering practitioner can lease a fully equipped computer and color printer system for $800 per month or purchase one for $8500 now and pay a $75 per month maintenance fee. If the nominal interest rate is 15% per year, determine the months of use necessary for the two to break even.

8.19 Paul is a chemical engineer at an east coast refinery that produces JP-4 jet fuel. He is evaluating two equivalent methods that may assist in the reduction of vapor emission during the processing cycle. If MARR is 12% per year, determine the annual breakeven gallonage for a study period of (*a*) 1 year, (*b*) 3 years, and (*c*) 5 years.

Method	A	B
Fixed cost, $ per year	400,000	750,000
Variable cost, $ per 1000 gallons	60	40 the first year, decreasing by 3 each year thereafter

8.20 Field painting on steel structures is performed as necessary by Vaksun International Construction. Over the last several years the equivalent annual cost at 8% per year has been $28,000. A subcontractor will perform all field painting for an initial fee of $100,000 for a 5-year contract and an add-on rate of $5.00 per 100 sq ft per coat (layer).

 a. Determine the breakeven square footage using the contract period of 5 years.

 b. Past yearly records indicate the total areas covered, including multiple coats if necessary. Determine the years in which the contract would have been an economic advantage for Vaksun.

Year	2003	2004	2005	2006	2007
100 sq ft	420	852	1004	339	571

8.21 The Ecology Group wishes to purchase a piece of equipment for various metals recycling. Machine 1 costs $123,000, has a life of 10 years, an annual cost of $5000, and requires one operator at a cost of $24 per hour. It can process 10 tons per hour. Machine 2 costs $70,000, has a life of 6 years, an annual cost of $2500, and requires two operators at a cost of $24 per hour each. It can process 6 tons per hour.

 a. Determine the breakeven tonnage of scrap metal at $i = 7\%$ per year and select the better machine for a processing level of 1000 tons per year.

 b. Calculate and plot the sensitivity of the breakeven tons per year to ±15% change in the hourly cost of an operator.

The following information is used for Problems 8.22 through 8.25.

Mid-Valley Industrial Extension Service, a state-sponsored agency, provides water quality sampling services to all business and industrial firms in a 10-county region. Just last month, the service purchased all necessary lab equipment for full inhouse testing and analysis. Now, an outsourcing agency has offered to take over this function on a per-sample basis. Data and quotes for the two options have been collected. The interest rate for government projects is 5% per year and a study period of 8 years is chosen.

 Inhouse: Equipment and supplies initially cost $125,000 for a life of 8 years and an AOC of $15,000 and annual salaries of $175,000. Sample costs average $25. There is no significant market value for the equipment and supplies currently owned.

 Outsourced: Contractors quote sample cost averages of $100 for the first 5 years, increasing to $125 per sample for years 6 through 8.

8.22 Determine the breakeven number of tests between the two options.

8.23 Graph the AW curves for both options for test loads between 0 and 4000 per year in increments of 1000 tests. What is the estimated breakeven quantity?

8.24 The service director has asked the outsource company to reduce the per sample costs by 25% across the board. What will this do to the breakeven point? (Hint: Look carefully at your graph from the previous problem before answering.) What is the new value?

8.25 Suppose the Service reduces its personnel costs to $100,000 per year and the per sample cost to $20. What will this do to the breakeven point? (Hint: Again, look carefully at your graph from the previous problem before answering.) What is the new annual breakeven test quantity?

8.26 Lorraine can select from two nutrient injection systems for her cottage-industry hydroponics tomato and lettuce greenhouses.

 a. Use an AW relation to determine the minimum number of hours per year to operate the pumps that will justify the Auto Green system, if the MARR is 10% per year.

 b. Which pump is economically better, if it operates 7 hours per day, 365 days per year?

	Nutra Jet (N)	Auto Green (A)
Initial cost, $	−4,000	−10,300
Life, years	3	6
Rebuild cost, $	−1,000	−2,200
Time before rebuild, annually or minimum hours	2,000	8,000
Cost to operate, $ per hour	1.00	0.90

8.27 Donny and Barbara want to join a sports and exercise club. The HiPro plan has no upfront charge and the first month is free. It then charges a total of $100 at the end of each subsequent month. Bally charges a membership fee of $100 per person now and $20 per person per month starting the first month. How many months will it take the two plans to reach a breakeven point?

8.28 Balboa Industries' Electronics Division is trying to reduce supply chain risk by making more responsible make/buy decisions through improved cost estimation. A high-use component (expected usage is 5000 units per year) can be purchased for $25 per unit with delivery promised within a week. Alternatively, Balboa can make the component inhouse and have it readily available at a cost of $5 per unit, if equipment costing $150,000 is purchased. Labor and other operating costs are estimated to be $35,000 per year over the study period of 5 years. Salvage is estimated at 10% of first cost and $i = 12\%$ per year. Neglect the element of availability (a) to determine the breakeven quantity, and (b) to recommend making or buying at the expected usage level.

8.29 Claris Water Company makes and sells filters for public water drinking fountains. The filter sells for $50. Recently a make/buy analysis was conducted based on the need for new manufacturing equipment. The equipment first cost of $200,000 and $25,000 annual operation cost comprise the fixed cost, while Claris's variable cost is $20 per filter. The equipment has a 5-year life, no salvage value, and the MARR is 6% per year. The decision to make the filter was based on the breakeven point and the historical sales level of 5000 filters per year.
  a. Determine the breakeven point.
  b. An engineer at Claris learned that an outsourcing firm offered to make the filters for $30

each, but this offer was rejected by the president as entirely too expensive. Perform the breakeven analysis of the two options and determine if the "make" decision was correct.
  c. Develop and use the profit relations for both options to verify the preceding answers.

Sensitivity Analysis of Estimates

8.30 An engineer collected average cost and revenue data for Arenson's FC1 handheld financial calculator.

  Fixed cost = $ 300,000 per year
  Cost per unit = $40
  Revenue per unit = $70

  a. (2 questions) What is the range of the breakeven quantity to variation in the fixed cost from $200,000 to $400,000 per year? Use $50,000 increments. What is the incremental change in the breakeven quantity for each $50,000 change in fixed cost?
  b. (2 questions) Show the sensitivity of profit to variation in revenue from $55 to $75 per unit using a $5 increment. Perform this analysis at two sales quantities: (1) the breakeven quantity for the collected data, and (2) 20% greater than this breakeven quantity. (Note: Be sure to use the original estimates for FC and cost per unit.)

The following information is used for Problems 8.31 through 8.34.
A new online patient diagnostics system for surgeons will cost $200,000 to install, $5000 annually to maintain, and have an expected life of 5 years. The revenue is estimated to be $60,000 per year. The MARR is 10% per year. Examine the sensitivity of present worth to variation in individual parameter estimates, while others remain constant.

8.31 Sensitivity to first cost variation: $150,000 to $250,000 (−25% to +25%)

8.32 Sensitivity to revenue variation: $45,000 to $75,000 (−25% to +25%)

8.33 Sensitivity to life variation: 4 years to 7 years (−20% to +40%)

8.34 Plot the results on a gr̶ and comment on the r̶ parameter.

8.35 Review the situation in ̶ ager at Hamilton wants dors R and T, since ren̶ viable option at this time. However, vendor R is not firm on the first cost, and the zero salvage value is likely wrong. Perform two AW-based sensitivity analyses for the following ranges of P and S for vendor R and determine if the selections change. Vendor T was selected using the original estimates.

a. P variation: $\$-60,000$ to $\$-90,000$ (-20% to $+20\%$)

b. S variation: up to $\$25,000$

c. If the most favorable estimates for P and S are realized ($P = \$-60,000$; $S = \$25,000$), does the selection change to vendor R?

8.36 Charlene plans to place an annual savings amount of $A = \$27,185$ into a retirement program at the end of each year for 20 years starting next year. She expects to retire and start to draw a total of $R = \$60,000$ per year one year after the twentieth deposit. Assume an effective earning rate of $i = 6\%$ per year on the retirement investments and a long life. Determine and comment on the sensitivity of the size of the annual withdrawal R for variations in A and i.

a. Variation of $\pm5\%$ in the annual deposit A.

b. Variation of $\pm1\%$ in the effective earning rate i, that is, ranging from 5% to 7% per year.

8.37 Ned Thompson Labs performs tests on superalloys, titanium, aluminum, and most metals. Tests on metal composites that rely upon scanning electron microscope results can be subcontracted or the labs can purchase new equipment. Evaluate the sensitivity of the economic decision to purchase the equipment over a range of $\pm20\%$ (in 10% increments) of the estimates for P, AOC, R, n, and MARR (range on MARR is 12% to 16%). Use the AW method and plot the results on a sensitivity graph (like Figure 8.7). For which parameter(s) is the AW most sensitive? Least sensitive?

First cost, $P = \$-180,000$
Salvage, $S = \$20,000$
Life, $n = 10$ years
Annual operating cost, AOC = $\$-30,000$ per year
Annual revenue, $R = \$70,000$ per year
MARR = 15% per year

8.38 Determine if the selection of system 1 or 2 is sensitive to variation in the return required by management. The corporate MARR ranges from % to 16% per year on different projects.

	System 1	System 2
First cost, $	−50,000	−100,000
AOC, $ per year	−6,000	−1,500
Salvage value, $	30,000	0
Rework at midlife, $	−17,000	−30,000
Life, years	4	12

8.39 Titan manufactures and sells gas-powered electricity generators. It can purchase a new line of fuel injectors from either of two companies. Cost and savings estimates are available, but the savings estimate is unreliable at this time. Use an AW analysis at 10% per year to determine if the selection between company A and company B changes when the savings per year may vary as much as $\pm40\%$ from the best estimates made thus far.

	Company A	Company B
First cost, $	−50,000	−37,500
AOC, $ per year	−7,500	−8,000
Savings best estimate, $ per year	15,000	13,000
Salvage, $	5,000	3,700
Life, years	5	5

Alternative Selection Using Multiple Estimates

8.40 In Problem 5.12b, Holly Farms evaluated two environmental chambers where the first cost of the D103 model could vary from $300,000 to $500,000. The selection changes if the lower first cost is correct. The AW values are:

$$AW_{490G} = \$-135,143$$

If $P_{D103} = \$-300,000$, $AW_{D103} = \$-115,571$
If $P_{D103} = \$-400,000$, $AW_{D103} = \$-152,761$
If $P_{D103} = \$-500,000$, $AW_{D103} = \$-189,952$

Because of the high profile of E. coli detection, the manager wants to do a "worst case analysis" to be sure that the expected life does not also significantly influence the selection. Use the three estimates that follow for P and n for D103 to determine which alternative to select. The 490G and other D103 estimates are considered firm. MARR is 10% per year.

	Pessimistic	Most Likely	Optimistic
First cost, $	−500,000	−400,000	−300,000
Life, years	1	3	5

8.41 Harmony Auto Group is considering three options for its warranty work. (This is Problem 5.20.) The marketing manager and service manager have now developed a range of possible income and cost estimates for a more thorough analysis. Use them and the remainder of the data provided in Problem 5.20 to determine the best option(s).

	Contract	License	Inhouse
Annual cost, $ million per year			
P	−2	−0.2	−7
ML	−2	−0.2	−5
O	−2	−0.2	−4
Annual income, $ million per year			
P	0.5	0.5	6
ML	2.5	1.5	9
O	4.0	2.0	10

8.42 The engineer in Example 8.6 is impressed with alternative C's design and expected flexibility in the workplace, even though alternative B is the most economical under all three sets of estimates (see Figure 8.8). Perform the two analyses outlined below to determine the changes necessary in first costs to make B and C just break even, thus giving C increased opportunities for selection.

a. Using the optimistic estimates for alternative C, what is the first cost of C necessary to break even with the optimistic estimates for B?

b. Using the pessimistic estimates for alternative B, what is the first cost of B necessary to break even with the optimistic estimates for C?

c. Prepare a graph of AW values versus first costs (similar to Figure 8.8) to demonstrate these results.

8.43 Doug has performed a complete three-estimate AW analysis of the equipment to manufacture wood or plastic kitchen cabinets. His spreadsheet (shown below) shows the pessimistic, most likely, and optimistic values for P, AOC, and n. It also presents AW values at the MARR of 8% per year for each alternative's three estimates, and a plot of AW versus n estimates indicating that plastic has the lower annual cost for ML and O estimates.

a. Plot the AW values for the other two varying parameters to indicate the better alternative(s).

b. Duplicate Doug's spreadsheet for yourself to display the estimates and AW values. Use GOAL SEEK to find optimistic values for the wood alternative for each of the three varying parameters (P, AOC, and n) that will cause the two alternatives to be equally acceptable.

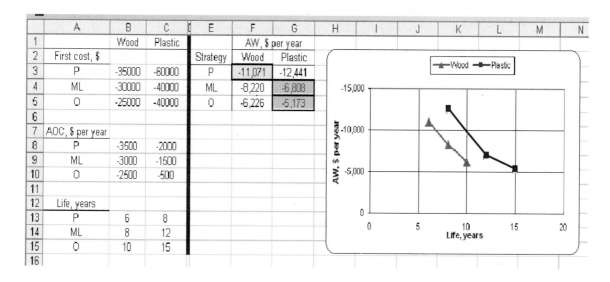

Payback Analysis

8.44 Laura's grandparents helped her purchase a small self-serve laundry business to make extra money during her five college years. When she completed her engineering management degree, she sold the business and her grandparents told her to keep the money as a graduation present. For the net cash flows listed, determine

    a. if they returned the investment in 5 years.

    b. the actual rate of return over the 5-year period.

    c. how long it took to pay back the $75,000 investment plus a 7% per year return.

Year	0	1	2	3	4	5
Net cash flow, $ per year	−75,000	−10,500	18,600	−2,000	28,000	105,000

8.45 Buhler Tractor sold a tractor for $45,000 to Tom Edwards 10 years ago. (*a*) What is the uniform net cash flow that Tom must make each year to realize payback and a return of 5% per year on his investment over a period of 3 years? 5 years? 8 years? All 10 years? (*b*) If the net cash flow was actually $5000 per year, what is the amount Tom should have paid for the tractor to realize payback plus the 5% per year return over these 10 years?

8.46 National Parcel Service has historically owned and maintained its own delivery trucks. Leasing is an option being seriously considered because costs for maintenance, fuel, insurance, and some liability issues will be transferred to Pacific Leasing, the truck leasing company. The study period is no more than 24 months for either alternative. The annual lease cost is paid at the beginning of each year and is not refundable for partially used years. Use the first cost and net cash flow estimates to determine the payback in months with a nominal 9% per year return for the (*a*) purchase and (*b*) lease option.

    Purchase: $P = \$-30,000$ now; monthly cost $= \$-1000$; monthly revenue $= \$4500$

    Lease: $P = \$-10,000$ at the beginning of each year (months 0 and 12); monthly cost $= \$-2500$; monthly revenue $= \$4500$

The following information is used for Problems 8.47 through 8.49.

Julian Browne, owner of Clear Interior Environments, purchased an air scrubber, HEPA vacuum, and other equipment for mold removal for $15,000 eight months ago. Net cash flows were $−2000$ for each of the first two months, followed by $1000 per month for months 3 and 4. For the last four months, a contract generated a net $6000 per month. Julian sold the equipment yesterday for $3000 to a friend.

8.47 Determine (*a*) the no-return payback period and (*b*) the 18% per year payback period.

8.48 (2 questions) What is the estimated annual (nominal) rate of return based only upon the net cash flows for the 8 months of ownership? What fallacy of payback analysis does this ROR value and the payback periods above demonstrate, if the owner had disposed of the equipment after 7 months (approximate payback time) for $S = 0$?

8.49 For the net cash flows experienced, what could Julian have paid for the equipment 8 months ago to have payback plus a return of exactly 18% per year over the ownership period?

The following information is used for Problems 8.50 through 8.53.

Darrell, an engineer with Hamilton Watches, is considering two alternative processes to waterproof the new line of scuba-wear watches. Estimates follow.

    Process 1: $P = \$-50,000$; $n = 5$ years ; NCF $= \$24,000$ per year; no salvage value.

    Process 2: $P = \$-120,000$; $n = 10$ years; NCF $= \$42,000$ for year 1, decreasing by $2500 per year thereafter; no salvage value.

8.50 Darrell first decided to use no-return payback to select the process, because his boss told him most investments at Hamilton must pay back in 3 to 4 years. Determine which process Darrell will select.

8.51 Next Darrell decided to use the AW method at the corporate MARR of 12% per year that he used previously on another evaluation. Now what process will he select?

8.52 (2 questions) Finally, Darrell decided to calculate the rate of return for the cash flows of each process over its respective life. What process does this analysis indicate as better? Explain the fundamental assumptions and errors made when this approach is used. (Hint: Before answering, review Sections 6.3 and 6.4 in the ROR chapter.)

8.53 Of the evaluations above, which is the correct method upon which to base the final economic decision? Why is this method the only correct one?

PROBLEMS FOR TEST REVIEW AND FE EXAM PRACTICE

8.54 When the variable cost is reduced for linear total cost and revenue lines, the breakeven point decreases. This is an economic advantage because
 a. the revenue per unit will increase.
 b. the two lines will now cross at zero.
 c. the profit will increase for the same revenue per unit.
 d. the total cost line becomes nonlinear.

8.55 The profit relation for the following estimates at a quantity that is 10% above breakeven is

$$\text{Fixed cost} = \$500,000 \text{ per year}$$
$$\text{Cost per unit} = \$200$$
$$\text{Revenue per unit} = \$250$$

 a. Profit $= 200(11,000) - 250(11,000) - 500,000$
 b. Profit $= 250(11,000) - 500,000 - 200(11,000)$
 c. Profit $= 250(11,000) - 200(11,000) + 500,000$
 d. Profit $= 250(10,000) - 200(10,000) - 500,000$

8.56 $AW_1 = -23,000(A/P,10\%,10)$
 $\qquad + 4000(A/F,10\%,10) - 3000 - 3X$
 $AW_2 = -8,000(A/P,10\%,4) - 2000 - 6X$

For these two AW relations, the breakeven point X in miles per year is closest to
 a. 1130.
 b. 1224.
 c. 1590.
 d. 655.

8.57 To make an item inhouse, equipment costing $250,000 must be purchased. It will have a life of 4 years, an annual cost of $80,000, and each unit will cost $40 to manufacture. Buying the item externally will cost $100 per unit. At $i = 15\%$ per year, it is cheaper to make the item inhouse if the number per year needed is
 a. above 1047 units.
 b. above 2793 units.
 c. equal to 2793 units.
 d. below 2793 units.

8.58 The sensitivity of two parameters (P and n) for one project is evaluated by graphing the AW values versus percentage variation from the most likely estimates. The AW curve for n has a slope very close to zero, while the P curve has a significant negative slope. One good conclusion from the graph is that
 a. both PW and AW values are more sensitive to variations in P than n.

 b. the project should be rejected, since AW values vary with P and n.
 c. a better estimate of P needs to be made.
 d. the ROR is equally sensitive for both parameters.

The following information is used for Problems 8.59 through 8.60.
Four mutually exclusive alternatives are evaluated using three estimates or strategies (pessimistic, most likely, and optimistic) for several parameters. The resulting PW values over the LCM are determined as shown.

	Alternative's PW Over LCM, $			
Strategy	**1**	**2**	**3**	**4**
Pessimistic (P)	4,500	−6,000	3,700	−1,900
Most likely (ML)	6,000	−500	5,000	−100
Optimistic (O)	9,500	2,000	10,000	3,500

8.59 The best alternative to select for a given strategy is
 a. pessimistic: select alternative 2.
 b. optimistic: select alternative 2.
 c. pessimistic: select alternative 1.
 d. optimistic: select alternative 4.

8.60 If none of the strategies is more likely than any other strategy, the alternative to select is
 a. 2.
 b. 1 and 2 are equally acceptable.
 c. 1.
 d. 3.

8.61 Which of the following is *not a concern* when using payback analysis to select between alternatives?
 a. No-return payback does not consider the time value of money.
 b. No net cash flows after the payback period are considered in the analysis.
 c. An incremental analysis must be performed over the least common multiple of lives.
 d. Short-lived alternatives may be favored over longer-lived alternatives that produce a higher rate of return.

8.62

Month	0	1	2	3	4	5	6
NCF, $1000 per month	−40	+5	+7	+9	+11	+13	+25

For a return of 12% per year on the investment and the net cash flows (NCF) shown, the payback period is
a. between 3 and 4 months.
b. between 4 and 5 months.
c. exactly 6 months.
d. between 4 and 5 years.

8.63 Tom evaluated these alternatives using AW-based payback analysis with a 15% per year return required. For the cash flows shown below (in $1000), the results indicate that A is better with a shorter payback period. If the AW method over the respective lives is used at MARR = 15% per year, the conclusion is
a. the same as the 15% payback selection.
b. to select neither alternative, since AW is negative for both.
c. inconclusive because the LCM must be used.
d. to select B.

8.64 Payback is a good screening method for project evaluation when
a. only projects that exceed a specified payback period are considered viable.
b. the estimates may vary over a narrow range of values.
c. the net cash flow is constant for the entire project life.
d. the breakeven point is considered too high or too low.

Year	Alternative A		Alternative B	
	Cash Flow	AW Payback Analysis	Cash Flow	AW Payback Analysis
0	$−150		$−200	
1	55	$−117.50	50	$−180.00
2	55	−37.27	50	−73.02
3	55	−10.70	50	−37.60
4	55	2.46	50	−20.05
5	55		100	−2.25
6			100	9.43
7			100	

Replacement and Retention Decisions

The McGraw-Hill Companies, Inc./ Ken Cavanagh

One of the most commonly performed engineering economy studies is that of replacement or retention of an asset or system that is currently installed. This differs from previous studies where all the alternatives are new. The fundamental question answered by a replacement study about a currently installed asset or system is, *Should it be replaced now or later?* When an asset is currently in use and its function is needed in the future, it will be replaced at some time. So, in reality, a replacement study answers the question of *when*, not *if*, to replace.

A replacement study is usually designed to first make the economic decision to retain or replace *now*. If the decision is to replace, the study is complete. If the decision is to retain, the cost estimates and decision will be revisited each year to ensure that the decision to retain is still economically correct. This chapter explains how to perform the initial year and follow-on year replacement studies.

A replacement study is an application of the AW method of comparing unequal-life alternatives, first introduced in Chapter 5. In a replacement study with no specified study period, the AW values are determined by a technique of cost evaluation called the *economic service life (ESL)* analysis. If a study period is specified, the replacement study procedure is different from that used when no study period is set. Both procedures are covered in this chapter.

Objectives

Purpose: Perform a replacement study between an in-place asset or system and a new one that could replace it.

1. Understand the fundamentals and terms for a replacement study.

2. Determine the economic service life of an asset that minimizes the total AW of costs.

3. Perform a replacement study between the defender and the best challenger.

4. Determine the replacement value to make the defender and challenger equally attractive.

5. Perform a replacement study over a specified number of years.

6. Use a spreadsheet to determine the ESL, to perform a replacement study, and to calculate the replacement value.

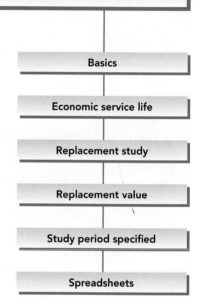

Basics

Economic service life

Replacement study

Replacement value

Study period specified

Spreadsheets

9.1 BASICS OF A REPLACEMENT STUDY

Up to this point, none of the two or more mutually exclusive alternatives compared are currently in place. It is very common to face the situation that the currently used asset (or system or service) could be either replaced with a more economical alternative or retained as is. This is called a *replacement study,* which may be necessary for several reasons—unacceptable performance or reliability, physical deterioration, competitive or technological obsolescence, or changed requirements. At some point in time, the function of every currently used asset must be replaced. A replacement study provides an answer to the question: Is replacement with a specified alternative economical at this point?

The in-place asset is referred to as the *defender,* and the replacement alternative is called the *challenger.* The replacement analysis is performed from the perspective (viewpoint) of a *consultant* or *outsider.* That is, the analysis assumes that neither of the alternatives is owned currently; selection is between the proposed challenger and the in-place defender.

To conduct the economic analysis, the estimates for the challenger are developed as presented in previous chapters. The challenger first cost is the actual investment needed for acquisition and installation. (A common temptation is to increase the challenger's initial cost by the defender's unrecovered depreciation. This amount is a *sunk cost* related to the defender and must not be borne by the challenging alternative.)

In a replacement study, estimates for the defender's n and P values are obtained as follows:

- The expected life n is the number of years at which the lowest AW of costs occurs. This is called the economic service life or ESL. (Section 9.2 details the technique to find ESL.)
- The defender's "initial investment" P is estimated by the *defender's current market value,* that is, the amount required to acquire the services provided by the in-place asset. (If mandatory additional capital investment is necessary for the defender to provide the current services, this amount is also included in the P value.) The equivalent annual capital recovery and costs for the defender must be based on the entire amount required *now* to continue the defender's services in the future. This approach is correct because all that matters in an economic analysis is what will happen from now on. Previous costs are sunk costs and are considered *irrelevant* to the replacement study. (A common temptation is to use the defender's current book value for P. Again, this figure is irrelevant to the replacement study as depreciation is for tax purposes, as discussed in Chapter 12.)

An annual worth analysis is most commonly used for the replacement analysis. The length of the replacement study period is either *unlimited* or *specified.* If the *period is not limited,* the assumptions of the AW method discussed in Section 5.1 are made—the service is needed for the indefinite future, and cost estimates are the same for future life cycles changing with the rate of inflation or deflation. If the *period is specified,* these assumptions are unnecessary, since estimates are made for only the fixed time period.

EXAMPLE 9.1   The Arkansas Division of ADM, a large agricultural products corporation, purchased a state-of-the-art ground-leveling system for rice field preparation 3 years
ago for $120,000. When purchased, it had an expected service life of 10 years,
an estimated salvage of $25,000 after 10 years, and AOC of $30,000. Current
account book value is $80,000. The system is deteriorating rapidly; 3 more years
of use and then salvaging it for $10,000 on the international used farm equip-
ment network are now the expectations. The AOC is averaging $30,000.

A substantially improved, laser-guided model is offered today for $100,000
with a trade-in of $70,000 for the current system. The price goes up next week
to $110,000 with a trade-in of $70,000. The ADM division engineer estimates
the laser-guided system to have a useful life of 10 years, a salvage of $20,000,
and an AOC of $20,000. A $70,000 market value appraisal of the current system was made today.

If no further analysis is made on the estimates, state the correct values to
include if the replacement study is performed today.

Solution

Take the consultant's viewpoint and use the most current estimates.

Defender	Challenger
$P = \$-70,000$	$P = \$-100,000$
$\text{AOC} = \$-30,000$	$\text{AOC} = \$-20,000$
$S = \$10,000$	$S = \$20,000$
$n = 3$ years	$n = 10$ years

The defender's original cost, AOC, and salvage estimates, as well as its current
book value, are all *irrelevant* to the replacement study. *Only the most current
estimates should be used.* From the consultant's perspective, the services that
the defender can provide could be obtained at a cost equal to the *defender market value* of $70,000.

9.2 ECONOMIC SERVICE LIFE

Until now the estimated life n of a project or alternative has been stated. In actuality,
this value is determined prior to an evaluation. An asset should be retained for a time
period that minimizes its cost to the owner. This time is called the economic service
life (ESL) or minimum cost life. *The smallest total AW of costs identifies the ESL value.*
This n value is used in all evaluations for the asset, including a replacement study.

Total AW of costs is the sum of the asset's annual capital recovery and AW
of annual operating costs, that is,

$$\textbf{Total AW} = \textbf{−capital recovery} - \textbf{AW of annual operating costs}$$
$$= \textbf{−CR} - \textbf{AW of AOC} \qquad\qquad \text{[9.1]}$$

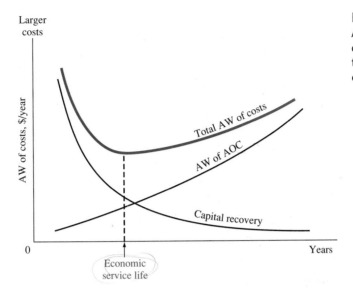

FIGURE 9.1

Annual worth curves
of cost elements
that determine the
economic service life.

These are cost estimates; all, except for salvage value, are negative numbers. Figure 9.1 shows the characteristic concave shape of the total AW curve. The CR component decreases with time and the AOC component increases. For k years of service, the components are calculated using the following formulas:

$$CR_k = -P(A/P,i,k) + S_k(A/F,i,k) \qquad [9.2]$$

$$\begin{aligned}(AW \text{ of } AOC)_k = [AOC_1(P/F,i,1) &+ AOC_2(P/F,i,2) + \ldots \\ &+ AOC_k(P/F,i,k)](A/P,i,k) \qquad [9.3]\end{aligned}$$

In Equation [9.2], the salvage S_k after k years of service is the estimated future market value were the asset (defender or challenger) purchased now. The ESL (best n value) is indicated by the smallest total AW in the series calculated using Equation [9.1]. Example 9.2 illustrates ESL calculations. Spreadsheet usage is presented in Section 9.6.

EXAMPLE 9.2

A device that monitors rotational vibration changes in turbines may be purchased for use in southern California wind farms. The first cost is $40,000 with a constant AOC of $15,000 over a maximum service period of 6 years. Use the decreasing future market values and $i = 20\%$ per year to find the best n value for an economic evaluation.

After k years of service	1	2	3	4	5	6
Estimated market value is	$32,000	30,000	24,000	20,000	11,000	0

Solution

Determine the total AW of costs using Equations [9.1] through [9.3] for years 1 through 6. The AW of AOC is constant at $15,000 in Equation [9.3]. For one year of retention, $k = 1$.

$$\text{Total AW}_1 = -40,000(A/P,20\%,1) + 32,000(A/F,20\%,1) - 15,000$$
$$= -16,000 - 15,000$$
$$= \$-31,000$$

For two years of retention, $k = 2$.

$$\text{Total AW}_2 = -40,000(A/P,20\%,2) + 30,000(A/F,20\%,2) - 15,000$$
$$= -12,546 - 15,000$$
$$= \$-27,546$$

Table 9.1 shows the AW values over all possible 6 years of service. The smallest total AW cost value is the ESL, which occurs at $-26,726 for $k = 4$. Use an estimated life of $n = 4$ years in the evaluation.

Comment: The CR component (Table 9.1) does not decrease every year, only through year 4. This is the effect of the changing future market values that are used for the salvage estimates in Equation [9.2].

TABLE 9.1   Calculation of Total AW of Costs Including Capital Recovery and AOC, Example 9.2

Years of Retention	1	2	3	4	5	6
Capital recovery, $/year	−16,000	−12,546	−12,395	−11,726	−11,897	−12,028
AW of AOC, $/year	−15,000	−15,000	−15,000	−15,000	−15,000	−15,000
Total AW, $/year	−31,000	−27,546	−27,395	−26,726	−26,897	−27,028

9.3 PERFORMING A REPLACEMENT STUDY

Replacement studies are performed in one of two ways: without a study period specified or with one defined. Figure 9.2 gives an overview of the approach taken for each situation. The procedure discussed in this section applies when no study period (planning horizon) is specified. If a specific number of years is identified for the replacement study, for example, over the next 5 years, with no continuation considered after this time period, the procedure in Section 9.5 is applied.

A replacement study determines when a challenger replaces the in-place defender. The complete study is finished if the challenger (C) is selected to replace the defender (D) now. However, if the defender is retained now, the study may extend over a number of years equal to the life of the defender n_D, after which a challenger replaces the defender. Use the annual worth and life values for C and D

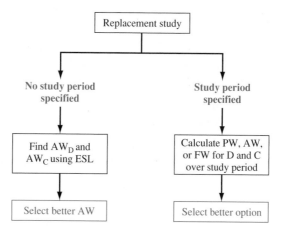

FIGURE 9.2
Overview of replacement study approaches.

determined in the ESL analysis to apply the following replacement study procedure. This assumes the services provided by the defender could be obtained at the AW_D amount.

New replacement study:
1. On the basis of the better AW_C or AW_D value, select the challenger alternative (C) or defender alternative (D). When the challenger is selected, replace the defender now, and expect to keep the challenger for n_C years. This replacement study is complete. If the defender is selected, plan to retain it for up to n_D more years, but next year, perform the following analysis.

One-year-later analysis:
2. Are all estimates still current for both alternatives, especially first cost, market value, and AOC? If no, proceed to step 3. If yes and this is year n_D, replace the defender. If this is not year n_D, retain the defender for another year and repeat this same step. This step may be repeated several times.
3. Whenever the estimates have changed, update them, perform new ESL analyses, and determine new AW_C and AW_D values. Initiate a new replacement study (step 1).

If the defender is selected initially (step 1), estimates may need updating after 1 year of retention (step 2). Possibly there is a new best challenger to compare with D. Either significant changes in defender estimates or availability of a new challenger indicates that a new replacement study is to be performed. In actuality, a replacement study can be performed each year to determine the advisability of replacing or retaining any defender, provided a competitive challenger is available.

EXAMPLE 9.3

Two years ago, Toshiba Electronics made a $15 million investment in new assembly line machinery. It purchased approximately 200 units at $70,000 each and placed them in plants in 10 different countries. The equipment sorts, tests, and performs insertion-order kitting on electronic components in preparation for special-purpose printed circuit boards. A new international industry standard requires a $16,000 additional cost next year (year 1 of retention) on each unit in addition to the expected operating cost. Due to the new standards, coupled with rapidly changing technology, a new system is challenging these 2-year-old machines. The chief engineer at Toshiba USA has asked that a replacement study be performed this year and each year in the future, if need be. At $i = 10\%$ and with the estimates below, do the following:

a. Determine the AW values and economic service lives necessary to perform the replacement study.

Challenger: First cost: $50,000
  Future market values: decreasing by 20% per year
  Estimated retention period: no more than 5 years
  AOC estimates: $5000 in year 1 with increases of $2000 per year thereafter

Defender: Current international market value: $15,000
  Future market values: decreasing by 20% per year
  Estimated retention period: no more than 3 more years
  AOC estimates: $4000 next year, increasing by $4000 per year thereafter, plus the extra $16,000 next year

b. Perform the replacement study now.
c. After 1 year, it is time to perform the follow-up analysis. The challenger is making large inroads into the market for electronic components assembly equipment, especially with the new international standards features built in. The expected market value for the defender is still $12,000 this year, but it is expected to drop to virtually nothing in the future—$2000 next year on the worldwide market and zero after that. Also, this prematurely outdated equipment is more costly to keep serviced, so the estimated AOC next year has been increased from $8000 to $12,000 and to $16,000 two years out. Perform the follow-up replacement study analysis.

Solution

a. The results of the ESL analysis, shown in Table 9.2, include all the market values and AOC estimates for the challenger in the top of the table. Note that $P = \$50,000$ is also the market value in year 0. The total AW of costs is shown by year, should the challenger be placed into service for that number of years. As an example, if the challenger is kept for 4 years, AW_4 is

$$\text{Total AW}_4 = -50,000(A/P,10\%,4) + 20,480(A/F,10\%,4)$$
$$-[5000 + 2000(A/G,10\%,4)]$$
$$= \$-19,123$$

The defender costs are analyzed in the same way in Table 9.2 (bottom) up to the maximum retention period of 3 years.

The lowest AW cost (numerically largest) values for the replacement study are

Challenger: $AW_C = \$-19,123$ for $n_C = 4$ years
Defender: $AW_D = \$-17,307$ for $n_D = 3$ years

TABLE 9.2 **Economic Service Life (ESL) Analysis of Challenger and Defender Costs, Example 9.3**

		Challenger		
Challenger Year k	Market Value	AOC	Total AW If Owned k Years	
0	$50,000	—	—	
1	40,000	$ −5,000	$−20,000	
2	32,000	−7,000	−19,524	
3	25,600	−9,000	−19,245	
4	20,480	−11,000	−19,123	ESL
5	16,384	−13,000	−19,126	

		Defender		
Defender Year k	Market Value	AOC	Total AW If Retained k Years	
0	$15,000	—	—	
1	12,000	$−20,000	$−24,500	
2	9,600	−8,000	−18,357	
3	7,680	−12,000	−17,307	ESL

b. To perform the replacement study now, apply only the first step of the procedure. Select the defender because it has the better AW of costs ($−17,307), and expect to retain it for 3 more years. Prepare to perform the one-year-later analysis 1 year from now.

c. One year later, the situation has changed significantly for the equipment Toshiba retained last year. Apply the steps for the one-year-later analysis:

   2. After 1 year of defender retention, the challenger estimates are still reasonable, but the defender market value and AOC estimates are substantially different. Go to step 3 to perform a new ESL analysis for the defender.

   3. The defender estimates in Table 9.2 (bottom) are updated below, and new AW values are calculated. There is now a maximum of 2 more years of retention, 1 year less than the 3 years determined last year.

Year k	Market Value	AOC	Total AW If Retained k More Years
0	$12,000	—	—
1	2,000	$−12,000	$−23,200
2	0	−16,000	−20,819

The defender ESL is 2 years. The AW and n values for the new replacement study are:

Challenger: unchanged at $AW_C = \$-19{,}123$ for $n_C = 4$ years
Defender: new $AW_D = \$-20{,}819$ for $n_D = 2$ more years

Now select the challenger based on its favorable AW value. Therefore, replace the defender now, not 2 years from now. Expect to keep the challenger for 4 years, or until a better challenger appears on the scene.

The use of the expected first cost P of the challenger and current market value of the defender is called the *conventional* or *opportunity-cost approach* to a replacement study. *This is the correct approach.* Another, incorrect method is entitled the cash-flow approach, in which the defender's market value is subtracted from the challenger's first cost and the defender's first cost set equal to zero. The same economic decision results using either approach, but a falsely low challenger capital recovery (CR) amount is calculated using cash flow since the challenger's P estimate is lowered for the evaluation. This, plus the fact that the equal service assumption is violated when defender and challenger lives are unequal, makes it essential that the approach illustrated in Example 9.3 be taken in all replacement studies.

9.4 DEFENDER REPLACEMENT VALUE

Oftentimes it is helpful to know the minimum defender market value that, if exceeded, will make the challenger the better alternative. This defender value, called its *replacement value (RV)*, yields a breakeven value between challenger and defender. RV is found by setting $AW_C = AW_D$ with the defender's first cost P replaced by the unknown RV. In Example 9.3b, $AW_C = \$-19{,}123$, which is larger than $AW_D = \$-17{,}307$. Using the estimates in Table 9.2, the alternatives are equally attractive at RV = \$22,341, determined from the breakeven relation.

$$
\begin{aligned}
-19{,}123 = {} & -RV(A/P,10\%,3) + 0.8^3 RV(A/F,10\%,3) - [20{,}000(P/F,10\%,1) \\
& + 8{,}000(P/F,10\%,2) + 12{,}000(P/F,10\%,3)](A/P,10\%,3) \\
RV = {} & \$22{,}341
\end{aligned}
$$

Any trade-in offer (market value) above this amount is an economic indication to replace the defender now. The current market value was estimated at \$15,000, thus the defender's selection in Example 9.3b.

9.5 REPLACEMENT STUDY OVER A SPECIFIED STUDY PERIOD

The right branch of Figure 9.2 applies when the time period for the replacement study is limited to a specified study period or planning horizon, for example, 3 years. In this case, the only relevant cash flows are those that occur within the 3-year period. Situations such as this often arise because of international competition and swift obsolescence of in-place technologies. Skepticism and uncertainty about the future are often reflected in management's desire to impose *abbreviated study periods* upon all economic evaluations, knowing that it may be necessary to consider yet another replacement in the near future. This approach, though reasonable from management's perspective, usually forces recovery of the initial investment and the required MARR over a shorter period of time than the ESL of the asset. In fixed study period analyses, the AW, PW, or FW is determined based on the estimates that apply only from the present time through the end of the study period.

For the data shown in Table 9.3 (partially developed from Table 9.2), determine which alternative is better at $i = 10\%$ per year, if the study period is (*a*) 1 year and (*b*) 3 years.

EXAMPLE 9.4

Solution

a. Use AW relations for a 1-year study period.

$$AW_C = -50{,}000(A/P{,}10\%{,}1) + 40{,}000(A/F{,}10\%{,}1) - 5000$$
$$= \$-20{,}000$$
$$AW_D = -15{,}000(A/P{,}10\%{,}1) + 12{,}000(A/F{,}10\%{,}1) - 20{,}000$$
$$= \$-24{,}500$$

Select the challenger.

TABLE 9.3 Challenger and Defender Estimates for Replacement Study, Example 9.4.

Challenger		
Challenger **Year *k***	**Market** **Value**	**AOC**
0	$50,000	—
1	40,000	$ −5,000
2	32,000	−7,000
3	25,600	−9,000
4	20,480	−11,000
5	16,384	−13,000
Defender		
Defender **Year *k***	**Market** **Value**	**AOC**
0	$15,000	—
1	12,000	$−20,000
2	9,600	−8,000
3	7,680	−12,000

b. For a 3-year study period, the AW equations are

$$AW_C = -50,000(A/P,10\%,3) + 25,600(A/F,10\%,3)$$
$$- [5000 + 2000(A/G,10\%,3)]$$
$$= \$-19,245$$

$$AW_D = -15,000(A/P,10\%,3) + 7680(A/F,10\%,3)$$
$$- [20,000(P/F,10\%,1) + 8000(P/F,10\%,2)$$
$$+ 12,000(P/F,10\%,3)](A/P,10\%,3)$$
$$= \$-17,307$$

Select the defender.

If there are several options for the number of years that the defender may be retained before replacement with the challenger, the first step is to develop the succession options and their AW values. For example, if the study period is 5 years, and the defender will remain in service 1 year, or 2 years, or 3 years, cost estimates must be made to determine AW values for each defender retention period. In this case, there are four options; call them W, X, Y, and Z.

Option	Defender Retained	Challenger Serves
W	3 years	2 years
X	2	3
Y	1	4
Z	0	5

The respective AW values for defender retention and challenger service define the cash flows for each option. Example 9.5 illustrates the procedure.

EXAMPLE 9.5   Amoco Canada has oil field equipment placed into service 5 years ago for which a replacement study has been requested. Due to its special purpose, it has been decided that the current equipment will have to serve for either 2, 3, or 4 more years before replacement. The equipment has a current market value of $100,000, which is expected to decrease by $25,000 per year. The AOC is constant now, and is expected to remain so, at $25,000 per year. The replacement challenger is a fixed-price contract to provide the same services at $60,000 per year for a minimum of 2 years and a maximum of 5 years. Use MARR of 12% per year to perform a replacement study over a 6-year period to determine when to sell the current equipment and purchase the contract services.

Solution

Since the defender will be retained for 2, 3, or 4 years, there are three viable options (X, Y, and Z).

Option	Defender Retained	Challenger Serves
X	2 years	4 years
Y	3	3
Z	4	2

The defender annual worth values are identified with subscripts D2, D3, and D4 for the number of years retained.

$$AW_{D2} = -100,000(A/P,12\%,2) + 50,000(A/F,12\%,2) - 25,000$$
$$= \$-60,585$$
$$AW_{D3} = -100,000(A/P,12\%,3) + 25,000(A/F,12\%,3) - 25,000$$
$$= \$-59,226$$
$$AW_{D4} = -100,000(A/P,12\%,4) - 25,000 = \$-57,923$$

For all options, the challenger has an annual worth of

$$AW_C = \$-60,000$$

Table 9.4 presents the cash flows and PW values for each option over the 6-year study period. A sample PW computation for option Y is

$$PW_Y = -59,226(P/A,12\%,3) - 60,000(F/A,12\%,3)(P/F,12\%,6)$$
$$= \$-244,817$$

Option Z has the lowest cost PW value ($-240,369). Keep the defender all 4 years, then replace it. Obviously, the same answer will result if the annual worth, or future worth, of each option is calculated at the MARR.

TABLE 9.4 Equivalent Cash Flows and PW Values for a 6-Year Study Period Replacement Analysis, Example 9.5

Option	Defender	Challenger	1	2	3	4	5	6	Option PW, $
	Time in Service, Years		AW Cash Flows for Each Option, $/Year						
X	2	4	−60,585	−60,585	−60,000	−60,000	−60,000	−60,000	−247,666
Y	3	3	−59,226	−59,226	−59,226	−60,000	−60,000	−60,000	−244,817
Z	4	2	−57,923	−57,923	−57,923	−57,923	−60,000	−60,000	−240,369

Comment: If the study period is long enough, it is possible that the ESL of the challenger should be determined and its AW value used in developing the options and cash flow series. An option may include more than one life cycle of the challenger for its ESL period. Partial life cycles of the challenger can be included. Regardless, any years beyond the study period must be disregarded for the replacement study, or treated explicitly, in order to ensure that equal-service comparison is maintained, especially if PW is used to select the best option.

9.6 USING SPREADSHEETS FOR A REPLACEMENT STUDY

This section includes two examples. The first illustrates the use of a spreadsheet to determine the ESL of an asset. The second demonstrates a one-worksheet replacement study, including ESL determination followed by a breakeven analysis to find the defender's replacement value using the GOAL SEEK tool.

Using the PMT function each year k to find the components in Equation [9.1]—CR and AW of AOC—and adding the two offers a rapid way to determine the ESL. The function formats for each component follow.

For CR component: $= -\text{PMT}(i\%,k,P,S)$

For AW of AOC component: $= -\text{PMT}(i\%,k,\text{NPV}(i\%,\text{AOC}_1:\text{AOC}_k))$

The minus sign retains the sign sense of the cash flow values. For the AOC component, the NPV function is embedded in the PMT function to calculate the AW value of all AOC estimates from years 1 through k in one operation. Example 9.6 demonstrates this technique.

EXAMPLE 9.6   The Navarro County District has purchased new $850,000 rugged trenching equipment for preparing utility cuts in very rocky areas. Because of physical deterioration and heavy wear and tear, market value will decrease 30% per year annually and AOC will increase 30% annually from the expected AOC of $13,000 in year 1. Capital equipment policy dictates a 5-year retention period. Benjamin, a young engineer with the county, wants to confirm that 5 years is the best life estimate. At $i = 5\%$ per year, help Benjamin perform the spreadsheet analysis.

Solution

Equations [9.2] and [9.3] are determined using the PMT functions. Figure 9.3, columns B and C, detail the annual S and AOC values over a 13-year period. The right side (columns D and E) shows the CR and AW of AOC values for 1 through k years of ownership.

Once the functions are set using cell reference formatting, they can be dragged down through the years. (As an example, when $k = 2$, the function $= -\text{PMT}(5\%,2,-850000,416500)$ displays CR $= \$-253,963$; and

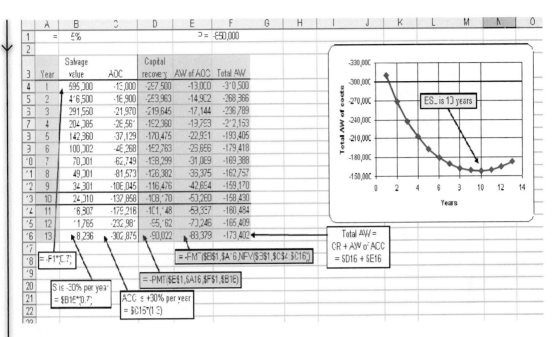

FIGURE 9.3 Determination of ESL using PMT functions, Example 9.6.

$= -PMT(5\%,2,NPV(5\%,2,C4:C5))$ displays AW of AOC as $\$-14{,}902$. The total AW is $\$-268{,}866$.) Cell tags detail the functions for year 13 using cell reference formatting.

Care is needed when inserting minus signs on estimates and functions in order to ensure correct answers. For example, the first cost P has a plus sign, so the (positive) salvage value (cell B4) is calculated using a plus sign. Likewise, the PMT functions are preceded by a minus to ensure that the answer remains a negative (cost) amount.

The lowest total AW value in the table and the chart indicate an ESL of 10 years. The differences in total AW are significantly lower for 10 than for 5 years ($\$-158{,}430$ vs. $\$-193{,}405$). Therefore, the 5-year retention period dictated by county policy will have an equivalently higher annual cost based on the estimates.

It is quite simple to use one spreadsheet to find the ESL values, perform the replacement study, and, if desired, find the RV value for the defender. The PMT functions determine the total AW values in the same way as the previous example. The resulting ESL and AW values are used to make the replace or retain decision. Then, the equivalent of setting up the RV relation of $AW_C = AW_D$ is accomplished by applying the GOAL SEEK tool. An illustration of this efficient use of a spreadsheet follows.

EXAMPLE 9.7 Revisit the situation in Example 9.3. Use these estimates and one spreadsheet (*a*) to determine the ESL values, (*b*) to decide to replace or retain the defender, and (*c*) to find the minimum defender market value to make the challenger more attractive economically.

Solution

a. Figure 9.4 includes all estimates and functions that determine the total AW and ESL values.

Challenger: minimum total AW = $-19,123; ESL = 4 years

Defender: minimum total AW = $-17,307; ESL = 3 years

Remember to use minus signs correctly so that costs and salvage values have the correct sign sense.

b. Retaining the defender is economically favorable since its total AW is lower.

c. Use GOAL SEEK to set the defender total AW for 3 years equal to $-19,123, the challenger total AW at its ESL of 4 years. This is normally accomplished on the same worksheet. However, for clarity, the defender portion from Figure 9.4 is duplicated in Figure 9.5 (top part) with the GOAL SEEK screen added. In short, this will force the defender market value/first cost higher (currently $-15,000) so its total AW goes from $-17,307 to the challenger AW, which is $-19,123. This is the point of breakeven between the two alternatives. Initiation of GOAL SEEK

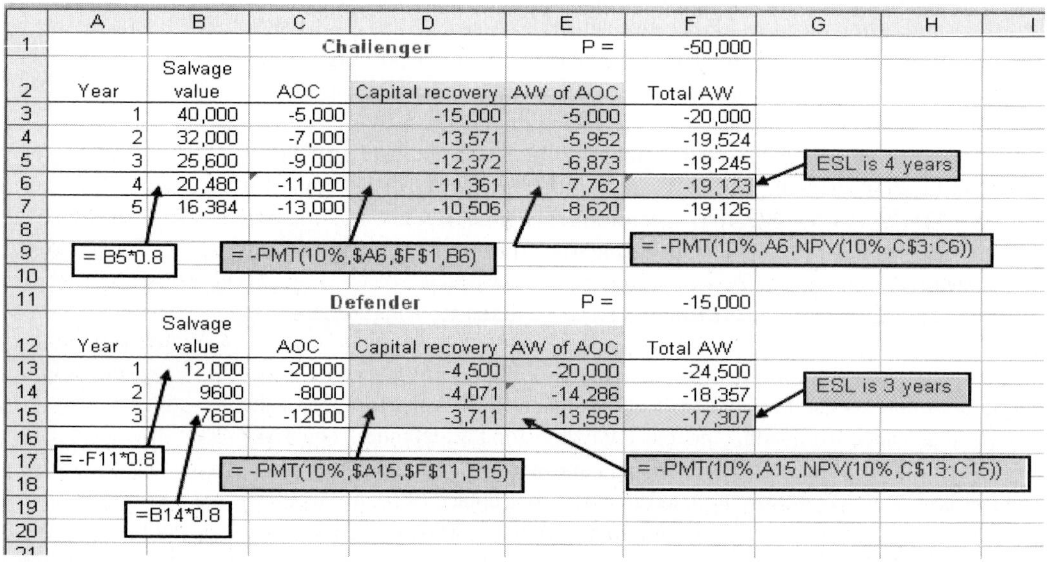

	A	B	C	D	E	F	G	H	I
1			Challenger		P =	-50,000			
2	Year	Salvage value	AOC	Capital recovery	AW of AOC	Total AW			
3	1	40,000	-5,000	-15,000	-5,000	-20,000			
4	2	32,000	-7,000	-13,571	-5,952	-19,524			
5	3	25,600	-9,000	-12,372	-6,873	-19,245			
6	4	20,480	-11,000	-11,361	-7,762	-19,123			
7	5	16,384	-13,000	-10,506	-8,620	-19,126			
8									
9									
10									
11			Defender		P =	-15,000			
12	Year	Salvage value	AOC	Capital recovery	AW of AOC	Total AW			
13	1	12,000	-20000	-4,500	-20,000	-24,500			
14	2	9600	-8000	-4,071	-14,286	-18,357			
15	3	7680	-12000	-3,711	-13,595	-17,307			
16									
17									
18									
19									
20									

FIGURE 9.4 Spreadsheet determination of minimum total AW of costs and ESL values, Example 9.7.

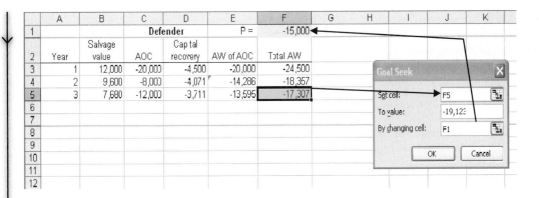

FIGURE 9.5   Use of GOAL SEEK to determine defender replacement value on the spreadsheet in Figure 9.4, Example 9.7.

displays the RV of $22,341 and the new total AW in Figure 9.5 (lower part). As before, a defender trade-in offer greater than RV indicates that the challenger is better.

SUMMARY

It is important in a replacement study to compare the best challenger with the defender. *Best (economic) challenger is described as the one with the lowest AW of costs for some period of years.* However, if the expected remaining life of the defender and the estimated life of the challenger are specified, the AW values over these years must be used in the replacement study.

The economic service life (ESL) analysis is designed to determine the best challenger's years of service and the resulting lowest total AW of costs. The resulting n_C and AW_C values are used in the replacement study procedure. The same analysis can be performed for the ESL of the defender.

Replacement studies in which no study period (planning horizon) is specified utilize the annual worth method of comparing two unequal-life alternatives. The better AW value determines how long the defender is retained before replacement.

When a study period is specified, it is vital that market value and cost estimates for the defender be as accurate as possible. All the viable time options for using the defender and challenger are enumerated, and their AW equivalent cash flows determined. For each option, the PW, AW, or FW value is used to select the best option. This option determines how long the defender is retained before replacement.

PROBLEMS

Foundations of Replacement

9.1 In a replacement study, the in-place asset is referred to as what?

9.2 What does it mean to take a consultant's viewpoint in a replacement analysis?

9.3 In a replacement study, what value is used for the in-place asset's "initial investment" amount, that is, its P value?

9.4 In conducting a replacement study of assets with different lives, can the annual worth values over the asset's own life cycle be used in the comparison, if the study period is (a) unlimited, (b) limited wherein the period *is not* an even multiple of asset lives, and (c) limited wherein the period *is* a multiple of asset lives? Explain your answers.

Economic Service Life

9.5 Determine the economic service life at $i = 10\%$ per year for a machine that has a first cost of $10,000 and estimated operating costs and year-end salvage values as shown.

Year	Operating Cost, $	Salvage Value, $
1	−1000	7000
2	−1200	5000
3	−1500	4200
4	−2000	3000
5	−3000	2000

9.6 To improve package tracking at a UPS transfer facility, conveyor equipment was upgraded with RFID sensors at a cost of $345,000. The operating cost is expected to be $148,000 per year for the first 3 years and $210,000 for the next 3 years. The salvage value of the equipment is expected to be $140,000 for the first 3 years, but due to obsolescence, it won't have a significant value after that. At an interest rate of 10% per year, determine the economic service life of the equipment and the associated annual worth.

9.7 A construction company bought a 180,000 metric ton earth sifter at a cost of $65,000. The company expects to keep the equipment a maximum of 7 years. The operating cost is expected to follow the series described by $40{,}000 + 10{,}000k$, where k is the number of years since it was purchased ($k = 1$ through 7). The salvage value is estimated to be $30,000 for years 1 and 2 and $20,000 for years 3 through 7. At $i = 10\%$ per year, determine the economic service life of the sifter and the associated annual worth.

9.8 A large heat treating oven (with appurtenances) for powder-coating automobile frames and large pieces of furniture was purchased for $60,000. The estimated operating costs, maintenance costs, and salvage values are shown below. Determine the economic service life and the associated annual worth.

Year	Operating Cost, $	Maintenance Cost, $	Salvage Value, $
1	−15,000	−3000	35,000
2	−17,000	−3000	30,000
3	−19,000	−3000	25,000
4	−21,000	−3000	20,000
5	−23,000	−3000	15,000

9.9 A bulldozer with a first cost of $70,000 may be used for a maximum of 6 years. Its salvage value, which decreases by 15% per year, is described by the equation $S = 70{,}000(1 - 0.15)^n$, where n is the number of years after purchase. The operating cost of the dozer will be constant at $75,000 per year. At an interest rate of 12% per year, what is the economic service life of the machine? What is the associated AW value?

Replacement Study

9.10 In a one-year later analysis, what action should be taken if (a) all estimates are still current and the year is n_D, (b) all estimates are still current and the year is not n_D, and (c) the estimates have changed?

9.11 A piece of equipment that was purchased two years ago for $50,000 was expected to have a useful life of 5 years with a $5000 salvage value. Since its performance was less than expected, it was upgraded for $20,000 one year ago. Increased demand now requires that the equipment be upgraded again for an additional $15,000 so that it can be used for 3 more years. Its annual operating cost will be $27,000 and it will have a $12,000 salvage after 3 years. Alternatively, it can be replaced with new equipment costing $65,000, with operating costs of $14,000 per year and a salvage value of $23,000 after 3 years. If replaced now, the existing equipment will be sold for $7000. Calculate the annual worth of the defender at an interest rate of 10% per year.

9.12 Based on estimates she obtained from vendor and company records, an engineer with Haliburton calculated the AW values shown for retaining a presently owned machine additional years.

Retention Period, Years	AW Value, $ per Year
1	−92,000
2	−81,000
3	−87,000
4	−89,000
5	−95,000

A challenger has an economic service life of 7 years with an AW of $−86,000 per year. Assuming all future costs remain as estimated for the analysis, when should the company purchase the challenger, if the MARR is 12% per year? Assume used machines like the one presently owned will always be available.

9.13 A recent college grad is trying to decide whether he should keep his presently owned car or purchase a new one. A new car will cost $26,000 and have annual operation and maintenance costs of $1200 per year with an $8000 salvage value in 5 years, which is its economic service life. The presently owned car has a resale value *now* of $5000; 1 year from now it will be $3000, 2 years from now $2500, and 3 years from now $2200. Its operating cost is expected to be $1900 this year, with costs increasing by $200 per year. The presently owned car will definitely not be kept longer than 3 more years. Assuming used cars like the one presently owned *will always* be available, should the presently owned car be sold now, 1 year from now, 2 years from now, or 3 years from now? Use annual worth calculations at $i = 10\%$ per year.

9.14 A biotech company planning a plant expansion is trying to determine whether it should upgrade the existing controlled-environment rooms or purchase new ones. The presently owned rooms were purchased 4 years ago for $250,000. They have a current "quick sale" value of $30,000. However, for an investment of $100,000 now, they can be adequate for another 4 years, after which they would be sold for $40,000. Alternatively, new controlled-environment rooms cost $300,000. They are expected to have a 10-year economic life with a $50,000 salvage value at that time. Determine whether the company should upgrade the existing controlled-environment rooms or purchase new ones. Use an MARR of 12% per year and assume that used controlled-environment rooms will always be available.

9.15 A pulp and paper company is evaluating whether it should retain the current bleaching process, which uses chlorine dioxide, or replace it with the proprietary "oxypure" process. The relevant information for each process is shown below. Use an interest rate of 15% per year to select a process.

	Current Process	Oxypure Process
Original cost 6 years ago, $	−450,000	—
Investment cost now, $	—	−700,000
Current market value, $	25,000	—
Annual operating cost, $/year	−180,000	−70,000
Remaining life, years	5	10
Salvage value, $	0	50,000

9.16 A critical machine in the Phelps-Dodge copper refining operation was purchased 7 years ago for $160,000. Last year a replacement study was performed with the decision to retain for 3 more years. The situation has changed. The equipment is estimated to have a value of $8000 if "scavenged" for parts now or anytime in the future. If kept in service, it can be minimally upgraded at a cost of $43,000 to make it usable for up to 2 more years. Its operating cost is estimated at $22,000 the first year and $25,000 the second year. Alternatively, the company can purchase a new system that will have an equivalent annual worth of $47,063 per year over its ESL. The company uses a MARR of 10% per year. Use annual worth analysis to determine when the company should replace the machine.

9.17 A tire retreading machine purchased 9 years ago for $45,000 is expected to have the salvage values and operating costs shown below for additional years of service. It can be sold immediately for $9000 by placing an ad in a trade magazine. A replacement machine will cost $125,000, have a $10,000 salvage value after its 10-year ESL, and have an AOC of $35,000 per year. At an interest rate of 12% per year, determine if the company should replace the defender with the challenger and, if so, when.

Year	Salvage Value at End of Year, $	Operating Cost, $
1	6000	−50,000
2	4000	−53,000
3	1000	−60,000

Defender Replacement Values

9.18 In 2006, Violet Rose Computer Corporation purchased a new quality inspection system for $550,000. The estimated salvage value was $50,000 after 10 years. Currently the expected remaining life is 7 years with an AOC of $27,000 per year and an estimated salvage value of $40,000. The new president has recommended early replacement of the system with one that costs $400,000 and has a 12-year economic service life, a $35,000 salvage value, and an estimated AOC of $50,000 per year. If the MARR for the corporation

is 12% per year, find the minimum trade-in value necessary now to make the president's replacement economically advantageous.

9.19 A lathe purchased by a machine tool company 10 years ago for $75,000 can be used for 3 more years. Estimates are an annual operating cost of $63,000 and a salvage value of $25,000. A challenger will cost $130,000 with an economic life of 6 years and an operating cost of $32,000 per year. Its salvage value will be $45,000. On the basis of these estimates, what market value for the existing asset will render the challenger equally attractive? Use an interest rate of 12% per year.

9.20 Hydrochloric acid, which fumes at room temperatures, creates a very corrosive work environment. A machine purchased 4 years ago for $80,000 can be used for only 2 more years, at which time it will be scrapped for no value. Its operating cost is $71,000 per year. A more corrosion-resistant challenger will cost $210,000 with an operating cost of $48,000 per year. It is expected to have a $60,000 salvage value after its 10-year ESL. At an interest rate of 15% per year, what minimum replacement value would render the challenger attractive?

9.21 Machine A was purchased 5 years ago for $90,000. Its operating cost is higher than expected, so it will be used for only 4 more years. Its operating cost this year will be $40,000, increasing by $2000 per year through the end of its useful life. The challenger, machine B, will cost $150,000 with a $50,000 salvage value after its 10-year ESL. Its operating cost is expected to be $10,000 for year 1, increasing by $500 per year thereafter. What is the market value for machine A that would make the two machines equally attractive at an interest rate of 12% per year?

Replacement Study Over a Study Period

9.22 Angstrom Technologies intends for the company to use the newest and finest equipment in its labs. Accordingly, a senior engineer has recommended that a 2-year-old piece of precision measurement equipment be replaced immediately. This engineer believes it can be shown that the proposed equipment is economically advantageous at a 15%-per-year return and a planning horizon of 5 years. Perform the replacement analysis using the annual worth method for a 5-year study period using the estimates shown.

	Current	Proposed
Original purchase price, $	−30,000	−40,000
Current market value, $	15,000	—
Remaining life, years	5	15
Estimated value in 5 years, $	7,000	10,000
Salvage value after 15 years, $	—	5,000
Annual operating cost, $ per year	−8,000	−3,000

9.23 An industrial engineer at a fiber optic manufacturing company is considering two robots to reduce costs in a production line. Robot X will have a first cost of $82,000, an annual maintenance and operation (M&O) cost of $30,000, and salvage values of $50,000, $42,000, and $35,000 after 1, 2, and 3 years, respectively. Robot Y will have a first cost of $97,000, an annual M&O cost of $27,000, and salvage values of $60,000, $51,000, and $42,000 after 1, 2, and 3 years, respectively. Which robot should be selected if a 2-year study period is used at an interest rate of 15% per year?

9.24 A machine purchased 3 years ago for $140,000 is now too slow to satisfy increased demand. The machine can be upgraded now for $70,000 or sold to a smaller company for $40,000. The current machine will have an annual operating cost of $85,000 per year and a $30,000 salvage value in 3 years. If upgraded, the presently owned machine will be retained for only 3 more years, then replaced with a machine to be used in the manufacture of several other product lines.

The replacement machine, which will serve the company now and for at least 8 years, will cost $220,000. Its salvage value will be $50,000 for years 1 through 5; $20,000 in year 6; and $10,000 thereafter. It will have an estimated operating cost of $65,000 per year. The company asks you to perform an economic analysis at 15% per year using a 3-year planning horizon. Should the company replace the presently owned machine now, or do it 3 years from now? What are the AW values?

9.25 Two processes can be used for producing a polymer that reduces friction loss in engines. Process K will have a first cost of $160,000, an operating cost of $7000 per month, and a salvage value of $50,000 after 1 year and $40,000 after its maximum 2-year life. Process L will have a first cost of $210,000, an operating cost of $5000 per month, and salvage values of $100,000 after 1 year, $70,000 after 2 years, $45,000 after 3 years, and $26,000 after its maximum 4-year life. You have been asked to determine which process is better when (a) a 2-year study period is used, and (b) a 3-year study period is used. The company's MARR is 12% per year compounded monthly.

PROBLEMS FOR TEST REVIEW AND FE EXAM PRACTICE

9.26 In a replacement study conducted last year, it was determined that the defender should be kept for 3 more years. In reviewing the current costs, it appears that the estimates for this year and next are still valid. The proper course of action is to

a. replace the existing asset now.

b. replace the existing asset 2 years from now, as was determined last year.

c. conduct a new replacement study using the new estimates.

d. conduct a new replacement study using last year's estimates.

9.27 At an interest rate of 10% per year, the economic service life of an asset that has a current market value of $15,000 and the expected cash flows shown is

a. 1 year. **b.** 2 years. **c.** 3 years. **d.** 4 years.

Year	Salvage Value at End of Year, $	Operating Cost, $
1	10,000	−50,000
2	8,000	−53,000
3	5,000	−60,000
4	0	−68,000

9.28 In trying to decide whether or not to replace a sorting/baling machine in a solid waste recycling operation, an engineer calculated the annual worth values for the in-place machine and a challenger. On the basis of these costs, the defender should be replaced

a. now.

b. 1 year from now.

c. 2 years from now.

d. 3 years from now.

Year	AW of Defender, $ per year	AW of Challenger, $ per year
1	−24,000	−31,000
2	−25,500	−28,000
3	−26,900	−25,000
4	−27,000	−25,900
5	−28,000	−27,500

9.29 In a replacement study, the correct value to use when *amortizing* the defender is
  a. its first cost when it was purchased.
  b. its book value.
  c. its market value plus the first cost of needed upgrades (if any).
  d. only the cost of needed upgrades.

9.30 A replacement analysis is most objectively conducted from the viewpoint of
  a. a consultant.
  b. the company that is selling the challenger.
  c. the company that originally sold the defender.
  d. the maintenance department.

9.31 A consulting engineering firm purchased field test equipment 2 years ago for $50,000 that was expected to have a useful life of 5 years with a $5000 salvage value. Its performance was less than expected and it was upgraded for $20,000 one year ago. Increased demand now requires that the equipment be upgraded again with another $15,000 capital investment so that it can be used for 3 more years. Its annual operating cost will be $27,000, and it will have a $12,000 salvage value after 3 years. If replaced now, the existing equipment will be sold for $7000. In conducting a replacement study, the value to use for *P* (i.e., the amount to be amortized) in calculating the annual worth of the defender is
  a. $−15,000.
  b. $−22,000.
  c. $−50,000.
  d. $−70,000.

9.32 An engineer determined that the equivalent annual worth of an existing machine over its remaining useful life of 3 years will be $−70,000 per year. It can be replaced now or later with a machine that will have an AW of $−80,000 if it is kept for 2 years or less, $−68,000 if it is kept between 3 and 4 years, and $−75,000 if it is kept for 5 to 10 years. The company wants an analysis of what it should do for a 3-year study period at an interest rate of 15% per year. You should recommend that the existing machine
  a. be replaced now.
  b. be replaced 1 year from now.
  c. be replaced 2 years from now.
  d. not be replaced.

Effects of Inflation

Digital Vision/Getty Images

This chapter concentrates upon understanding and calculating the effects of inflation in time value of money computations. Inflation is a reality that we deal with nearly every day in our professional and personal lives.

The annual inflation rate is closely watched and historically analyzed by government units, businesses, and industrial corporations. An engineering economy study can have different outcomes in an environment in which inflation is a serious concern compared to one in which it is of minor consideration. The inflation rate is sensitive to real, as well as perceived, factors of the economy. Factors such as the cost of energy, interest rates, availability and cost of skilled people, scarcity of materials, political stability, and other less tangible factors have short-term and long-term impacts on the inflation rate. In some industries, it is vital that the effects of inflation be integrated into an economic analysis. The basic techniques to do so are covered here.

Objectives

1. Determine the difference inflation makes between money now and money in the future.

   Impact of inflation

2. Calculate present worth with an adjustment for inflation.

   PW with inflation

3. Determine the real interest rate and the inflation-adjusted MARR, and calculate a future worth with an adjustment for inflation.

   FW with inflation

4. Calculate an annual amount in future dollars that is equivalent to a specified present or future sum.

   AW with inflation

5. Use a spreadsheet to perform inflation-adjusted equivalence calculations.

   Spreadsheets

10.1 UNDERSTANDING THE IMPACT OF INFLATION

All of us are very aware that $20 now does not purchase the same amount as $20 did in 1999 or 2000, and significantly less than in 1990. This is primarily because of inflation. *Inflation is the increase in the amount of money necessary to purchase the same amount of a product or service over time.* It occurs because the value of the currency has decreased, so it takes more money to obtain the same amount of goods or services. Associated with inflation is an increase in the money supply, that is, the government prints more dollars, while the supply of goods does not increase.

To consider inflation when making comparisons between monetary amounts that occur at different time periods, the *different-value* dollars must be converted to *constant-value* dollars so that they represent the same purchasing power. This is an important consideration for alternative evaluation since all estimates are for the future.

There are two ways to make meaningful economic calculations when the currency is changing in value, that is, when inflation is considered:

- Convert the amounts that occur in different time periods into equivalent amounts that force the currencies to have the same value. This is accomplished *before* any time value of money calculation is made.
- Change the interest rate used in the economic evaluation to account for the changing currencies (inflation) *plus* the time value of money.

All the computations that follow work for any country's currency; dollars are used here.

The first method above is referred to as making calculations in constant-value (CV) dollars. Money in one time period is brought to the same value as money in a later period.

$$\text{Dollars in period } t_1 = \frac{\text{Dollars in period } t_2}{(1 + \text{Inflation rate between } t_1 \text{ and } t_2)} \qquad \textbf{[10.1]}$$

where:

Dollars in period t_1 = Constant-value (CV) dollars, also called today's dollars

Dollars in period t_2 = Future dollars, also called inflated or then-current dollars

If f represents the inflation rate *per period* (year) and n is the number of periods (years) between t_1 and t_2. Equation [10.1] may be used to express future dollars in CV dollar terms, or vice versa.

$$\textbf{CV dollars} = \frac{\textbf{Future dollars}}{(1 + f)^n} \qquad \textbf{[10.2]}$$

$$\textbf{Future dollars} = \textbf{CV dollars } (1 + f)^n \qquad \textbf{[10.3]}$$

Equations [10.2] and [10.3] are used to find the average inflation rate over a given period of time by substituting the actual monetary amounts and solving for f. This is how the Consumer Price Index (CPI) is determined. (Note that the average inflation rate cannot be determined by taking an arithmetic average.) As an example, use the price of 87 octane (regular) gasoline. From 1986 to 2006 the average pump price in the United States went from $0.92 per gallon to $2.86. Solving Equation [10.3] for f yields an average price increase of 5.83% per year over these 20 years.

$$2.86 = 0.92(1 + f)^{20}$$
$$(1 + f) = 3.1087^{0.05}$$
$$f = 5.83\% \text{ per year}$$

Assume, for a minute, that gas price increases continue at an average of 5.83% per year. Predicted average prices are:

In 2007: $2.86(1.0583)^1 = \$3.03$ per gallon
In 2008: $2.86(1.0583)^2 = \$3.20$ per gallon
In 2010: $2.86(1.0583)^4 = \$3.59$ per gallon

A 5.83% annual increase causes a 25.5% increase from 2006 to 2010. In some areas of the world, hyperinflation may average 40 to 50% per year, with some spikes as high as 1200% for short periods of time. Hyperinflation is discussed further in Section 10.3.

Placed into an industrial or business context, even at a reasonably low inflation rate averaging, say, 4% per year, equipment or services with a first cost of $209,000 will increase by 48% to $309,000 over a 10-year span. This is before any consideration of the rate of return requirement is placed upon the equipment's revenue-generating ability. Clearly, inflation must be taken into account.

There are actually three different inflation-related rates that are important: the real interest rate (i), the market interest rate (i_f), and the inflation rate (f). Only the first two are interest rates.

Real or inflation-free interest rate i. This is the rate at which interest is earned when the effects of changes in the value of currency (inflation) have been removed. Thus, the real interest rate presents an actual gain in purchasing power. (The equation used to calculate i, with the influence of inflation removed, is derived later in Section 10.3.) The real rate of return that generally applies for individuals is approximately 3.5% per year. This is the "safe investment" rate. The required real rate for corporations (and many individuals) is set above this safe rate when a MARR is established without an adjustment for inflation.

Inflation-adjusted interest rate i_f. As its name implies, this is the interest rate that has been adjusted to take inflation into account. The *market interest rate,* which is the one we hear everyday, is an inflation-adjusted rate. This rate is a combination of the real interest rate i and the inflation rate f, and, therefore, it changes as the inflation rate changes.

Inflation rate f. As described above, this is a measure of the rate of change in the value of the currency.

A company's MARR adjusted for inflation is referred to as the inflation-adjusted MARR. The determination of this value is discussed in Section 10.3.

Deflation is the opposite of inflation in that when deflation is present, the purchasing power of the monetary unit is greater in the future than at present. That is, it will take fewer dollars in the future to buy the same amount of goods or services as it does today. In deflationary economic conditions, the market interest rate is always less than the real interest rate.

Temporary price deflation may occur in specific sectors of the economy due to the introduction of improved products, cheaper technology, or imported materials or products that force current prices down. In normal situations, prices equalize at a competitive level after a short time. However, deflation over a short time in a specific sector of an economy can be orchestrated through *dumping.* An example of dumping may be the importation of materials, such as steel, cement, or cars, into one country from international competitors at very low prices compared to current market prices in the targeted country. The prices will go down for the consumer, thus forcing domestic manufacturers to reduce their prices in order to compete for business. If domestic manufacturers are not in good financial condition, they may fail, and the imported items replace the domestic supply. Prices may then return to normal levels and, in fact, become inflated over time, if competition has been significantly reduced.

On the surface, having a moderate rate of deflation sounds good when inflation has been present in the economy over long periods. However, if deflation occurs at a more general level, say nationally, it is likely to be accompanied by the lack of money for new capital. Another result is that individuals and families have less money to spend due to fewer jobs, less credit, and fewer loans available; an overall "tighter" money situation prevails. As money gets tighter, less is available to be committed to industrial growth and capital investment. In the extreme case, this can evolve over time into a deflationary spiral that disrupts the entire economy. This has happened on occasion, notably in the United States during the Great Depression of the 1930s.

Engineering economy computations that consider deflation use the same relations as those for inflation. Equations [10.2] and [10.3] are used, except the deflation rate is a $-f$ value. For example, if deflation is estimated to be 2% per year, an asset that costs $10,000 today would have a first cost 5 years from now determined by Equation (10.3).

$$10,000(1 - f)^n = 10,000(0.98)^5 = 10,000(0.9039) = \$9039$$

10.2 PW CALCULATIONS ADJUSTED FOR INFLATION

Present worth calculations may be performed using either of the two methods described at the beginning of Section 10.1—constant-value (CV) dollars and the real interest rate i, or future dollars and the inflation-adjusted rate i_f. By way of introduction to these two methods, consider an asset that may be purchased now or in any of the next four years at the future dollar equivalent. Table 10.1 (columns 2 and 3) indicates a first cost of $5000 now, increasing each future year at 4% per year inflation. The estimated cost four years from now is $5849. However, the CV dollar cost is always $5000 (column 4).

TABLE 10.1 **Inflation Calculations and Present Worth Calculation Using Constant-Value Dollars ($f = 4\%$, $i = 10\%$)**

Year t (1)	Cost Increase Due to 4% Inflation (2)	Cost in Future Dollars (3)	Future Cost in Constant-Value Dollars (4) = (3)(P/F,4%,t)	Present Worth at Real $i = 10\%$ (5) = (4)(P/F,10%,t)
0		$5000	$5000	$5000
1	$5000(0.04) = $200	5200	5000	4545
2	5200(0.04) = 208	5408	5000	4132
3	5408(0.04) = 216	5624	5000	3757
4	5624(0.04) = 225	5849	5000	3415

By the first method, use CV dollars and a real interest rate i to find the PW now of the first cost for any year in the future. By this method, all effects of inflation are removed before the real interest rate is applied. If $i = 10\%$ per year, the equivalent cost now for each future year t is PW $= 5000(P/F,10\%,t)$, as calculated in column 5. In CV dollar terms, $3415 now will buy an asset worth $5000 four years hence. Similarly, $3757 now will buy the same asset if it is worth $5000 three years hence.

Figure 10.1 shows the differences over a 4-year period of the constant-value amount of $5000, the future-dollar costs at 4% inflation, and the present worth at 10% real interest with inflation considered. The effect of compounded inflation and interest rates increases rapidly, as indicated by the shaded area.

FIGURE 10.1

Comparison of constant-value dollars, future dollars, and their present worth values.

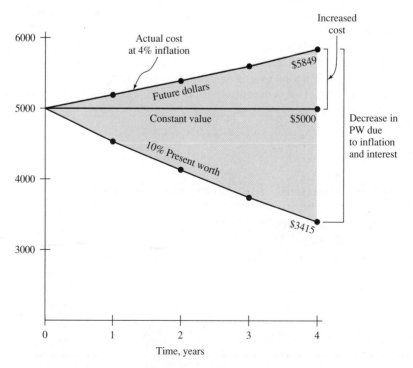

The second method to calculate PW is more commonly used since it utilizes future-dollars estimates and adjusts the interest rate for inflation. Consider the P/F formula, where i is the real interest rate.

$$P = F\frac{1}{(1 + i)^n}$$

The F, which is a future-dollar amount with inflation built in, can be converted into CV dollars by using Equation [10.2].

$$P = \frac{F}{(1 + f)^n}\frac{1}{(1 + i)^n}$$

$$= F\frac{1}{(1 + i + f + if)^n} \qquad [10.4]$$

If the term $i + f + if$ is defined as i_f, the equation becomes

$$P = F\frac{1}{(1 + i_f)^n} = F(P/F, i_f, n) \qquad [10.5]$$

The *inflation-adjusted interest rate* i_f is defined as

$$i_f = i + f + if \qquad [10.6]$$

For a real interest rate of $i = 10\%$ per year and an inflation rate of $f = 4\%$ per year, Equation [10.6] yields an inflated interest rate of 14.4%.

$$i_f = 0.10 + 0.04 + 0.10(0.04) = 0.144$$

Table 10.2 illustrates the use of $i_f = 14.4\%$ in PW calculations for $5000 now, which inflates to $5849 in future dollars 4 years hence. As shown in column 5, the present worth for each year is the same as column 5 of Table 10.1.

The present worth of any series of cash flows—equal, arithmetic gradient, or geometric gradient—can be found similarly. That is, either i or i_f is introduced into the P/A, P/G, or P_g factors, depending upon whether the cash flow is expressed in CV dollars or future dollars.

TABLE 10.2 **Present Worth Calculation Using an Inflation-Adjusted Interest Rate**

Year n (1)	Cost Increase at $f = 4\%$ (2)	Cost in Future Dollars (3)	$(P/F,14.4\%,n)$ (4)	PW (5) = (3)(4)
0	—	$5000	1	$5000
1	$200	5200	0.8741	4545
2	208	5408	0.7641	4132
3	216	5624	0.6679	3757
4	225	5849	0.5838	3415

EXAMPLE 10.1   Joey is one of several winners who shared a lottery ticket. There are three plans offered to receive the after-tax proceeds.

> **Plan 1:** $100,000 now
>
> **Plan 2:** $15,000 per year for 8 years beginning 1 year from now. Total is $120,000.
>
> **Plan 3:** $45,000 now, another $45,000 four years from now, and a final $45,000 eight years from now. Total is $135,000.

Joey, a quite conservative person financially, plans to invest all of the proceeds as he receives them. He expects to make a real return of 6% per year. Use the 8-year time frame and an average inflation of 4% per year to determine which plan provides the best deal.

Solution

Select the plan with the largest PW now with inflation considered. Either of the two methods discussed earlier may be used. In the first method, all amounts are converted to CV dollars at 4% inflation, and the PW value is determined at the real rate of 6% per year. For the second method, find the PW of the quoted future amounts at the inflation-adjusted rate of 10.24% determined by Equation [10.6].

$$i_f = 0.06 + 0.04 + (0.06)(0.04) = 0.1024$$

Using the i_f method to obtain PW values for each plan,

> $PW_1 = \$100,000$ since all proceeds are taken immediately
>
> $PW_2 = 15,000(P/A,10.24\%,8) = 15,000(5.2886) = \$79,329$
>
> $PW_3 = 45,000[1 + (P/F,10.24\%,4) + (P/F,10.24\%,8)]$
> $= 45,000(2.1355) = \$96,099$

It is economically best to take the money now (plan 1).

Comment: This result supports the principle that it is better to take the money as early as possible because it is usually possible to make a better return investing it yourself. In this case, plan 1 is marginally the best only if all proceeds are invested and they earn at a real rate of 6% per year. Plan 3 is a close contender; if i_f goes down to 8.811% per year, plans 1 and 3 are economically equivalent with PW values of $100,000.

EXAMPLE 10.2   A self-employed chemical engineer is on contract with Dow Chemical, currently working in a relatively high-inflation country. She wishes to calculate a project's PW with estimated costs of $35,000 now and $7000 per year for ↓

5 years beginning 1 year from now with increases of 12% per year thereafter for the next 8 years. Use a real interest rate of 15% per year to make the calculations (*a*) without an adjustment for inflation and (*b*) considering inflation at a rate of 11% per year.

Solution

a. Figure 10.2 presents the cash flows. The PW without an adjustment for inflation is found using $i = 15\%$ and $g = 12\%$ in Equation [2.7] for the geometric series.

$$PW = -35{,}000 - 7000\,(P/A,15\%,4)$$

$$-\left\{ \dfrac{7000\left[1 - \left(\dfrac{1.12}{1.15}\right)^9\right]}{0.15 - 0.12} \right\}(P/F,15\%,4)$$

*— must use 9 & 4
not 8 & 5*

$$= -35{,}000 - 19{,}985 - 28{,}247$$
$$= \$-83{,}232$$

In the P/A factor, $n = 4$ because the $7000 cost in year 5 is the A_1 term in Equation [2.7].

b. To adjust for inflation, calculate the inflated interest rate by Equation [10.6].

$$i_f = 0.15 + 0.11 + (0.15)(0.11) = 0.2765$$
$$PW = -35{,}000 - 7000(P/A,27.65\%,4)$$

$$-\left\{ \dfrac{7000\left[1 - \left(\dfrac{1.12}{1.2765}\right)^9\right]}{0.2765 - 0.12} \right\}(P/F,27.65\%,4)$$

$$= -35{,}000 - 7000(2.2545) - 30{,}945(0.3766)$$
$$= \$-62{,}436$$

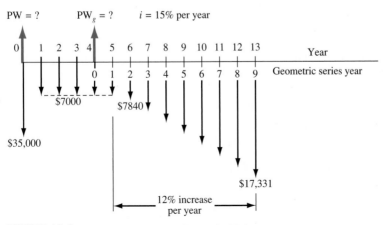

FIGURE 10.2   Cash flow diagram, Example 10.2.

This result demonstrates that in a high-inflation economy, when negotiating the amount of the payments to repay a loan, it is economically advantageous for the borrower to use future (inflated) dollars whenever possible to make the payments. The present value of future inflated dollars is significantly less when the inflation adjustment is included. And the higher the inflation rate, the larger the discounting because the P/F and P/A factors decrease in size.

The last example seems to add credence to the "buy now, pay later" philosophy of financial management. However, at some point, the debt-ridden company or individual will have to pay off the debts and the accrued interest with the inflated dollars. If cash is not readily available, the debts cannot be repaid. This can happen, for example, when a company unsuccessfully launches a new product, when there is a serious downturn in the economy, or when an individual loses a salary. In the longer term, this "buy now, pay later" approach must be tempered with sound financial practices now, and in the future.

10.3 FW CALCULATIONS ADJUSTED FOR INFLATION

In future worth calculations, a future amount F can have any one of four different interpretations:

Case 1. The *actual amount* of money that will be accumulated at time n.

Case 2. The *purchasing power* of the actual amount accumulated at time n, but stated in today's (constant-value) dollars.

Case 3. The number of *future dollars required* at time n to maintain the same purchasing power as a given amount today; that is, inflation is considered, but interest is not.

Case 4. The number of dollars required at time n to *maintain purchasing power and earn a stated real interest rate.*

Depending upon which interpretation is intended, the F value is calculated differently, as described below. Each case is illustrated.

Case 1: Actual Amount Accumulated

It should be clear that F, the actual amount of money accumulated, is obtained using the inflation-adjusted (market) interest rate.

$$F = P(1 + i_f)^n = P(F/P,i_f,n) \qquad [10.7]$$

For example, when a market rate of 10% is quoted, the inflation rate is included. Over a 7-year period, $1000 will accumulate to

$$F = 1000(F/P,10\%,7) = \$1948$$

Case 2: Constant-Value with Purchasing Power

The purchasing power of future dollars is determined by first using the market rate i_f to calculate F and then deflating the future amount through division by $(1 + f)^n$.

$$F = \frac{P(1 + i_f)^n}{(1 + f)^n} = \frac{P(F/P,i_f,n)}{(1 + f)^n} \qquad [10.8]$$

This relation, in effect, recognizes the fact that inflated prices mean $1 in the future purchases less than $1 now. The percentage loss in purchasing power is a measure of how much less. As an illustration, consider the same $1000 now, a 10% per year market rate, and an inflation rate of 4% per year. In 7 years, the purchasing power has risen, but only to $1481.

$$F = \frac{1000(F/P,10\%,7)}{(1.04)^7} = \frac{\$1948}{1.3159} = \$1481$$

This is $467 (or 24%) less than the $1948 actually accumulated at 10% (case 1). Therefore, we conclude that 4% inflation over 7 years reduces the purchasing power of money by 24%.

Also for case 2, the future amount of money accumulated with today's buying power could equivalently be determined by calculating the real interest rate and using it in the F/P factor to compensate for the decreased purchasing power of the dollar. This *real interest rate* is the i in Equation [10.6].

$$i_f = i + f + if$$
$$= i(1 + f) + f$$
$$i = \frac{i_f - f}{1 + f} \qquad [10.9]$$

The real interest rate i represents the rate at which today's dollars expand with their *same purchasing power* into equivalent future dollars. An inflation rate larger than the market interest rate leads to a negative real interest rate. The use of this interest rate is appropriate for calculating the future worth of an investment (such as a savings account or money market fund) when the effect of inflation must be removed. For the example of $1000 in today's dollars from Equation [10.9]

$$i = \frac{0.10 - 0.04}{1 + 0.04} = 0.0577, \text{ or } 5.77\%$$
$$F = 1000(F/P,5.77\%,7) = \$1481$$

The market interest rate of 10% per year has been reduced to a real rate that is less than 6% per year because of the erosive effects of inflation.

Case 3: Future Amount Required, No Interest

This case recognizes that prices increase when inflation is present. Simply put, future dollars are worth less, so more are needed. No interest rate is considered at all in this case. This is the situation present if someone asks, How much will a car cost in 5 years if its current cost is $20,000 and its price will increase by 6% per year? (The answer is $26,765.) No interest rate, only inflation, is involved. To find the future cost, substitute f for the interest rate in the F/P factor.

$$F = P(1 + f)^n = P(F/P,f,n) \qquad [10.10]$$

Reconsider the $1000 used previously. If it is escalating at exactly the inflation rate of 4% per year, the amount 7 years from now will be

$$F = 1000(F/P,4\%,7) = \$1316$$

Case 4: Inflation and Real Interest

This is the case applied when a MARR is established. Maintaining purchasing power and earning interest must account for both increasing prices (case 3) and the time value of money. If the growth of capital is to keep up, funds must grow at a rate equal to or above the real interest rate i plus a rate equal to the inflation rate f. Thus, to make a *real rate of return of 5.77%* when the inflation rate is 4%, i_f is the market (inflation-adjusted) rate that must be used. For the same $1000 amount,

$$i_f = 0.0577 + 0.04 + 0.0577(0.04) = 0.10$$
$$F = 1000(F/P,10\%,7) = \$1948$$

This calculation shows that $1948 seven years in the future will be equivalent to $1000 now with a real return of $i = 5.77\%$ per year and inflation of $f = 4\%$ per year.

Table 10.3 summarizes which rate is used in the equivalence formulas for the different interpretations of F. The calculations made in this section reveal that:

- Case 1: $1000 now at a market rate of 10% per year accumulates to $1948 in 7 years.

TABLE 10.3   Calculation Methods for Various Future Worth Interpretations

Future Worth Desired	Method of Calculation	Example for $P = \$1000$, $n = 7$, $i_f = 10\%$, $f = 4\%$
Case 1: Actual dollars accumulated	Use stated market rate i_f in equivalence formulas	$F = 1000(F/P,10\%,7)$ $= \$1948$
Case 2: Purchasing power of accumulated dollars in terms of today's dollars	Use market rate i_f in equivalence and divide by $(1 + f)^n$ or Use real i	$F = \dfrac{1000(F/P,10\%,7)}{(1.04)^7}$ or $F = 1000(F/P,5.77\%,7)$ $= \$1481$
Case 3: Dollars required for same purchasing power	Use f in place of i in equivalence formulas	$F = 1000(F/P,4\%,7)$ $= \$1316$
Case 4: Future dollars to maintain purchasing power and to earn a return	Calculate i_f and use in equivalence formulas	$F = 1000(F/P,10\%,7)$ $= \$1948$

- Case 2: The $1948 has the purchasing power of $1481 of today's dollars if $f = 4\%$ per year.
- Case 3: An item with a cost of $1000 now will cost $1316 in 7 years at an inflation rate of 4% per year.
- Case 4: It takes $1948 of future dollars to be equivalent to the $1000 now at a real interest rate of 5.77% with inflation considered at 4%.

Most corporations evaluate alternatives at a MARR large enough to cover inflation plus some return greater than their cost of capital, and significantly higher than the safe investment return of approximately 3.5% mentioned earlier. Therefore, for case 4, the resulting MARR will normally be higher than the market rate i_f. Define the symbol $MARR_f$ as the inflation-adjusted MARR, which is calculated in a fashion similar to i_f.

$$MARR_f = i + f + i(f) \qquad [10.11]$$

The real rate of return i used here is the required rate for the corporation relative to its cost of capital. (Cost of capital was introduced in Section 1.3 and is detailed in Section 13.5.) Now the future worth F, or FW, is calculated as

$$F = P(1 + MARR_f)^n = P(F/P,MARR_f,n) \qquad [10.12]$$

For example, if a company has a cost of capital of 10% per year and requires that a project return 3% per year, the real return is $i = 13\%$. The inflation-adjusted MARR is calculated by including the inflation rate of, say, 4% per year. Then, the project PW, AW, or FW will be determined at the rate obtained from Equation [10.11].

$$MARR_f = 0.13 + 0.04 + 0.13(0.04) = 17.52\%$$

EXAMPLE 10.3

Abbott Mining Systems wants to determine whether it should buy now or buy later for upgrading a piece of equipment used in deep mining operations. If the company selects plan N, the equipment will be purchased now for $200,000. However, if the company selects plan L, the purchase will be deferred for 3 years when the cost is expected to rise rapidly to $340,000. Abbott is ambitious; it expects a real MARR of 12% per year. The inflation rate in the country has averaged 6.75% per year. From only an economic perspective, determine whether the company should purchase now or later (*a*) when inflation is not considered and (*b*) when inflation is considered.

Solution

a. *Inflation not considered:* The real rate, or MARR, is $i = 12\%$ per year. The cost of plan L is $340,000 three years hence. Calculate the FW value for plan N three years from now, and select N.

$$FW_N = -200,000(F/P,12\%,3) = \$-280,986$$
$$FW_L = \$-340,000$$

b. *Inflation considered:* This is case 4; there is a real rate (12%), and inflation is 6.75%. First, compute the inflation-adjusted MARR by Equation [10.11].

$$\text{MARR}_f = 0.12 + 0.0675 + 0.12(0.0675) = 0.1956$$

Compute the FW value for plan N in future dollars.

$$\text{FW}_N = -200,000(F/P,19.56\%,3) = \$-341,812$$
$$\text{FW}_L = \$-340,000$$

Purchasing later is now selected, because it requires fewer equivalent future dollars. The inflation rate of 6.75% per year has raised the equivalent future worth of costs by 21.6% to \$341,812. This is the same as an increase of 6.75% per year, compounded over 3 years, or $(1.0675)^3 - 1 = 21.6\%$.

Most countries have inflation rates in the range of 2% to 8% per year, but *hyperinflation* is a problem in countries where political instability, overspending by the government, weak international trade balances, etc., are present. Hyperinflation rates may be very high—10% to 100% *per month*. In these cases, the government may take drastic actions: redefine the currency in terms of the currency of another country, control banks and corporations, or control the flow of capital into and out of the country in order to decrease inflation.

In a hyperinflated environment, people usually spend all their money immediately since the cost will be so much higher the next month, week, or day. To appreciate the disastrous effect of hyperinflation on a company's ability to keep up, we can rework Example 10.3b using an inflation rate of 10% per month, that is, a nominal 120% per year (not considering the compounding of inflation). The FW_N amount skyrockets and plan L is a clear choice. Of course, in such an environment the \$340,000 purchase price for plan L 3 years hence would obviously not be guaranteed, so the entire economic analysis is unreliable. Good economic decisions in a hyperinflated economy are very difficult to make, since the estimated future values are totally unreliable and the future availability of capital is uncertain.

10.4 AW CALCULATIONS ADJUSTED FOR INFLATION

It is particularly important in capital recovery calculations used for AW analysis to include inflation because current capital dollars must be recovered with future inflated dollars. Since future dollars have less buying power than today's dollars, it is obvious that more dollars will be required to recover the present investment. This suggests the use of the inflated interest rate in the A/P formula. For example, if \$1000 is invested today at a real interest rate of 10% per year when the inflation rate is 8% per year, the equivalent amount that must be recovered each year for 5 years in future dollars is

$$A = 1000(A/P,18.8\%,5) = \$325.59$$

On the other hand, the decreased value of dollars through time means that investors can spend fewer present (higher-value) dollars to accumulate a specified amount of future (inflated) dollars. This suggests the use of a higher interest rate, that is, the i_f rate, to produce a lower A value in the A/F formula. The annual equivalent (with adjustment for inflation) of $F = \$1000$ five years from now in future dollars is

$$A = 1000(A/F,18.8\%,5) = \$137.59$$

This method is illustrated in the next example.

For comparison, the equivalent annual amount to accumulate $F = \$1000$ at a real $i = 10\%$, (without adjustment for inflation) is $1000(A/F,10\%,5) = \$163.80$. Thus, when F is fixed, uniformly distributed future costs should be spread over as long a time period as possible so that the leveraging effect of inflation will reduce the payment ($\$137.59$ versus $\$163.80$ here).

EXAMPLE 10.4

a. What annual deposit is required for 5 years to accumulate an amount of money with the same purchasing power as $680.58 today, if the market interest rate is 10% per year and inflation is 8% per year?
b. What is the real interest rate?

Solution

a. First, find the actual number of future (inflated) dollars required 5 years from now. This is case 3, as described earlier.

$$F = (\text{present buying power})(1 + f)^5 = 680.58(1.08)^5 = \$1000$$

The actual amount of the annual deposit is calculated using the market (inflated) interest rate of 10%. This is case 4 using A instead of P.

$$A = 1000(A/F,10\%,5) = \$163.80$$

b. Using Equation [10.9], i is calculated.

$$i = (0.10 - 0.08)/(1.08)$$
$$= 0.0185 \text{ or } 1.85\% \text{ per year}$$

Comment: To put these calculations into perspective, consider the following: If the inflation rate is zero when the real interest rate is 1.85%, the future amount of money with the same purchasing power as $680.58 today is obviously $680.58. Then the annual amount required to accumulate this future amount in 5 years is $A = 680.58(A/F,1.85\%,5) = \$131.17$. This is $32.63 lower than the $163.80 calculated above for $f = 8\%$. This difference is due to the fact that during inflationary periods, dollars deposited at the beginning of a period have more buying power than the dollars returned at the end of the period. To make up the buying power difference, more lower-value dollars are required. That is, to maintain equivalent purchasing power at $f = 8\%$ per year, an extra $32.63 per year is required.

This logic explains why, in times of increasing inflation, lenders of money (credit card companies, mortgage companies, and banks) tend to further increase their market interest rates. People tend to pay off less of their incurred debt at each payment because they use any excess money to purchase other items before the price is further inflated. Also, the lending institutions must have more dollars in the future to cover the expected higher costs of lending money. All this is due to the spiraling effect of increasing inflation. Breaking this cycle is difficult for the individual and family to do, and much more difficult to alter at a national level.

10.5 USING SPREADSHEETS TO ADJUST FOR INFLATION

When using a spreadsheet, adjust for inflation in PW, AW, and FW calculations by entering the correct percentage rate into the appropriate Excel function. For example, suppose PW is required for a receipt of $A = \$15,000$ per year for 8 years at a real rate of $i = 5\%$ per year. To adjust for the expected inflation rate $f = 3\%$ per year, first determine the inflation-adjusted rate by using Equation [10.6] or by entering its equivalent relation $= i\% + f\% + i\%*f\%$ into a spreadsheet cell.

$$i_f = 5\% + 3\% + 5\%(3\%) = 8.15\%$$

The single-cell PV functions used to determine the PW value with and without inflation considered follow. The minus keeps the sign of the answer positive.

With inflation: $= -PV(8.15\%,8,15000)$ PW = \$85,711
Without inflation: $= -PV(5\%,8,15000)$ PW = \$96,948

The next example uses different Excel functions when inflation is considered and the timing of the cash flows varies between plans.

EXAMPLE 10.5   This is a reconsideration of the three plans presented in Example 10.1. To recap, the plans are:

Plan 1: Receive $100,000 now.

Plan 2: Receive $15,000 per year for 8 years beginning 1 year from now.

Plan 3: Receive 45,000 now, another $45,000 four years from now, and a final $45,000 eight years from now.

If $i = 6\%$ per year and $f = 4\%$ per year, (*a*) find the best plan based on PW, (*b*) determine the worth in future dollars of each plan 8 years from now in inflated dollars, and (*c*) determine the future value of the plans in terms of today's purchasing power.

Solution

a. Figure 10.3: Of course, $PW_1 = \$100,000$. The value $PW_2 = \$79,329$ can be displayed using the function $= -PV(10.24\%,8,15000)$. However, for PW_3 the cash flows must be entered with intervening zeros and the NPV function applied. The spreadsheet shows the computation of $i_f = 10.24\%$ and all PW values using NPV functions. Plan 1 has the largest PW value.

b. Row 16 of Figure 10.3 shows the equivalent future dollars of each plan 8 years from now at 10.24% per year using the FV function. The minus sign maintains the positive sign.

c. To determine FW while retaining today's purchasing power (case 2), use the real rate of 6% in the FV function. Row 18 shows the significant decrease from the previous FW values with inflation considered. The impact of inflation is large. For plan 3, as an example, the winner could have the equivalent of over \$209,000 in 8 years. However, this will purchase only approximately \$153,000 worth of goods in terms of today's dollars.

	A	B	C	D	E	F	G	H
1		Inflated-adjusted Rate =		10.24%				
2								
3	Year	Plan 1	Plan 2	Plan 3	= 6% + 4% + 6%*4%			
4	0	100,000		45000				
5	1		15000	0				
6	2		15000	0				
7	3		15000	0				
8	4		15000	45000				
9	5		15000	0			= NPV(D1,D5:D12) + D4	
10	6		15000	0				
11	7		15000	0				
12	8		15000	45000			= NPV(D1,C5:C12)	
13	PW with inflation	$ 100,000	$ 79,329	$ 96,099				
14								
15						= -FV(D1,8,,D13)		
16	FW with inflation	$218,129	$173,041	$209,619				
17						= -FV(6%,8,,D13)		
18	FW without inflation	$159,385	$126,439	$153,167				
19								

FIGURE 10.3 PW and FW calculations with inflation considered and with today's purchasing power maintained, Example 10.5.

SUMMARY

Inflation, treated computationally as an interest rate, makes the cost of the same product or service increase over time due to the decreased value of money. The chapter covers two ways to consider inflation in engineering economy computations: (1) in terms of today's (constant-value) dollars, and (2) in terms of future dollars. Some important relations are:

Inflated interest rate: $i_f = i + f + if$

Real interest rate: $i = (i_f - f)/(1 + f)$

PW of a future amount with inflation considered: $P = F(P/F,i_f,n)$

Future worth of a present amount in constant-value dollars with the same purchasing power: $F = P(F/P,i,n)$

Future amount to cover a current amount with no interest: $F = P(F/P,f,n)$

Future amount to cover a current amount with interest: $F = P(F/P,i_f,n)$

Annual equivalent of a future dollar amount: $A = F(A/F,i_f,n)$

Annual equivalent of a present amount in future dollars: $A = P(A/P,i_f,n)$

PROBLEMS

Understanding Inflation

10.1 There are two ways to account for inflation in economic calculations. What are they?

10.2 In an inflationary period, what is the difference between inflated dollars and "then-current" future dollars?

10.3 At a time when the inflation rate is 6% per year, $100,000 fifteen years from now (that is, inflated dollars) is the same as how many of today's constant-value dollars?

10.4 When the inflation rate is 5% per year, how many inflated dollars will be required 10 years from now to buy the same things that $10,000 buys now?

10.5 In 1950, the cost of a loaf of bread was $0.14. If the cost in 2007 was $1.29, determine the average inflation rate per year between 1950 and 2007.

10.6 In 2006, the average annual salary of petroleum engineers was $74,400. Predict the average salary in 2015 if salaries increase only by the inflation rate. Assume the inflation rate over the time period is constant at 2.5% per year.

10.7 If the inflation rate is 7% per year, how many years will it take for the cost of something to double, if the price increases only by the inflation rate?

10.8 Assume that you want to retire 30 years from now with an amount of money that will have the same value (same buying power) as $1.5 million today. If you estimate the inflation rate will be 4% per year, how many then-current dollars will you need?

10.9 A certain website allows users to determine how many inflated dollars have the same buying power as a given amount of money at any time in the past between the years of 1774 and 2006. When the values 1920, 2005, and $1000 are entered for the initial year, desired year, and initial amount, respectively, the result is $9745.51. On the basis of these numbers, what was the average inflation rate per year between 1920 and 2005?

10.10 The inflation rate over a 10-year period for an item that now costs $1000 is shown in the following table. (a) What will be the cost at the end of year 10? (b) Do you get the same cost using an average inflation rate of 5% per year through the 10-year period? Why?

Year	Inflation Rate
1	10%
2	0%
3	10%
4	0%
5	10%
6	0%
7	10%
8	0%
9	10%
10	0%

10.11 An engineer who is now 65 years old began planning for retirement 40 years ago. At that time, he thought that if he had $1 million when he retired, he would have more than enough money to live his remaining life in luxury. If the inflation rate over the 40-year time period averaged a constant 4% per year, what is the constant-value dollar amount of his $1 million? Use the day he started 40 years ago as the base year.

Present Worth Calculations with Inflation

10.12 The president of a medium-sized oil company wants to buy a private plane to reduce the total travel time between cities where refineries are located. The company can buy a used Lear jet now or wait for a new very light jet (VLJ) that will be available 3 years from now. The cost of the VLJ will be $1.5 million, payable when the plane is delivered in 3 years. The president has asked you to determine the present worth of the plane so that he can decide whether to buy the used Lear now or wait for the VLJ. If the company's MARR is 15% per year and the inflation rate is projected to be 3% per year, what is the present worth of the VLJ with inflation considered?

10.13 A piece of heavy equipment can be purchased for $85,000 now or for an estimated $130,000 five years from now. At a real interest rate of 10% per

year and an inflation rate of 6% per year, determine if the company should buy now or later with inflation considered.

10.14 An environmental testing company needs to purchase $50,000 worth of equipment 2 years from now. At a real interest rate of 10% per year and inflation rate of 4% per year, what is the present worth of the equipment? Assume the $50,000 are "then-current" dollars.

10.15 Find the present worth of the cash flows shown below. Some are expressed as constant-value (CV) dollars and some are inflated (inf) dollars. Assume a real interest rate of 8% per year and an inflation rate of 6% per year.

Year	1	2	3	4	5
Cash Flow, $	3000 (CV)	6000 (inf)	8000 (inf)	4000 (CV)	5000 (CV)

10.16 A salesman from vendor A, who is trying to get his foot in the door of a large account, offered water desalting equipment for $2.1 million. This is $400,000 more than the price that a competing saleswoman from vendor B offered. However, vendor A said the company won't have to pay for the equipment until the warranty runs out. If the equipment has a 2-year warranty, determine which offer is better. The company's real MARR is 12% per year and the inflation rate is 4% per year.

10.17 How much can the manufacturer of superconducting magnetic energy storage systems afford to spend now on new equipment in lieu of spending $75,000 four years from now? The company's real MARR is 12% per year and the inflation rate is 3% per year.

10.18 A chemical engineer is considering two sizes of pipes—small (S) and large (L)—for moving distillate from a refinery to the tank farm. A small pipeline will cost less to purchase (including valves and other appurtenances) but will have a high head loss and, therefore, a higher pumping cost. In writing the report, the engineer compared the alternatives on the basis of future worth values, but the company president wants the costs expressed as present dollars. Determine present worth values if future worth values are $FW_S = \$2.3$ million and $FW_L = \$2.5$ million. The company uses a real interest rate of 1% per month and an inflation rate of 0.4% per month. Assume the future worth values were for a 10-year project period.

Future Worth Calculations with Inflation

10.19 A pulp and paper company is planning to set aside $150,000 now for possibly replacing its large synchronous refiner motors. If the replacement isn't needed for 5 years, how much will the company have in the account if it earns the market interest rate of 10% per year and the inflation rate is 4% per year?

10.20 In order to encourage its employees to save money for retirement, a large pharmaceutical company offers a guaranteed rate of return of 10% per year, without regard to the market interest rate. If a new engineer invests $5000 per year for 10 years at a time when the inflation rate averages 3% per year, how much will be in the account at the end of the 10-year period?

10.21 Well-managed companies set aside money to pay for emergencies that inevitably arise in the course of doing business. If a commercial solid waste recycling and disposal company puts 0.5% of its after-tax income into such an account, how much will the company have after 7 years? The company's annual after-tax income averages $15.2 million and inflation and market interest rates are 5% per year and 9% per year, respectively?

10.22 What is the real interest rate when the market rate is 10% per year and the inflation rate is 6% per year?

10.23 What inflation rate can be implied from a market interest rate of 9% per year and a real interest rate of 3% per year?

10.24 If you deposit $10,000 now into an account that pays interest at 8% per year for 10 years, what will be the purchasing power (with respect to today's dollars) of the accumulated amount? Assume the inflation rate averages 3% per year over the 10 years?

10.25 A mechanical consulting company is examining its cash flow requirements for the next 6 years. The company expects to replace office machines and computer equipment at various times over the 6-year planning period. Specifically, the company expects to spend $6000 two years from now, $9000 three years from now, and $5000 six years from now. What is the buying power (with respect to today's dollars) of each expenditure in its respective year, if the inflation rate over the 6-year period is 4% per year?

10.26 Harmony Corporation plans to set aside $60,000 per year beginning 1 year from now for replacing equipment 5 years from now. What will be the buying power of the amount accumulated (with respect to today's dollars) if the investment grows by 10% per year, but inflation averages 4% per year?

10.27 If a company deposits $100,000 into an account that earns a market interest rate of 10% per year at a time when the *deflation* rate is 1% per year, what will be the buying power of the accumulated amount (with respect to today's dollars) at the end of 15 years?

10.28 The strategic plan of a solar energy company that manufactures high-efficiency solar cells includes an expansion of its physical plant in 4 years. The engineer in charge of planning estimates the expenditure required now to be $8 million, but in 4 years, the cost will be higher by an amount equal to the inflation rate. If the company sets aside $7,000,000 now into an account that earns interest at 7% per year, what will the inflation rate have to be in order for the company to have exactly the right amount of money for the expansion?

10.29 Construction equipment has a cost today of $40,000. If its cost has increased by only the inflation rate of 6% per year when the market interest rate has been 10% per year, what was its cost 10 years ago?

10.30 A Toyota Tundra can be purchased today for $32,350. A civil engineering firm is going to need 3 more trucks in 2 years because of a land development contract it just won. If the price of the truck increases exactly in accordance with an estimated inflation rate of 3.5% per year, determine how much the 3 trucks will cost in 2 years.

10.31 Timken Roller Bearing is a manufacturer of seamless tubes for drill bit collars. The company is planning to add larger capacity robotic arms to one of its assembly lines 3 years from now. If done now, the cost of the equipment with installation is $2.4 million. If the company's real MARR is 15% per year, determine the equivalent amount the company can spend 3 years from now in then-current dollars? The inflation rate is 2.8% per year.

10.32 The data below shows two patterns of inflation that are exactly the opposite of each other over a 20-year time period. (*a*) If each machine costs $10,000 in year 0 and they both increase in cost exactly in accordance with the inflation rate, how much will each machine cost at the end of year 20? (*b*) What is the average inflation rate over the time period for machine A (that is, what single inflation rate would result in the same final cost for machine A?) (*c*) In which years will machine A cost *more than* machine B?

Year	Machine A	Machine B
1	10%	2%
2	10%	2%
3	2%	10%
4	2%	10%
5	10%	2%
6	10%	2%
7	2%	10%
8	2%	10%
.	.	.
.	.	.
.	.	.
19	2%	10%
20	2%	10%

10.33 If the inflation rate is 6% per year and a person wants to earn a true (real) interest rate of 10% per year, determine the number of "then-current" dollars he has to receive 10 years from now if the present investment is $10,000.

Capital Recovery with Inflation

10.34 Johnson Thermal Products used austenitic nickel-chromium alloys to manufacture resistance heating wire. The company is considering a new annealing-drawing process to reduce costs. The new process will cost $3.1 million dollars now. How much must be saved each year to recover the investment in 5 years if the company's MARR is a real 10% per year and the inflation rate is 3% per year?

10.35 The costs associated with a small X-ray inspection system are $40,000 now and $24,000 per year, with a $6000 salvage value after 3 years. Determine the equivalent annual cost of the system if the real interest rate is 10% per year and the inflation rate is 4% per year.

10.36 MetroKlean LLC, a hazardous waste soil cleaning company, borrowed $2.5 million to finance start-up costs for a site reclamation project. How much must the company receive each year in revenue to earn a real rate of return of 20% per year for the 5-year project period? Assume inflation is 5% per year.

10.37 Maintenance costs for pollution control equipment on a pulverized coal cyclone furnace are expected to be $80,000 now and another $90,000 three years from now. The CFO of Monongahela Power wants to know the equivalent annual cost of the equipment in years 1 through 5. If the company uses a real interest rate of 12% per year and the inflation rate averages 4% per year, what is the equivalent annual cost of the equipment?

10.38 In wisely planning for your retirement, you invest $12,000 per year for 20 years into a 401k account. How much will you be able to withdraw each year for 10 years, starting one year after your last deposit, if you earn a real return of 10% per year and the inflation rate averages 2.8% per year?

PROBLEMS FOR TEST REVIEW AND FE EXAM PRACTICE

10.39 For a market interest rate of 12% per year and an inflation rate of 7% per year, the real interest rate per year is closest to
a. 4.7%.
b. 7%.
c. 12%.
d. 19.8%.

10.40 Inflation occurs when
a. productivity increases.
b. the value of the currency decreases.
c. the value of the currency increases.
d. the price of gold decreases.

10.41 In order to calculate how much something will cost if you expect its cost to increase by exactly the inflation rate, you should
a. multiply by $(1 + f)^n$.
b. multiply by $(1 + i)^n$.
c. divide by $(1 + f)^n$.
d. multiply by $(1 + i_f)^n$.

10.42 When future cash flows are expressed in *constant-value dollars,* the interest rate to use in calculating the present worth of the cash flows is
a. i.
b. i_f.
c. $1 + i$.
d. $i - f$.

10.43 A machine has a cost today of $20,000. The *future worth* (in then-current dollars) of the machine in year 10, if the inflation rate is 4% per year and the interest rate is 9% per year, is closest to
a. $29,000.
b. $47,300.
c. $67,900.
d. $70,086.

10.44 For a real interest rate of 1% per month and an inflation rate of 1% per month, the *nominal inflated* interest rate *per year* is closest to
a. 1%.
b. 2.0%.
c. 24%.
d. 24.12%.

10.45 Construction equipment has a cost today of $40,000. Because of high demand for this type of equipment, its cost is expected to increase by *double* the inflation rate. If the inflation rate is 6% per year and the market interest rate is 10% per year, the cost of the equipment 4 years from now is expected to be
a. $62,941.
b. $58,564.
c. $50,499.
d. $47,359.

10.46 When the inflation rate is equal to 0% per year,
  a. the market interest rate is greater than the real interest rate.
  b. the market interest rate is less than the real interest rate.
  c. the market interest rate is equal to the real interest rate.
  d. the inflated interest rate is greater than the real interest rate.

10.47 If the market interest rate is 12% per year and the inflation rate is 5% per year, the number of future dollars in year 7 that will have the *same buying power* as $2000 now is determined by equation

  a. Future dollar amount = $2000(1 + 0.05)^7$
  b. Future dollar amount = $2000/(1 + 0.05)^7$
  c. Future dollar amount = $2000(1 + 0.12)^7$
  d. Future dollar amount = $2000/(1 + 0.12)^7$

10.48 Geraldine invests $5000 now. The inflation rate is 6% per year and she wants to earn a real interest rate of 10% per year. The amount of then-current dollars she will get back 10 years from now is closest to

  a. $8,955.
  b. $12,970.
  c. $18,150.
  d. $23,225.

Estimating Costs

C. Borland/PhotoLink/Getty Images

U p to this point, cost and revenue cash flow values have been stated or assumed as known. In reality, they are not; they must be estimated. This chapter explains what cost estimation involves, and applies cost estimation techniques. *Cost estimation* is important in all aspects of a project, but especially in the stages of project conception, preliminary design, detailed design, and economic analysis. In engineering practice, the estimation of costs receives much more attention than revenue estimation; costs are the topic of this chapter.

Unlike direct costs for labor and materials, indirect costs are not easily traced to a specific function, department, machine, or processing line. Therefore, *allocation of indirect costs* for functions such as IT, utilities, safety, management, purchasing, and quality is made using some rational basis. Both the traditional method of allocation and the Activity-Based Costing (ABC) method are summarized in this chapter.

Objectives

1. Describe different approaches to cost estimation.

   | Approaches |

2. Use a unit cost factor to estimate preliminary cost.

   | Unit method |

3. Use a cost index to estimate present cost based on historic data.

   | Cost indexes |

4. Estimate the cost of a component, system, or plant by using a cost-capacity equation.

   | Cost-capacity equations |

5. Estimate total plant cost using the factor method.

   | Factor method |

6. Estimate time and cost using the learning curve relationship.

   | Learning curve |

7. Allocate indirect cost using traditional indirect cost rates and the Activity-Based Costing (ABC) method.

   | Indirect cost rates and ABC allocation |

11.1 HOW COST ESTIMATES ARE MADE

Cost estimation is a major activity performed in the initial stages of virtually every effort in industry, business, and government. In general, most cost estimates are developed for either a *project* or a *system;* however, combinations of these are very common. A *project* usually involves physical items, such as a building, bridge, manufacturing plant, or offshore drilling platform, to name just a few. A *system* is usually an operational design that involves processes, software, and other non-physical items. Examples might be a purchase order system, a software package to design highways, or an Internet-based remote-control system. The cost estimates are usually made for the initial development of the project or system, with the life-cycle costs of maintenance and upgrade estimated as a percentage of first cost. Much of the discussion that follows concentrates on physical-based projects. However, the logic is widely applicable to cost estimation in all areas.

Costs are comprised of *direct costs* (largely humans, machines, and materials) and *indirect costs* (mostly support functions, utilities, management, taxes, etc.). Normally direct costs are estimated with some detail, then the indirect costs are added using standard rates and factors. Direct costs in many industries have become a relatively small percentage of overall product cost, while indirect costs have become much larger. Accordingly, many industrial settings require some estimating for indirect costs as well. Indirect cost allocation is discussed in Section 11.7. Primarily, direct costs are discussed here.

Because cost estimation is a complex activity, the following questions form a structure for our discussion.

- What cost components must be estimated?
- What approach to cost estimation will be applied?
- How accurate should the estimates be?
- What estimation techniques will be utilized?

Costs to Estimate

If a project revolves around a single piece of equipment, for example, a multistory building, the *cost components* will be significantly simpler and fewer than the components for a complete system, such as the design, manufacturing, and testing of a new commercial aircraft. Therefore, it is important to know up front how much the cost estimation task will involve. Examples of direct cost components are the first cost P and the annual operating cost (AOC), also called the M&O costs (maintenance and operating). Each component will have several *cost elements,* some that are directly estimated, others that require examination of records of similar projects, and still others that must be modeled using an estimation technique. Following are sample elements of the first cost and AOC components.

> *First cost component P:*
>> Elements: Equipment cost
>>> Delivery charges
>>> Installation cost

Insurance coverage

Initial training of personnel for equipment use

Delivered-equipment cost is the sum of the first two elements; installed-equipment cost adds the third element.

AOC component, a part of the equivalent annual cost A:

Elements: Direct labor cost for operating personnel

Direct materials

Maintenance (daily, periodic, repairs, etc.)

Rework and rebuild

Some of these elements, such as equipment cost, can be determined with high accuracy; others, such as maintenance costs, may be harder to estimate. However, a wide variety of data are available, such as McGraw-Hill Construction, R. S. Means cost books, Marshall & Swift, NASA Cost Estimating website, and many others. When costs for an entire system must be estimated, the number of cost components and elements is likely to be in the hundreds. It is then necessary to prioritize the estimation tasks.

For familiar projects (houses, office buildings, highways, and some chemical plants) there are standard cost estimation packages available in paper or software form. For example, state highway departments utilize software packages that prompt for the correct cost components (bridges, pavement, cut-and-fill profiles, etc.) and estimate costs with time-proven, built-in relations. Once these components are estimated, exceptions for the specific project are added. One such exception is location cost adjustments. (R. S. Means has city index values for over 700 cities in the United States and Canada.)

Cost Estimation Approach

Traditionally in industry, business, and the public sector, a "bottom-up" approach to cost estimation is applied. For a simple rendition of this approach, see Figure 11.1 (left). The progression is as follows: cost components and their elements are identified, cost elements are estimated, and estimates are summed to obtain total direct cost. The price is then determined by adding indirect costs and the profit margin, which is usually a percentage of the total cost. This approach works well when competition is not the dominant factor in pricing the product or service.

The bottom-up approach treats the required price as an output variable and the cost estimates as input variables.

Figure 11.1 (right) shows a simplistic progression for the design-to-cost, or top-down, approach. The competitive price establishes the target cost.

The design-to-cost, or top-down, approach treats the competitive price as an input variable and the cost estimates as output variables.

This approach places greater emphasis on the accuracy of the price estimation activity. The target cost must be realistic, or it becomes very difficult to make sensible, realizable cost estimates for the various components.

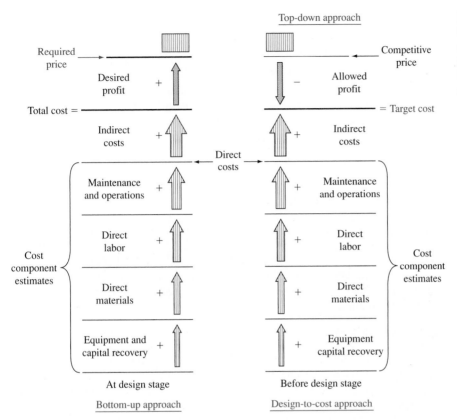

FIGURE 11.1
Simplified cost estimation processes for bottom-up and top-down approaches.

The design-to-cost approach is best applied in the early stages of a product design. This approach is useful in encouraging innovation, new design, process improvement, and efficiency. These are some of the essentials of *value engineering*.

Usually, the resulting approach is some combination of these two cost estimation philosophies. However, it is helpful to understand up front what approach is to be emphasized. Historically, the bottom-up approach has been more predominant in Western engineering cultures. The design-to-cost approach is considered routine in Eastern engineering cultures.

Accuracy of the Estimates

No cost estimates are expected to be exact; however, they are expected to be reasonable and accurate enough to support economic scrutiny. The accuracy required increases as the project progresses from preliminary design to detailed design and on to economic evaluation. Cost estimates made before and during the preliminary design stage are expected to be good "first-cut" estimates that serve as input to the project budget.

When utilized at early and conceptual design stages, estimates are referred to as *order-of-magnitude* estimates and generally range within ±20% of actual cost. At the detailed design stage, cost estimates are expected to be accurate enough to support economic evaluation for a go–no go decision. Every project

FIGURE 11.2

Characteristic curve of estimate's accuracy versus time spent to estimate construction cost of a building.

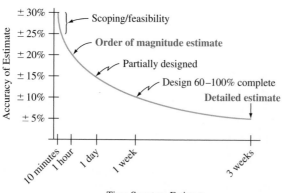

setting has its own characteristics, but a range of ±5% of actual costs is expected at the detailed design stage. Figure 11.2 shows the general range of estimate accuracy for the construction cost of a building versus time spent in preparing the estimate. Obviously, the desire for better accuracy has to be balanced against the cost of obtaining it.

Cost Estimation Techniques

Methods such as expert opinion and comparison with comparable installations serve as excellent estimators. The use of the *unit method* and *cost indexes* base the present estimate on past cost experiences, with inflation considered. Models such as *cost-capacity equations,* the *factor method,* and the *learning curve* are simple mathematical techniques applied at the preliminary design stage. They are called cost-estimating relationships (CER). There are many additional methods discussed in the handbooks and publications of different industries.

Most cost estimates made in a professional setting are accomplished in part or wholly using *software packages.* These are usually linked to an updated database that contains cost indexes and rates for the locations and product or process types being studied. There are a large number and a wide variety of estimators, cost trackers, and cost compliance software systems, most of them developed for specific industries. Corporations usually standardize on one or two packages to ensure consistency over time and projects.

11.2 UNIT METHOD

The *unit method* is a popular preliminary estimation technique applicable to virtually all professions. The total estimated cost C_T is obtained by multiplying the number of units N by a per unit cost factor U.

$$C_T = U \times N \qquad [11.1]$$

Unit cost factors must be updated frequently to remain current with changing costs, areas, and inflation. Some sample unit cost factors (and values) are

Total average cost of operating an automobile (43¢ per mile)

Cost to bury fiber cable in a suburban area ($30,000 per mile)

Cost to construct a parking space in a parking garage ($4500 per space)

Cost of constructing interstate highway ($6.2 million per mile)

Cost of house construction per livable area ($225 per square foot)

Applications of the unit method to estimate costs are easily found. If house construction costs average $225 per square foot, a preliminary cost estimate for an 1800-square-foot house, using Equation [11.1], is $405,000. Similarly, a 200-mile trip should cost about $86 for the car only at 43¢ per mile.

When there are several components to a project or system, the unit cost factors for each component are multiplied by the amount of resources needed and the results are summed to obtain the total cost C_T. This is illustrated in the next example.

Justin, an ME with Dynamic Castings, has been asked to make a preliminary estimate of the total cost to manufacture 1500 sections of high pressure gas pipe using an advanced centrifugal casting method. Since a ±20% estimate is acceptable at this preliminary stage, a unit method estimate is sufficient. Use the following resource and unit cost factor estimates to help Justin.

EXAMPLE 11.1

    Materials: 3000 tons at $45.90 per ton

    Machinery and tooling: 1500 hours at $120 per hour

    Direct labor in-plant:

        Casting and treating: 3000 hours at $55 per hour

        Finishing and shipping: 1200 hours at $45 per hour

    Indirect labor: 400 hours at $75 per hour

Solution

Apply Equation [11.1] to each of the five areas and sum the results to obtain the total cost estimate of $566,700. Table 11.1 provides the details.

TABLE 11.1 **Total Cost Estimate Using Unit Cost Factors for Several Resource Areas, Example 11.1**

Resource	Amount, N	Unit Cost Factor, U	Cost Estimate, $U \times N$
Materials	3000 tons	$ 45.90 per ton	$137,700
Machinery, tooling	1500 hours	$120 per hour	180,000
Labor, casting	3000 hours	$ 55 per hour	165,000
Labor, finishing	1200 hours	$ 45 per hour	54,000
Labor, indirect	400 hours	$ 75 per hour	30,000
Total cost estimate			**$566,700**

11.3 COST INDEXES

A *cost index* is a ratio of the cost of something today to its cost sometime in the past. As such, the index is a dimensionless number that shows the relative cost change over time. One such index that most people are familiar with is the Consumer Price Index (CPI), which shows the relationship between present and past costs for many of the things that "typical" consumers must buy. This index includes such items as rent, food, transportation, and certain services. Other indexes track the costs of equipment, and goods and services that are more pertinent to the engineering disciplines. Table 11.2 is a listing of some common indexes.

TABLE 11.2   Types and Sources of Various Cost Indexes

Type of Index	Source
Overall prices	
Consumer (CPI)	Bureau of Labor Statistics
Producer (wholesale)	U.S. Department of Labor
Construction	
Chemical plant overall	*Chemical Engineering*
Equipment, machinery, and supports	
Construction labor	
Buildings	
Engineering and supervision	
Engineering News Record overall	*Engineering News Record (ENR)*
Construction	
Building	
Common labor	
Skilled labor	
Materials	
EPA treatment plant indexes	Environmental Protection Agency; *ENR*
Large-city advanced treatment (LCAT)	
Small-city conventional treatment (SCCT)	
Federal highway	
Contractor cost	
Equipment	
Marshall and Swift (M&S) overall	Marshall & Swift
M&S specific industries	
Labor	
Output per man-hour by industry	U.S. Department of Labor

The general equation for updating costs through the use of a cost index over a period from time $t = 0$ (base) to another time t is

$$C_t = C_0\left(\frac{I_t}{I_0}\right) \qquad [11.2]$$

where C_t = estimated cost at present time t

C_0 = cost at previous time t_0

I_t = index value at time t

I_0 = index value at base time 0

Generally, the indexes for equipment and materials are made up of a mix of components that are assigned certain weights, with the components sometimes further subdivided into more basic items. For example, the equipment, machinery, and support component of the chemical plant cost index is subdivided into process machinery, pipes, valves and fittings, pumps and compressors, and so forth. These subcomponents, in turn, are built up from even more basic items such as pressure pipe, black pipe, and galvanized pipe. Table 11.3 presents several years' values of

TABLE 11.3 Values for Selected Indexes

Year	Chem. Eng. Plant Cost Index	*ENR* Construction Cost Index	M&S Equipment Cost Index
1985	325.3	4195	789.6
1986	318.4	4295	797.6
1987	323.8	4406	813.6
1988	342.5	4519	852.0
1989	355.4	4615	895.1
1990	357.6	4732	915.1
1991	361.3	4835	930.6
1992	358.2	4985	943.1
1993	359.2	5210	964.2
1994	368.1	5408	993.4
1995	381.1	5471	1027.5
1996	381.8	5620	1039.2
1997	386.5	5826	1056.8
1998	389.5	5920	1061.9
1999	390.6	6059	1068.3
2000	394.1	6221	1089.0
2001	394.3	6343	1093.9
2002	395.6	6538	1104.2
2003	402.0	6695	1123.6
2004	444.2	7115	1178.5
2005	468.2	7446	1244.5
2006	499.6	7751	1302.3
2007 (midyear)	517.7	8007	1362.7

the *Chemical Engineering* plant cost index, the *Engineering News Record (ENR)* construction cost index, and the Marshall and Swift (M&S) equipment cost index (also called installed-equipment index). The base period of 1957 to 1959 is assigned a value of 100 for the *Chemical Engineering* plant cost index, 1913 = 100 for the *ENR* index, and 1926 = 100 for the M&S equipment cost index.

Current and past values of several of the indexes may be obtained from the Internet (some for a fee). For example, the *CE* plant cost index is available at www. che.com/pindex. The *ENR* construction cost index is found at www.construction. com. This latter site offers a comprehensive series of construction-related resources, including several *ENR* cost indexes and cost estimation systems. A website used by many engineering professionals in the form of a "technical chat room" for all types of topics, including estimation, is www.eng-tips.com.

EXAMPLE 11.2   In evaluating the feasibility of a major construction project, an engineer is interested in estimating the cost of skilled labor for the job. The engineer finds that a project of similar complexity and magnitude was completed 5 years ago at a skilled labor cost of $360,000. The *ENR* skilled labor index was 3496 then and is now 4038. What is the estimated skilled labor cost?

Solution

The base time t_0 is 5 years ago. Using Equation [11.2], the present cost estimate is

$$C_t = 360{,}000\left(\frac{4038}{3496}\right)$$
$$= \$415{,}812$$

In the manufacturing and service industries, tabulated cost indexes may be difficult to find. The cost index will vary, perhaps with the region of the country, the type of product or service, and many other factors. The development of the cost index requires the actual cost at different times for a prescribed quantity and quality of the item. The *base period* is a selected time when the index is defined with a basis value of 100 (or 1). The index each year (period) is determined as the cost divided by the base-year cost and multiplied by 100. Future index values may be forecast using simple extrapolation or more refined mathematical techniques.

EXAMPLE 11.3   An engineer with Hughes Industries is in the process of estimating costs for a plant expansion. Two important items used in the process are a subcontracted 4 gigahertz microchip and a preprocessed platinum alloy. Spot checks on the contracted prices through the Purchasing Department at 6-month intervals show the following historical costs. Make January 2007 the base period, and determine the cost indexes using a basis of 100.

| Year | 2005 | | 2006 | | 2007 | | 2008 |
Month	Jan	Jul	Jan	Jul	Jan	Jul	Jan
Chip, $/unit	57.00	56.90	56.90	56.70	56.60	56.40	56.25
Platinum alloy, $/ounce	446.00	450.00	455.00	575.00	610.00	625.00	635.00

Solution

For each item, the index (I_t/I_0) is calculated with January 2007 cost used for the I_0 value. As indicated by the cost indexes shown, the index for the chip is stable, while the platinum alloy index is steadily rising.

| Year | 2005 | | 2006 | | 2007 | | 2008 |
Month	Jan	Jul	Jan	Jul	Jan	Jul	Jan
Chip cost index	100.71	100.53	100.53	100.17	100.00	99.65	99.38
Platinum alloy cost index	73.11	73.77	74.59	94.26	100.00	102.46	104.10

Cost indexes are sensitive over time to technological change. The predefined quantity and quality used to obtain cost values may be difficult to retain through time, so "index creep" may occur. Updating of the index and its definition is necessary when identifiable changes occur.

11.4 COST-ESTIMATING RELATIONSHIPS: COST-CAPACITY EQUATIONS

Design variables (speed, weight, thrust, physical size, etc.) for plants, equipment, and construction are determined in the early design stages. Cost-estimating relationships (CER) use these design variables to predict costs. Thus, a CER is generically different from the cost index method, because the index is based on the cost history of a defined quantity and quality of a variable.

One of the most widely used CER models is a *cost-capacity equation*. As the name implies, an equation relates the cost of a component, system, or plant to its capacity. This is also known as the *power law and sizing model*. Since many cost-capacity equations plot as a straight line on log-log paper, a common form is

$$C_2 = C_1 \left(\frac{Q_2}{Q_1}\right)^x \qquad [11.3]$$

where C_1 = cost at capacity Q_1
   C_2 = cost at capacity Q_2
    x = correlating exponent

The value of the exponent for various components, systems, or entire plants can be obtained from a number of sources, including *Plant Design and Economics for Chemical Engineers, Preliminary Plant Design in Chemical Engineering, Chemical Engineers' Handbook,* technical journals, the U.S. Environmental Protection Agency, professional and trade organizations, consulting firms, handbooks, and equipment companies. Table 11.4 is a partial listing of typical values of the exponent for various units. When an exponent value for a particular unit is not known, it is common practice to use the average value of 0.6. In fact, in the chemical processing industry, Equation [11.3] is referred to as the six-tenth model. Commonly, $0 < x \le 1$. For values $x < 1$, the economies of scale are taken advantage of; if $x = 1$, a linear relationship is present. When $x > 1$, there are diseconomies of scale in that a larger size is expected to be more costly than that of a purely linear relation.

It is especially powerful to combine the time adjustment of the cost index (I_t/I_0) from Equation [11.2] with a cost-capacity equation to estimate costs that change over time. If the index is embedded into the cost-capacity computation in

TABLE 11.4   Sample Exponent Values for Cost-Capacity Equations

Component/System/Plant	Size Range	Exponent
Activated sludge plant	1–100 MGD	0.84
Aerobic digester	0.2–40 MGD	0.14
Blower	1000–7000 ft/min	0.46
Centrifuge	40–60 in	0.71
Chlorine plant	3000–350,000 tons/year	0.44
Clarifier	0.1–100 MGD	0.98
Compressor, reciprocating (air service)	5–300 hp	0.90
Compressor	200–2100 hp	0.32
Cyclone separator	20–8000 ft^3/min	0.64
Dryer	15–400 ft^2	0.71
Filter, sand	0.5–200 MGD	0.82
Heat exchanger	500–3000 ft^2	0.55
Hydrogen plant	500–20,000 scfd	0.56
Laboratory	0.05–50 MGD	1.02
Lagoon, aerated	0.05–20 MGD	1.13
Pump, centrifugal	10–200 hp	0.69
Reactor	50–4000 gal	0.74
Sludge drying beds	0.04–5 MGD	1.35
Stabilization pond	0.01–0.2 MGD	0.14
Tank, stainless	100–2000 gal	0.67

NOTE: MGD = million gallons per day; hp = horsepower; scfd = standard cubic feet per day.

Equation [11.3], the cost at time t and capacity level 2 may be written as the product of two independent terms.

$$C_2 = C_1\left(\frac{Q_2}{Q_1}\right)^x\left(\frac{I_t}{I_0}\right)$$

[11.4]

EXAMPLE 11.4

The total design and construction cost for a digester to handle a flow rate of 0.5 million gallons per day (MGD) was $1.7 million in 2000. Estimate the cost today for a flow rate of 2.0 MGD. The exponent from Table 11.3 for the MGD range of 0.2 to 40 is 0.14. The cost index in 2000 of 131 has been updated to 225 for this year.

Solution

Equation [11.3] estimates the cost of the larger system in 2000, but it must be updated by the cost index to today's dollars. Equation [11.4] performs both operations at once. The estimated cost in current-value dollars is

$$C_2 = 1{,}700{,}000\left(\frac{2.0}{0.5}\right)^{0.14}\left(\frac{225}{131}\right)$$

$$= 1{,}700{,}000(1.214)(1.718) = \$3{,}546{,}178$$

11.5 COST-ESTIMATING RELATIONSHIPS: FACTOR METHOD

Another widely used CER model for preliminary cost estimates of process plants is the *factor method*. While the methods discussed previously can be used to estimate the costs of major items of equipment, processes, and the total plant costs, the factor method was developed specifically for total plant costs. The method is based on the premise that fairly reliable total plant costs can be obtained by multiplying the cost of the major equipment by certain factors. These factors are commonly referred to as Lang factors after Hans J. Lang, who first proposed the method.

In its simplest form, the factor method relation has the same form as the unit method.

$$C_T = h \times C_E$$

[11.5]

where C_T = total plant cost
  h = overall cost factor or sum of individual cost factors
  C_E = total cost of major equipment

The h may be one overall cost factor or, more realistically, the sum of individual cost components such as construction, maintenance, direct labor, materials, and indirect cost elements. This follows the cost estimation approaches presented in Figure 11.1.

In his original work, Lang showed that direct cost factors and indirect cost factors can be combined into one overall factor for some types of plants as follows: solid process plants, 3.10; solid-fluid process plants, 3.63; and fluid process plants, 4.74. These factors reveal that the total installed-plant cost is many times the first cost of the major equipment. Many consulting engineering firms use $h = 4$ in Equation [11.5] for overall cost estimating because equipment costs are easy to obtain and are highly accurate.

EXAMPLE 11.5 An engineer with Phillips Petroleum has learned that an expansion of the solid-fluid process plant is expected to have a delivered equipment cost of $1.55 million. If a specific overall cost factor for this type of plant is not known now, make a preliminary estimate of the plant's total cost.

Solution

At this early stage, the total plant cost is estimated by Equation [11.5] using $h = 4$.

$$C_T = 4.0(1,550,000) = \$6.2 \text{ million}$$

Subsequent refinements of the factor method have led to the development of separate factors for direct and indirect cost components. Direct costs (as discussed in Section 11.1) are specifically identifiable with a product, function, or process. Indirect costs are not directly attributable to a single function but are shared by several. Examples of indirect costs are administration, computer services, quality, safety, taxes, security, and a variety of support functions. For indirect costs, some of the factors apply to *equipment costs only,* while others apply to the total direct cost. In the former case, the simplest procedure is to *add the direct and indirect cost factors* before multiplying by the equipment cost C_E. In Equation [11.5], the overall cost factor h is now

$$h = 1 + \sum_{i=1}^{n} f_i \qquad [11.6]$$

where f_i = factor for each cost component, including indirect cost
         i = 1 to n components

If the *indirect cost factor* is applied to the *total direct cost,* only the direct cost factors are added to obtain h. Therefore, Equation [11.5] is rewritten.

$$C_T = \left[C_E \left(1 + \sum_{i=1}^{n} f_i \right) \right] (1 + f_I) \qquad [11.7]$$

where f_I = indirect cost factor
         f_i = factors for direct cost components only

EXAMPLE 11.6

A small activated sludge wastewater treatment plant is expected to have a delivered-equipment first cost of $273,000. The cost factor for the installation of piping, concrete, steel, insulation, supports, etc., is 0.49. The construction factor is 0.53, and the indirect cost factor is 0.21. Determine the total plant cost if the indirect cost is applied to (*a*) the cost of the delivered equipment and (*b*) the total direct cost.

Solution

a. Total equipment cost is $273,000. Since both the direct and indirect cost factors are applied to only the equipment cost, the overall cost factor from Equation [11.6] is

$$h = 1 + 0.49 + 0.53 + 0.21 = 2.23$$

The total plant cost estimate is

$$C_T = 2.23(273,000) = \$608,790$$

b. Now the total direct cost is calculated first, and Equation [11.7] is used to estimate the total plant cost.

$$h = 1 + \sum_{i=1}^{n} f_i = 1 + 0.49 + 0.53 = 2.02$$

$$C_T = [273,000(2.02)](1.21) = \$667,267$$

Comment: Note the decrease in estimated plant cost when the indirect cost is applied to the equipment cost only in part (*a*). This illustrates the importance of determining exactly what the factors apply to before they are used.

11.6 COST-ESTIMATING RELATIONSHIPS: LEARNING CURVE

Observation of completion times of repetitive operations in the aerospace industry several decades ago led to the conclusion that efficiency and improved performance do occur with more units. This fact is used to estimate the time to completion and cost of future units. The resulting CER is the *learning curve,* primarily used to predict the time to complete a specific repeated unit. The model incorporates a constant decrease in completion time every time the production is *doubled.* As an illustration, assume that Shell's Division of Offshore Rig Operations requires 32 PCs of the same, ruggedized configuration. The time to assemble and test the first unit was 60 minutes. If learning generates a reduction of 10% in completion time, 90% is the learning rate. Therefore, each time 2X units are finished, the estimated completion times are 90% of the previous doubling. Here, the second unit takes 60(0.90) = 54 minutes, the fourth takes 48.6 minutes, and so on.

The constant reduction of estimated time (and cost) for doubled production is expressed as an *exponential* model.

$$T_N = T_1 N^s \qquad [11.8]$$

where: N = unit number
T_1 = time or cost for the first unit
T_N = time or cost for the Nth unit
s = learning curve slope parameter, decimal

This equation estimates the time to complete a specific unit (1st or 2nd, . . . , or Nth), not total time or average time per unit. The slope s is a negative number, since it is defined as

$$s = \frac{\log \text{ (learning rate)}}{\log 2} \qquad [11.9]$$

Equation [11.8] plots as an exponentially decreasing curve on xy coordinates. Take the logarithm and the plot is a straight line on log-log paper.

$$\log T_N = \log T_1 + s \log N \qquad [11.10]$$

Figure 11.3 shows the per unit completion time estimates (numerical in the table and graphical on arithmetic and log-log scales) for the 32 PCs described previously. As production doubles, the time to produce each unit is 10% less than that of the previous time. From Equation [11.9], this learning rate of 90% results in a slope s of -0.152.

$$s = \frac{\log 0.90}{\log 2} = \frac{-0.04576}{0.30103} = -0.152$$

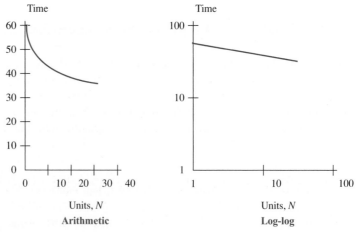

Unit, N	Time, minutes
1	60.0
2	54.0
4	48.6
8	43.7
16	39.4
32	35.4

FIGURE 11.3   Plots of time estimates for a learning rate of 90% for 32 units with T_1 = 60 minutes.

Now cost estimation is possible. To estimate the cost C_N for the Nth unit, or the total cost C_T for all N units, multiply the cost per unit c by the appropriate time estimate.

$$C_N = \text{(cost per unit) (time for Nth unit)} = c \times T_N \qquad \textbf{[11.11]}$$

$$C_T = \text{(cost per unit)(total time for all N units)} = c(T_1 + T_2 + \cdots + T_N) \;\textbf{[11.12]}$$

The learning curve offers good time estimates for larger-scale projects where the batch size is relatively small, say, 5 to 100 or 200. When the production is a routine process and repeated a very large number of times, continued learning is clearly not present, and the model may underestimate costs.

The learning curve has several other names, depending upon the application area, but the relationship is the same as Equation [11.8]. Some of these names include *manufacturing progress function* (commonly used in production and manufacturing sectors), *experience curve* (often used in process industries), and *improvement curve.*

EXAMPLE 11.7

FEMA (Federal Emergency Management Agency) has ordered 25 specialized test units capable of field checking 15 separate elements in potable water in emergency situations. Thompson Water Works, Inc., the contractor, took 200 hours to build the first unit. If direct and indirect labor costs average $50 per hour, and an 80% learning rate is assumed, estimate (*a*) the time needed to complete units 5 and 25, and (*b*) the total labor cost for the 25 units.

Solution

a. The 80% learning rate and Equation [11.9] are used to find the learning curve slope $s = \log 0.80/\log 2 = -0.322$. Equation [11.8] estimates the time for a specific unit.

$$T_5 = 200(5^{-0.322}) = 119.1 \text{ hours}$$
$$T_{25} = 200(25^{-0.322}) = 70.9 \text{ hours}$$

b. In order to apply Equation [11.12] for the total cost estimate, the individual time estimates for all 25 units must be made then multiplied by the $50 per hour labor cost. If T_1 through T_{25} are added, the total is 2461.4 hours. The total labor cost is estimated at 2461.4(50) = $123,070.

11.7 INDIRECT COST (OVERHEAD) ESTIMATION AND ALLOCATION

It is vitally important to estimate both direct *and* indirect costs for a project, process, system, or product prior to its economic evaluation. Previous sections have dealt with direct costs; this section discusses indirect costs, those that are not directly connected with a specific project, machine, or product. *Indirect costs* are also commonly referred to as *overhead.* These costs include support and infrastructure expenses such as maintenance, human resources, quality, safety, supervision and administration, planning and scheduling, taxes, legal, payroll, accounting, utilities, and a host of other costs. Indirect costs are too difficult to track in detail; some allocation method must be developed and applied. At the end of each fiscal period (quarter or year) or when a project is complete, the cost accounting system uses this method to charge overhead to the appropriate cost center. However, when preliminary and detailed cost estimates are developed, a method to allocate expected overhead is necessary. Some common methods are:

- Make indirect costs an implicit element in the estimate. The factor method, discussed earlier, does this by including an indirect cost factor. Process industries, especially chemical and petroleum related, do this commonly using the Lang factors (see Section 11.5).
- Apply established indirect cost rates that are calculated using some reasonable basis. This traditional method is presented next.
- Utilize Activity-Based Costing (ABC), which is especially applicable to high indirect cost industries. ABC is summarized later in this section.

Using the traditional method, indirect costs are estimated using an *indirect cost rate* that is allocated on some *basis.* Table 11.5 includes possible bases and sample cost categories that may be allocated using each basis. The rate is calculated using the relation

$$\text{Overhead or indirect cost rate} = \frac{\text{estimated total indirect costs}}{\text{estimated basis level}} \quad [11.13]$$

The next example illustrates how standard overhead rates may be established.

TABLE 11.5   Indirect Cost Allocation Bases and Sample Costs

Allocation Basis	Indirect Cost Category
Direct labor hours	Machine shop, human resources, supervision
Direct labor cost	Machine shop, supervision, accounting
Machine hours	Utilities, IT network servers
Cost of materials	Purchasing, receiving, inspection
Space occupied	Taxes, utilities, building maintenance
Amount consumed	Utilities, food services
Number of items	Purchasing, receiving, inspection
Number of accesses	Software
Number of inspections	Quality assurance

EXAMPLE 11.8

A make/buy decision is necessary for several steel components in the new, heavy duty transmission gearbox on off-road recreation vehicles. For the make alternative, accounting provided the engineer, Geraldine, the following standard rates for her use in estimating indirect costs:

Machine	Indirect Cost Rate
Vertical milling cutter	$ 1.00 per labor dollar
Turret lathe	$25.00 per labor hour
Numerical control drill press	$ 0.20 per material dollar

Geraldine had intended to apply a single (blanket) rate per direct labor hour until accounting showed her the wide-ranging rates above and the bases and usage data in Table 11.6. Explain how the indirect cost rates were derived.

TABLE 11.6 **Indirect Cost Bases and Activity, Example 11.8**

Machine	Indirect Cost Basis	Expected Annual Activity	Budgeted Indirect to Be Charged for the Year
Vertical milling cutter	Direct labor costs	$100,000	$100,000
Turret lathe	Direct labor hours	2000 hours	$ 50,000
Numerical control drill press	Materials cost	$250,000	$ 50,000

Solution

The rates would be determined at the beginning of the year using the total indirect costs budgeted for each machine. Equation [11.13] is used to determine the rates provided to Geraldine.

Vertical milling cutter: Rate = 100,000/100,000 = $1.00 per
                                        direct labor dollar

Turret lathe: Rate = 50,000/2000 = $25 per direct
                                        labor hour

Numerical control drill press: Rate = 50,000/250,000 = $0.20 per
                                        material dollar

When the same basis applies to allocate overhead to multiple cost centers, a *blanket rate* may be used. For example, if materials cost is a reasonable basis for four separate projects,

$$\text{Blanket indirect cost rate} = \frac{\text{total expected indirect costs}}{\text{total estimated materials cost}}$$

Blanket rates are easy to calculate and apply, but they do not account for the different accomplishments and functions of assets or people in the same department.

Consider a chemical processing line in which automated equipment is mixed with nonautomated (lower value added) methods. A blanket rate will over-accumulate indirect costs for the lower-value contribution. The correct approach is to apply different rates for the different machines and methods. When the rate is sensitive to the value added, that is, not an average or blanket rate, it is called the *productive hour rate method*. Realization that more than one basis should be normally used to allocate overhead has led to the ABC method, discussed below.

The sum of direct and indirect costs is called *cost of goods sold* or *factory cost*. These are presented in financial statements as shown in Appendix B. When indirect costs are estimated, they are included in the economic evaluation along with direct costs. The next example illustrates this for a make/buy evaluation.

EXAMPLE 11.9   For several years the Cuisinart Corporation has purchased the carafe assembly of its major coffee-maker line at an annual cost of $1.5 million. The suggestion to make the component inhouse has been made. For the three departments involved, the annual indirect cost rates and hours, plus estimated direct material and direct labor costs are found in Table 11.7. The allocated hours column is the time necessary to produce the carafes for a year.

Equipment must be purchased with the following estimates: first cost of $2 million, salvage value of $50,000, and life of 10 years. Perform a make/buy analysis at market MARR = 15% per year.

TABLE 11.7   Production Cost Estimates for Example 11.9

Department	Indirect Costs Basis, Hours	Rate Per Hour	Allocated Hours	Direct Material Cost	Direct Labor Cost
A	Labor	$10	25,000	$200,000	$200,000
B	Machine	5	25,000	50,000	200,000
C	Labor	15	10,000	50,000	100,000
				$300,000	$500,000

Solution

For making the components inhouse, the AOC is comprised of direct labor, direct material, and indirect costs. Use the data of Table 11.7 to distribute indirect costs.

$$\text{Department A:} \quad 25,000(10) = \$250,000$$
$$\text{Department B:} \quad 25,000(5) \ = \ 125,000$$
$$\text{Department C:} \quad 10,000(15) = \ \underline{150,000}$$
$$\$525,000$$

AOC = direct labor + direct material + indirect costs
= 500,000 + 300,000 + 525,000 = $1,325,000

The Make alternative annual worth is the total of capital recovery and AOC.

$$AW_{make} = -P(A/P,i,n) + S(A/F,i,n) - AOC$$
$$= -2,000,000(A/P,15\%,10) + 50,000(A/F,15\%,10)$$
$$- 1,325,000$$
$$= \$-1,721,037$$

Currently, the carafes are purchased at $AW_{buy} = \$-1,500,000$. It is cheaper to purchase outside, because the AW of costs is less.

As automation, software, and manufacturing technologies have advanced, the number of direct labor hours necessary to manufacture a product has decreased substantially. Where once as much as 35% to 45% of the final cost was labor, now it is commonly 5% to 15%. However, the indirect cost may represent as much as 35% to 45% of the total cost. The use of traditional bases such as direct labor hours to determine indirect cost rates is not accurate enough for automated and technologically advanced environments. As a result, a product or service that by traditional overhead allocation methods may appear to contribute to bottom-line profit may actually be a loser when indirect costs are allocated more realistically. This has led to the development of distribution methods that supplement (or replace) traditional ones, and allocation bases different from those traditionally used.

The best method for high-overhead industries is *Activity-Based Costing (ABC)*. Its approach is to identify *activities* and *cost drivers*. Implementing ABC includes several steps.

1. Identify each activity and its total cost.
2. Identify the cost drivers and their usage volumes.
3. Calculate the overhead rate for each activity.

$$\text{ABC overhead rate} = \frac{\textbf{total cost of activity}}{\textbf{total volume of cost driver}} \qquad [11.14]$$

This formula has the same format as Equation [11.13] for the traditional method.

4. Use the rate to allocate overhead to cost centers (products, departments, etc.)

Activities are usually support departments or functions—purchasing, quality, engineering, IT. Cost drivers are usually expressed in volumes—number of purchase orders, number of construction approvals, number of machine setups, or cost of engineering changes. As an illustration, suppose a company produces two models of industrial lasers (cost centers) and has three primary support departments (activities)—purchasing, quality, and personnel. Purchasing is an overhead activity with a cost driver of number of purchase orders issued. The ABC allocation rate, in $ per purchase order, is used to distribute budgeted indirect costs to the two cost centers.

EXAMPLE 11.10   A multinational aerospace firm with four plants in Europe uses traditional methods to distribute the annual business travel allocation on the basis of workforce size. Last year $500,000 in travel expenses were distributed, according to Equation [11.13], at a rate of 500,000/29,100 = $17.18 per employee.

City	Employees	Allocation
Paris	12,500	$214,777
Florence	8,600	147,766
Hamburg	4,200	72,165
Athens	3,800	65,292
	29,100	$500,000

A switch to ABC allocates the $500,000 in travel expenses on the basis of the number of travel vouchers, categorized by travelers working on each product line. In ABC terminology, travel is the activity and a travel voucher is the cost driver. Table 11.8 details the distribution of 500 vouchers to each product line by plant. Not all products are produced at each plant.

Use the ABC method to allocate travel expenses to each product line and each plant. Compare plant-by-plant allocations based on workforce size (traditional) and number of travel vouchers (ABC).

TABLE 11.8   Travel Vouchers Submitted by Five Product Lines at Four Plants

Plant	Product line					Total
	1	**2**	**3**	**4**	**5**	
Paris	50	25				75
Florence	80		30		30	140
Hamburg	100	25		20		145
Athens					140	140
Totals	230	50	30	20	170	500

Solution

The ABC allocation is to products, not to plants. Plant allocation is determined after allocation to products, based on where a product line is manufactured. Equation [11.14] determines the ABC allocation rate per voucher.

ABC allocation rate = 500,000/500 = $1000 per voucher

With the round-number rate of $1000, allocations to product-plant combinations are the same as the values in Table 11.8 times $1000. For example, total allocation to Product 1 is $230,000 and total allocation to the Paris plant is $75,000. ↓

Products 1 and 5 dominate the ABC allocation. Comparison of by-plant allocations for ABC with the respective traditional method indicates a substantial difference in the amounts for all plants, except Florence.

Plant	ABC Allocation	Traditional Allocation
Paris	$ 75,000	$214,777
Florence	140,000	147,766
Hamburg	145,000	72,165
Athens	140,000	65,292

This comparison supports the hypotheses that product lines, not plants, drive travel expenses and that travel vouchers are a good cost driver for ABC allocation.

The traditional method is better and easier to use for preliminary and detailed cost estimates, whereas ABC provides more detailed information once the project or fiscal period is completed. Some proponents of ABC recommend discarding the traditional method completely. However, from the viewpoints of cost estimation on one hand, and cost tracking and control on the other, the two methods work well together. The traditional method is good for estimation and allocation, while the ABC method traces costs more closely. ABC is more costly to implement and operate, but it assists in understanding the impact of management decisions and in controlling selected indirect costs.

SUMMARY

Cost estimates are not expected to be exact, but they should be accurate enough to support a thorough economic analysis using an engineering economy approach. There are bottom-up and top-down approaches; each treats price and cost estimates differently.

Costs can be estimated using the unit method or updated via a cost index, which is a ratio of costs for the same item at two separate times. The Consumer Price Index (CPI) is an often-quoted example of cost indexing.

Cost estimating may also be accomplished with a variety of models called Cost-Estimating Relationships. Three of them are

Cost-capacity equation. Good for estimating costs from design variables for equipment, materials, and construction.

Factor method. Good for estimating total plant cost.

Learning curve. Estimates time and cost for specific units produced. Good for manufacturing industries.

Traditional indirect cost rates are determined for a machine, department, product line, etc. Bases such as direct labor cost or direct material cost are used. For high automation and information technology environments, the Activity-Based Costing (ABC) method is an excellent alternative.

PROBLEMS

Cost Estimation Approaches and Unit Costs

11.1 Using an automobile as a *cost center* and miles traveled as the *cost driver,* classify the following cost components as either direct or indirect costs: gasoline, garage rental, license plate fee, insurance, tires, inspection fee, oil change.

11.2 In the bottom-up approach, identify the output and input variables.

11.3 In the top-down approach, what is the starting point?

11.4 In the early and conceptual design stages of a project, what are the cost estimates called and approximately how close should they be to the actual cost?

11.5 A preliminary estimate for the cost of a parking garage can be made using the unit cost method. If the cost per parking space is $4500, what is the estimated cost for a 500-space garage?

11.6 Preliminary cost estimates for jails can be made using costs based on either unit area (square feet) or unit volume (cubic feet). If the unit area cost is $185 per square foot and the average height of the ceilings is 10 feet, what is the unit volume cost?

11.7 The unit area and unit volume total project costs for a library are $114 per square foot and $7.55 per cubic foot, respectively. On the basis of these numbers, what is the average height of library rooms?

Cost Indexes

11.8 From historical data, you discover that the national average construction cost of middle schools is $10,500 per student. If the city index for El Paso is 76.9 and for Los Angeles it is 108.5, estimate the total construction cost for a middle school of 800 students in each city.

11.9 A consulting engineering firm is preparing a preliminary cost estimate for a design/construct project of a brackish groundwater desalting plant. The firm, which completed a similar project in 2001 with a construction cost of $30 million, wants to use the *ENR* construction cost index to update the cost. Estimate the cost of construction for a similar-size plant in mid-2006 when the index value was 7700.

11.10 If the editors at *ENR* decide to redo the construction cost index so that the year 2000 has a base value of 100, determine the value for the year (*a*) 1990; (*b*) 2005.

11.11 An engineer who owns a construction company that specializes in large commercial projects noticed that material costs increased at a rate of 1% *per month* over the past 12 months. If a material cost index is created for that year with the value of the index set at 100 for the beginning of the year, what is the value of the index at the end of the year? Express your answer to two decimal places.

11.12 Omega electropneumatic general-purpose pressure transducers convert supply pressure to regulated output pressure in direct proportion to an electrical input signal. If the cost of a certain model was $328 in 2006 when the M&S equipment index was at 1333.4, what was the cost in 2001 when the M&S index value was 1093.9? Assume the cost changed in proportion to the index.

11.13 The *ENR* construction cost index for New York City had a value of 12,381.40 in February 2007. For Pittsburgh and Atlanta, the values were 7341.32 and 4874.06, respectively. If a general contractor in Atlanta won construction jobs totaling $54.3 million, determine their equivalent total value in New York City.

11.14 The *ENR* construction cost index (CCI) for January 2007 had a value of 7879.58 when the base year was 1913 with a value of 100. If the base year is 1967, the CCI for January 2007 is 733.55. What is the CCI for 1967 when the base year is 1913?

11.15 Using 1913 as the base year with a value of 100, the *ENR* construction cost index (CCI) for October 2005 was 7562.50. For October 2006, the CCI value was 7882.53. (*a*) What was the inflation rate for construction for that one-year period? (*b*) The index value for January 2007 was 7879.58. What was the inflation rate over the period October 2006 and January 2007?

11.16 The *ENR* materials cost index (MCI) had a value of 2583.52 in January 2007. In the same month, the cost for cement was $95.90 per ton. If cement increased in price exactly in accordance with the MCI, what was the cost of cement per ton in 1913 when the MCI index value was 100?

11.17 In January 2007, the *ENR* 20-city construction common labor index had a value of 16,520.53 and the wage rate was $31.39 per hour. Assuming that labor rates increased in proportion to the common labor index, what was the common labor wage rate per hour in 1913 when the common labor index had a value of 100?

11.18 The cost of lumber per million board feet (MBF) in January 2007 was $464.49 when the value of the *ENR* materials cost index (MCI) was 2583.52. If the cost of lumber increased in proportion to the MCI, what was the value of the index when the cost of lumber was $400 per MBF?

Cost-Estimating Relationships

11.19 A 0.75 million gallon per day (MGD) induced draft packed tower for air-stripping trihalomethanes from drinking water costs $58,890. Estimate the cost of a 2 MGD tower if the exponent in the cost-capacity equation is 0.58.

11.20 The variable frequency drive (VFD) for a 300 hp motor costs $20,000. How much will the VFD for a 100 hp motor cost if the exponent in the cost-capacity equation is 0.61?

11.21 The cost of a 68 m^2 falling-film evaporator was 1.52 times the cost of the 30 m^2 unit. What exponent value in the cost-capacity equation yielded these results?

11.22 Reinforced concrete pipe (RCP) that is 12 inches in diameter had a cost of $12.54 per linear foot in Dallas, Texas in 2007. The cost for 24-inch RCP was $27.23 per foot. If the cross-sectional area of the pipe is considered the "capacity" in the cost-capacity equation, determine the value of the exponent in the cost-capacity equation that exactly relates the two pipe sizes.

11.23 A 100,000 barrel per day (BPD) fractionation tower cost $1.2 million in 2001 when the *Chemical Engineering* plant cost index value was 394.3. How much would a 300,000 BPD plant cost when the index value is 575.8, provided the exponent in the cost-capacity equation is 0.67?

11.24 A mini-wind tunnel for calibrating vane or hot-wire anemometers cost $3750 in 2002 when the M&S equipment index value was 1104.2. If the index value is now 1520.6, estimate the cost of a tunnel twice as large. The cost-capacity equation exponent is 0.89.

11.25 In 2005, an engineer estimated the cost for laser-based equipment to be $376,900. The engineer used the M&S equipment index for the years 1998 and 2005 and the cost-capacity equation with an exponent value of 0.61. If the original equipment had only one-fourth the capacity of the new equipment, what was the cost of the original equipment in 1998?

11.26 The equipment cost for removing arsenic from a well that delivers 800 gallons per minute is $1.8 million. If the overall cost factor for this type of treatment system is 2.25, what is the total plant cost expected to be?

11.27 A closed-loop filtration system for waterjet cutting industries eliminates the cost of makeup water treatments (water softeners, reverse osmosis, etc.) while maximizing orifice life and machine performance. If the equipment cost is $225,000 and the total plant cost is $1.32 million, what is the overall cost factor for the system?

11.28 A chemical engineer at Western Refining has estimated that the total cost for a diesel fuel desulfurization system will be $2.3 million. If the direct cost factor is 1.35 and the indirect cost factor is 0.41, what is the total equipment cost? Both factors apply to delivered equipment cost.

11.29 A mechanical engineer estimated that the equipment cost for a multitube cyclone system with a capacity of 60,000 cfm would be $400,000. If the direct cost factor is 3.1 and the indirect cost factor is 0.38, what is the estimated total plant cost? The indirect cost factor applies to the total direct cost.

11.30 The Pavonka family is contracted to frame 32 wooden houses of the same design in a new subdivision of 300 homes. The fourth unit took 400 hours and the father knows they cut about 10% off the time for each replication of the same design. Use the learning curve assumption of a constant decrease with each doubling of production to estimate past (units 1 and 2) and future completion times.

11.31 Determine the learning curve slope if the cost per unit is reduced by (*a*) 8% and (*b*) 14% each time processing output is doubled.

11.32 An engineer wants to predict the completion time per unit for KBR Construction to build 175 field-ready office/storage units in their plant. The Red Cross uses the units in international relief efforts. Plot the learning curves on arithmetic and log-log scales for a 90% learning rate if the first-unit time to completion is 90 hours.

11.33 An engineer at KBR wants to predict the cost per unit for KBR Construction to build 175 field-ready office/storage units in their plant. The Red Cross uses the units in international relief efforts. Plot the learning curves on arithmetic and log-log scales for a 95% learning rate applied to costs if the first unit costs $3000.

11.34 1st Class, which customizes the interiors of limousines, has an order for 100 vehicles. Since the design is new, the upholstery work will take extra time to cut, sew, and install. The first and fourth vehicles took 160 and 130 hours, respectively, to complete. A constant decrease of 12% per unit for the first 50 units is expected, with the remaining units taking the same amount of time. Use the learning curve approach to answer the following:

a. How close to the 130 hours is the predictive model?

b. What is the estimated completion time for the 50th and later vehicles?

c. What is the average completion time for the first 10 vehicles?

d. What is the amount of the invoice sent after the first 5 vehicles are customized, if the total material and labor rate is a flat $100 per hour?

Indirect Costs

11.35 The director of public works needs to distribute the indirect cost allocation of $1.2 million to the three branches around the city. She will use the information from last year (shown below) to determine the rates for this year.

Branch	Miles Driven	Direct Labor Hours	Basis	Allocation Last Year
North	350,000	40,000	Miles	$300,000
South	200,000	20,000	Labor	$200,000
Midtown	500,000	64,000	Labor	$450,000

a. Determine this year's indirect cost rates for each branch.

b. Use the rates to distribute the allocation for this year. What percentage of the $1.2 million is actually distributed this year?

	Records for This Year	
Branch	Miles Driven	Direct Labor Hours
North	275,000	38,000
South	247,000	31,000
Midtown	395,000	55,500

11.36 The Mechanical Components Division manager asks you to recommend a make/buy decision on a major automotive subassembly that is currently purchased externally for a total of $3.9 million this year. This cost is expected to continue rising at a rate of $300,000 per year.

Your manager asks that both direct and indirect costs be included when inhouse manufacturing (make alternative) is evaluated. New equipment will cost $3 million, have a salvage of $0.5 million and a life of 6 years. Estimates of materials, labor costs, and other direct costs are $1.5 million per year. Typical indirect rates, bases, and expected usage are shown. Perform the AW evaluation at MARR = 12% per year over a 6-year study period.

Department	Basis	Rate	Expected Usage
M	Direct labor in $	$2.40 per $	$450,000
P	Materials in $	$0.50 per $	$850,000
Q	Number of inspections	$25 per inspection	3,500

11.37 A project in digital imaging is detailed at H-P, but an economic evaluation has yet to be completed. Perform a PW analysis at 10% per year for the estimates shown. The AOC constitutes direct costs only. Indirect costs are concentrated primarily in two departments, for which annual standard rates and usage estimates are shown. The finance director wants the recommendation with indirect costs included, but the engineering director wants indirect costs left out. Does this make any difference in the recommendation?

Capital Equipment	Revenue and Direct Costs
P = $−1.5 million	Revenue, R = $1 million per year
S = nil	Direct AOC = $−500,000 per year
n = 5 years	

Indirect Costs				
Dept	**Basis**	**Year**	**Rate**	**Usage/Year**
A	Labor cost	1 to 5	$1.30/$	$100,000
B	IT time	1	$9.00/min	5,000 min
		2	$10.00/min	5,500 min
		3	$11.00/min	6,000 min
		4	$12.00/min	6,500 min
		5	$13.00/min	7,000 min

11.38 What is one primary reason that the ABC method of indirect cost analysis has gained popularity?

11.39 Factory Direct manufactures and sells manufactured homes. Traditionally, it has distributed indirect costs to its three construction plants based on materials cost. Each plant builds different models and floor plans. Advances in weight and shape of plastic and wood composite components have decreased cost and time to produce a unit. The president plans to use build time per unit as the new basis. However, he initially wants to determine what the allocation would have been this year had build time been the basis prior to incorporation of the new materials. The data shown represents average costs and times. Use this data and the three bases to determine the allocation rates and indirect cost distribution of $900,000 for this year.

Plant	**Texas**	**Oklahoma**	**Kansas**
Direct material cost, $ per unit	20,000	12,700	18,600
Previous build time per unit, person-hours	400	415	580
New build time per unit, person-hours	425	355	480

The following information is used in problems 11.40 through 11.42.

Blue Sky Airways traditionally distributes the indirect cost of lost baggage to its three major hubs using a basis of annual number of flights in and out of each airport. Last year $667,500 was distributed as follows:

Hub Airport	**Flights**	**Rate**	**Allocation**
DFW	110,000	$3/flight	$330,000
YYZ	62,500	$3/flight	187,500
MEX	75,000	$2/flight	150,000

The head of Baggage Management suggests that an allocation on the basis of baggage traffic, not flights, will be better at representing the distribution of lost bag costs. Total number of bags handled during the year are: 2,490,000 at DFW; 1,582,400 at YYZ, and 763,500 at MEX.

11.40 What is the activity and the cost driver for the baggage traffic basis?

11.41 Using the baggage traffic basis, determine the allocation rate using last year's allocation of $667,500 and distribute this amount to the hubs.

11.42 What are the percentage changes in allocation at each hub using the two different bases?

PROBLEMS FOR TEST REVIEW AND FE EXAM PRACTICE

11.43 A detailed cost estimate that takes several weeks to prepare should be accurate to within what percentage of the actual cost?
   a. 1%   **b.** 5%   **c.** 10%   **d.** 15%

11.44 The total mechanical and electrical unit costs for a library are listed as $34 per square foot, while total project cost is estimated at $114 per square foot. Based on these values, the percentage of total costs represented by mechanical and electrical costs is closest to
   a. 21%.   **b.** 29%.   **c.** 34%.   **d.** 38%.

11.45 A cost index is a
   a. guide to the types of costs that should be included in an economic analysis.

b. list of the costs that are included in a completed economic analysis report.

c. ratio of the cost today to the cost in the past.

d. ratio of the investment costs to the M&O costs.

11.46 Corrugated steel pipe with a diameter of 60 inches cost $49.20 per foot in Philadelphia in January 2007 when the *ENR* materials cost index (MCI) had a value of 2583.52. If the local MCI value is 2150, the estimated cost per foot of the 60-inch pipe is closest to

    a. $40.94. **b.** $45.94. **c.** $49.36. **d.** $59.12.

11.47 The *ENR* 20-city building cost skilled labor index had a value of 7458.80 in January 2007 when the skilled labor wage rate was $41.40 per hour. When the skilled labor wage rate was $30.65 per hour, the skilled labor index value was closest to

    a. 10,074.86.   **c.** 5522.03.

    b. 7436.12.   **d.** 4027.83.

11.48 The cost for a 200-horsepower (hp) pump, with controller, is $22,000. The cost estimate of a similar pump of 500 hp capacity, provided the exponent in the cost-capacity equation is 0.64, is closest to

    a. $12,240.   **c.** $33,780.

    b. $28,650.   **d.** $39,550.

11.49 If the cost of a certain high-speed assembly line robot is $80,000 and the cost for one with twice the capacity is $120,000, the value of the exponent in the cost-capacity equation is closest to

    a. 0.51. **b.** 0.58. **c.** 0.62. **d.** 0.69.

11.50 The equipment for applying specialty coatings that provide a high angle of skid for the paperboard and corrugated box industries has a delivered cost of $390,000. If the *overall* cost factor for the complete system is 2.96, the total plant cost is approximately

    a. $954,400.   **c.** $1,154,400.

    b. $1,054,400.   **d.** $1,544,400.

11.51 The delivered equipment cost for setting up a production and assembly line for high-sensitivity, gas-damped accelerometers is $650,000. If the direct cost and indirect cost factors are 1.82 and 0.31, respectively, and both factors apply to delivered equipment cost, the total plant cost estimate is approximately

    a. $2,034,500.   **c.** $1,384,500.

    b. $1,734,500.   **d.** $1,183,000.

11.52 If the processing cost decreases by a constant 4% every time the output doubles, the slope parameter of the learning curve is closest to

    a. −0.009.   **c.** 0.991.

    b. −0.059.   **d.** −1.699.

11.53 The first golf-cart chassis off a new fabrication line took 24.5 seconds to paint. If during the first 100 units, a learning rate of 95% is assumed, the paint time for unit 10 is closest to

    a. 20.7 seconds.

    b. 0.46 minutes.

    c. 2.75 seconds.

    d. 15.5 seconds.

11.54 The learning curve plots as a straight line on

    a. arithmetic scales.

    b. logarithmic scale for the *x*-axis.

    c. log-log paper.

    d. circle log paper.

11.55 A police department wants to allocate the indirect cost of speed monitoring to the three toll roads around the city. An allocation basis that may *not* be reasonable is

    a. miles of toll road monitored.

    b. average number of cars patrolling per hour.

    c. amount of car traffic per section of toll road.

    d. cost to operate a patrol car.

11.56 The IT department allocates indirect costs to user departments on the basis of CPU time at the rate of $2000 per second. For the first quarter, the two heaviest use departments logged 900 and 1300 seconds, respectively. If the IT indirect budget for the year is $8.0 million, the percentage of this year's allocation consumed by these departments is closest to

    a. 32%.

    b. 22.5%.

    c. 55%.

    d. Not enough information to determine.

11.57 If engineering change order is the activity for an application of the ABC method of overhead allocation, the most reasonable cost driver(s) may be

    1 – number of changes processed

    2 – size of the work force

    3 – management cost to process the change order

    a. 1. **b.** 2. **c.** 3. **d.** 1 and 3.

Chapter 12

Depreciation Methods

Simon Fell/Getty Images

The capital investments of a corporation in tangible assets—equipment, computers, vehicles, buildings, and machinery—are commonly recovered on the books of the corporation through *depreciation*. Although the depreciation amount is not an actual cash flow, the process of depreciating an asset, also referred to as *capital recovery* or *amortizing,* accounts for the decrease in an asset's value because of age, wear, and obsolescence.

Why is depreciation important to engineering economy? Depreciation is a *tax-allowed deduction* included in tax calculations in virtually all industrialized countries. Depreciation lowers income taxes via the relation

$$\text{Taxes} = (\text{income} - \text{expenses} - \text{depreciation})(\text{tax rate})$$

Income taxes are discussed further in Chapter 13.

This chapter concludes with an introduction to two methods of *depletion,* which are used to recover capital investments in deposits of natural resources such as minerals, ores, and timber.

> *Important note:* **To consider depreciation and after-tax analysis early in a course, cover this chapter and the next one (After-Tax Economic Analysis) after Chapter 5 (AW), Chapter 7 (B/C), or Chapter 9 (Replacement Analysis).**

Objectives

Purpose: Use a specific method to reduce the value of capital invested in an asset or natural resource.

1. Understand and use the basic terminology of depreciation. — **Depreciation terms**

2. Apply the straight line method. — **Straight line**

3. Apply the declining balance method. — **Declining balance**

4. Apply the Modified Accelerated Cost Recovery System (MACRS). — **MACRS**

5. Switch from one classical method to another; explain how MACRS uses switching. — **Switching**

6. Utilize percentage depletion and cost depletion methods for natural resource investments. — **Depletion**

7. Use various Excel functions to determine depreciation and book value schedules. — **Spreadsheets**

12.1 DEPRECIATION TERMINOLOGY

Primary terms used in depreciation are defined here.

Depreciation is the reduction in value of an asset. The method used to depreciate an asset is a way to account for the decreasing value of the asset to the owner *and* to represent the diminishing value (amount) of the capital funds invested in it. The annual depreciation amount D_t does not represent an actual cash flow, nor does it necessarily reflect the actual usage pattern.

Book depreciation and **tax depreciation** are terms used to describe the purpose for reducing asset value. Depreciation may be performed for two reasons:

1. Use by a corporation or business for internal financial accounting. This is book depreciation.
2. Use in tax calculations per government regulations. This is tax depreciation.

The methods applied for these two purposes may or may not utilize the same formulas, as is discussed later. *Book depreciation* uses a formula to indicate the reduced investment in an asset throughout its expected useful life. The amount of *tax depreciation* is important in an after-tax engineering economy study because the annual tax depreciation is usually tax deductible; that is, it is subtracted from income when calculating the amount of income taxes.

Tax depreciation may be calculated and referred to differently in countries outside the United States. For example, in Canada the equivalent is CCA (capital cost allowance), which is calculated based on the undepreciated value of all corporate properties that form a particular class of assets, whereas in the United States, depreciation may be determined for each asset separately.

First cost or **basis** is the delivered and installed cost of the asset including purchase price, installation fees, and any other depreciable direct costs.

Book value represents the remaining, undepreciated investment after the total amount of depreciation charges to date have been removed. The book value is determined at the end of each year, which is consistent with the end-of-year convention.

Recovery period is the depreciable life n in years. Often there are different n values for book and tax depreciation. Both of these values may be different from the asset's estimated productive life.

Market value is the estimated amount realizable if the asset were sold on the open market. Because of the structure of depreciation laws, the book value and market value may be substantially different. For example, a commercial building tends to increase in market value, but the book value will decrease with time. However, IT equipment usually has a market value much lower than its book value due to rapidly changing technology.

Salvage value is the estimated trade-in or market value at the end of the asset's depreciable life. The salvage value S, expressed as a dollar amount or a percentage of first cost, may be positive, zero, or negative (due to carry-away costs).

Depreciation rate or **recovery rate** d_t is the fraction of the first cost removed by depreciation each year.

Personal property, one of the two types of property for which depreciation is allowed, is the income-producing, tangible possessions of a corporation. Examples are vehicles, manufacturing equipment, computer equipment, chemical processing equipment, and construction assets.

Real property includes real estate and all improvements—office buildings, factories, warehouses, apartments, and other structures. *Land itself is considered real property, but it is not depreciable.*

Half-year convention assumes that assets are placed in service or disposed of in midyear, regardless of when these events actually occur. This convention is utilized in most U.S.-approved tax depreciation methods.

As mentioned before, there are several methods for depreciating assets. The straight line (SL) model is used, historically and internationally. Accelerated models, such as the declining balance (DB) model, decrease the book value to zero (or to the salvage value) more rapidly than the straight line method, as shown by the general book value curves in Figure 12.1. Accelerated methods defer some of the income tax burden to later in the asset's life; they do not reduce the total tax burden. In the 1980s the U.S. government standardized accelerated methods for *federal tax depreciation* purposes. In 1981, all classical methods (straight line, declining

FIGURE 12.1
General shape of book value curves for different depreciation models.

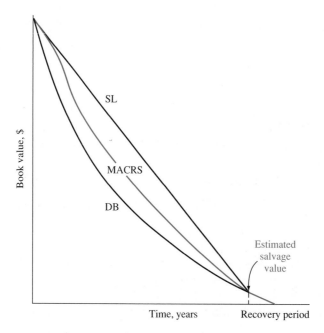

balance, and sum-of-year digits depreciation) were disallowed as acceptable procedures for income tax purposes and replaced by the Accelerated Cost Recovery System (ACRS). In a second round of standardization, MACRS (Modified ACRS) was
made the required tax depreciation method in 1986. To this date, the following is
the law in the United States: *tax depreciation must be calculated using MACRS;*
book depreciation may be calculated using any classical method or MACRS. This
chapter discusses the straight line, declining balance, and MACRS methods.

Tax law revisions occur often, and depreciation rules are changed from time to
time, but the basic principles and relations are always applicable. The U.S. Department of the Treasury, Internal Revenue Service (IRS) website at www.irs.gov has all
pertinent publications. Publication 946, *How to Depreciate Property,* is especially
applicable.

12.2 STRAIGHT LINE (SL) DEPRECIATION

Straight line is considered the standard against which any depreciation method is
compared. It derives its name from the fact that the book value decreases linearly
with time. For *book depreciation* purposes, it offers an excellent representation of
book value for any asset that is used regularly over a number of years.

The annual SL depreciation is determined by multiplying the first cost minus
the salvage value by the depreciation rate.

$$D_t = (B - S)d$$
$$= \frac{B - S}{n} \qquad [12.1]$$

where D_t = depreciation charge for year t ($t = 1, 2, \ldots, n$)
  B = first cost
  S = estimated salvage value
  n = recovery period
  d = depreciation rate = $1/n$

Since the asset is depreciated by the same amount each year, the book value after
t years of service, denoted by BV_t, is the first cost B minus the annual depreciation times t.

$$BV_t = B - tD_t \qquad [12.2]$$

The depreciation rate for a specific year is d_t. However, the SL model has the same
rate for all years.

$$d = d_t = \frac{1}{n} \qquad [12.3]$$

The format for the Excel function to display annual depreciation is

$$SLN(B,S,n)$$

EXAMPLE 12.1   If an asset has a first cost of $50,000 with a $10,000 estimated salvage value after 5 years, calculate the annual SL depreciation and plot the yearly book value.

Solution

The depreciation each year for 5 years is

$$D_t = \frac{B - S}{n} = \frac{50,000 - 10,000}{5} = \$8000$$

The book values, computed using Equation [12.2], are plotted in Figure 12.2. For year 5, for example,

$$BV_5 = 50,000 - 5(8000) = \$10,000 = S$$

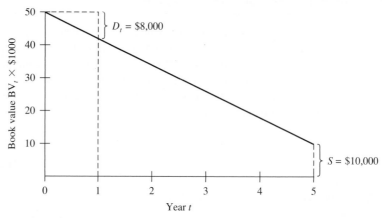

FIGURE 12.2   Book value of an asset depreciated using the straight line method, Example 12.1.

12.3 DECLINING BALANCE DEPRECIATION

Declining balance is also known as the fixed percentage or uniform percentage method. DB depreciation accelerates the write-off of asset value because the annual depreciation is determined by multiplying the *book value at the beginning of a year* by a fixed percentage d, expressed in decimal form. If $d = 0.2$, then 20% of the book value is removed each year. The amount of depreciation decreases each year.

The maximum annual depreciation rate for the DB method is twice the straight line rate.

$$d_{max} = 2/n \qquad\qquad [12.4]$$

This is called *double declining balance (DDB)*. If $n = 5$ years, the DDB rate is 0.4; so 40% of the book value is removed annually. Another commonly used percentage for the DB method is 150% of the SL rate, where $d = 1.5/n$.

The depreciation for year t is the fixed rate d times the book value at the end of the previous year.

$$D_t = (d)\text{BV}_{t-1} \qquad \text{[12.5]}$$

Book value in year t is determined by

$$\text{BV}_t = B(1 - d)^t \qquad \text{[12.6]}$$

The actual depreciation rate for each year t, relative to the first cost, is

$$d_t = d(1 - d)^{t-1} \qquad \text{[12.7]}$$

If BV_{t-1} is not known, the depreciation in year t can be calculated using B and d_t from Equation [12.7].

$$D_t = dB(1 - d)^{t-1} \qquad \text{[12.8]}$$

It is important to understand that the DB book value never goes to zero, because the book value is always decreased by a fixed percentage. The implied salvage value after n years is the BV_n amount.

$$\text{Implied } S = \text{BV}_n = B(1 - d)^n \qquad \text{[12.9]}$$

If a salvage value is initially estimated, this value is *not used* in the DB or DDB method. However, if the implied $S <$ estimated S, it is correct to stop charging further depreciation when the book value is at or below the estimated salvage value.

The Excel functions DDB and DB are used to display depreciation amounts for specific years. The formats are

$$\text{DDB}(B,S,n,t,d)$$

$$\text{DB}(B,S,n,t)$$

The d is a number between 1 and 2. If omitted, it is assumed to be 2 for DDB. The DDB function automatically checks to determine when the book value equals the estimated S value. No further depreciation is charged when this occurs. Consult Section 12.7 and Appendix A for further details on using the functions.

EXAMPLE 12.2

Albertus Natural Stone Quarry purchased a computer-controlled face-cutter saw for $80,000. The unit has an anticipated life of 5 years and a salvage value of $10,000. (*a*) Compare the schedules for annual depreciation and book value using two methods: DB at 150% of the straight line rate and at the DDB rate. (*b*) How is the estimated $10,000 salvage value used?

Solution

a. The DB depreciation rate is $d = 1.5/5 = 0.30$ while the DDB rate is $d_{\max} = 2/5 = 0.40$. Table 12.1 and Figure 12.3 present the comparison of

TABLE 12.1   Annual Depreciation and Book Value, Example 12.2

Year, t	Declining Balance, $d = 0.30$		Double Declining Balance, $d = 0.40$	
	D_t	BV_t	D_t	BV_t
0		$80,000		$80,000
1	$24,000	56,000	$32,000	48,000
2	16,800	39,200	19,200	28,800
3	11,760	27,440	11,520	17,280
4	$8,232	19,208	6,912	10,368
5	5,762	13,446	368	10,000

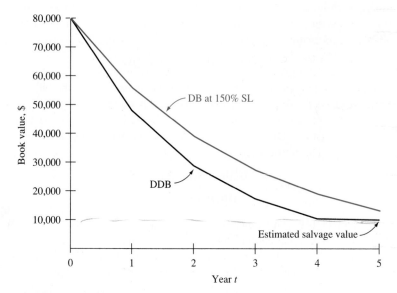

FIGURE 12.3   Plot of book values for two declining balance methods, Example 12.2.

depreciation and book value. Example calculations of depreciation and book value for each method follow.

150% DB for year 2 by Equation [12.5]
with $d = 0.30$ $D_2 = 0.30(56,000) = \$16,800$
                                by Equation [12.6] $BV_2 = 80,000(0.70)^2 = \$39,200$

DDB for year 3 by Equation [12.5]
with $d = 0.40$ $D_3 = 0.40(28,800) = \$11,520$
    by Equation [12.6] $BV_3 = 80,000(0.60)^3 = \$17,280$

The DDB depreciation is considerably larger during the first years, causing the book values to decrease faster, as indicated in Figure 12.3.

b. The $10,000 salvage value is not utilized by the 150% DB method since the book value is not reduced this far. However, the DDB method reduces book value to $10,368 in year 4. Therefore, not all of the calculated depreciation for year 5, $D_5 = 0.40(10,368) = \$4147$, can be removed; only the $368 above *S* can be written off.

12.4 MODIFIED ACCELERATED COST RECOVERY SYSTEM (MACRS)

In the 1980s, the U.S. introduced MACRS as the required *tax depreciation* method for all depreciable assets. MACRS rates take advantage of the accelerated DB and DDB methods. Corporations are still free to apply any of the classical methods for *book depreciation*.

MACRS determines annual depreciation amounts using the relation

$$D_t = d_t B \qquad [12.10]$$

where the depreciation rate is tabulated in Table 12.2. (Tab this page for future reference.) The book value in year *t* is determined by either subtracting the annual depreciation from the previous year's book value, or by subtracting the accumulated depreciation from the first cost.

$$BV_t = BV_{t-1} - D_t \qquad [12.11]$$

$$= B - \sum_{j=1}^{j=t} D_j \qquad [12.12]$$

The first cost is always completely depreciated, since MACRS assumes that $S = 0$, even though there may be an estimated positive salvage.

The MACRS recovery periods are standardized to the values of 3, 5, 7, 10, 15, and 20 years for personal property. Note that all MACRS depreciation rates (Table 12.2) are presented for 1 year longer than the recovery period, and that the extra-year rate is one-half of the previous year's rate. This is because the built-in *half-year convention* imposed by MACRS assumes that all property is placed in service at the midpoint of the tax year. Therefore, only 50% of the first-year DB depreciation applies for tax purposes. This removes some of the accelerated depreciation advantage and requires that leftover depreciation be taken in year $n + 1$.

TABLE 12.2 MACRS Depreciation Rates Applied to the First Cost

[handwritten: N = 4]

	Depreciation Rate (%)					
Year	n = 3	n = 5	n = 7	n = 10	n = 15	n = 20
1	33.33	20.00	14.29	10.00	5.00	3.75
2	44.45	32.00	24.49	18.00	9.50	7.22
3	14.81	19.20	17.49	14.40	8.55	6.68
4	7.41	11.52	12.49	11.52	7.70	6.18
5		11.52	8.93	9.22	6.93	5.71
6		*half →* 5.76	8.92	7.37	6.23	5.29
7			8.93	6.55	5.90	4.89
8			*half →* 4.46	6.55	5.90	4.52
9				6.56	5.91	4.46
10				6.55	5.90	4.46
11			*half life →* 3.28		5.91	4.46
12					5.90	4.46
13					5.91	4.46
14					5.90	4.46
15					5.91	4.46
16					2.95	4.46
17–20						4.46
21						2.23

MACRS depreciation rates incorporate the DDB method ($d = 2/n$) and switch to SL depreciation during the recovery period as an inherent component for *personal property* depreciation. The MACRS rates start at the DDB rate ($n = 3, 5, 7,$ and 10) or the 150% DB rate ($n = 15$ and 20) and switch when the SL method offers faster write-off.

For *real property (buildings)*, MACRS utilizes the SL method for $n = 39$ throughout the recovery period, and forces partial-year recovery in years 1 and 40. The MACRS real property rates in percentage amounts are

$$\text{Year 1} \qquad d_1 = 1.391\%$$
$$\text{Years 2–39} \qquad d_t = 2.564\%$$
$$\text{Year 40} \qquad d_{40} = 1.177\%$$

There is no specific Excel function for the MACRS method; however, the VDB function can be altered slightly to work for MACRS. See Example 12.7 or Appendix A.

Baseline, a nationwide franchise for environmental engineering services, has
acquired new workstations and 3-D modeling software for its 100 affiliate sites
at a cost of $4000 per site. The estimated salvage for each system after 3 years
is expected to be 5% of the first cost. The franchise manager in the home
office in San Francisco wants to compare the depreciation for a 3-year
MACRS model (tax depreciation) with that for a 3-year DDB model (book
depreciation). To help,

EXAMPLE 12.3

a. determine which model offers the larger total depreciation after 2 years, and
b. determine the book value for each method at the end of 3 years.

Solution

The basis is $B = \$400,000$, and the estimated $S = 0.05(400,000) = \$20,000$.
The MACRS rates for $n = 3$ are taken from Table 12.2, and the depreciation
rate for DDB is $d_{max} = 2/3 = 0.6667$. Table 12.3 presents the depreciation and
book values. Depreciation is calculated using Equations [12.10] (MACRS) and
[12.5] (DDB). Year 3 depreciation for DDB would be $\$44,444(0.6667) =$
$\$29,629$, except that this would make $BV_3 < \$20,000$. Only the remaining
amount of $24,444 can be removed.

a. The 2-year accumulated depreciation values are

$$\text{MACRS:} \quad D_1 + D_2 = \$133,320 + 177,800 = \$311,120$$
$$\text{DDB:} \quad D_1 + D_2 = \$266,667 + 88,889 = \$355,556$$

The DDB depreciation is larger. (Remember that for U.S. tax purposes,
Baseline does not have the choice of the DDB model.)
b. After 3 years, the MACRS book value is $29,640, while the DDB model
indicates $BV_3 = \$20,000$. This occurs because MACRS removes the
entire first cost, regardless of the estimated salvage value, but takes
an extra year to do so. This is usually a tax advantage of the MACRS
method.

TABLE 12.3 **Comparing MACRS and DDB Depreciation, Example 12.3**

Year	Rate	MACRS Tax Depreciation	MACRS Book Value	DDB Book Depreciation	DDB Book Value
0			$400,000		$400,000
1	0.3333	$133,320	266,680	$266,667	133,333
2	0.4445	177,800	88,880	88,889	44,444
3	0.1481	59,240	29,640	24,444	20,000
4	0.0741	29,640	0		

TABLE 12.4 Example MACRS Recovery Periods

| | MACRS n Value, Years | |
Asset Description	GDS	ADS Range
Special manufacturing and handling devices, tractors, racehorses	3	3–5
Computers and peripherals, oil and gas drilling equipment, construction assets, autos, trucks, buses, cargo containers, some manufacturing equipment	5	6–9.5
Office furniture; some manufacturing equipment; railroad cars, engines, tracks; agricultural machinery; petroleum and natural gas equipment; *all property not in another class*	7	10–15
Equipment for water transportation, petroleum refining, agriculture product processing, durable-goods manufacturing, ship building	10	15–19
Land improvements, docks, roads, drainage, bridges, landscaping, pipelines, nuclear power production equipment, telephone distribution	15	20–24
Municipal sewers, farm buildings, telephone switching buildings, power production equipment (steam and hydraulic), water utilities	20	25–50
Residential rental property (house, mobile home)	27.5	40
Nonresidential real property attached to the land, but not the land itself	39	40

All depreciable property is classified into *property classes,* which identify their MACRS-allowed recovery periods. Table 12.4, a summary of material from IRS Publication 946, gives examples of assets and the MACRS n values. This table provides two MACRS n values for each property. The first is the *general depreciation system (GDS)* value, which we use in examples and problems. The depreciation rates in Table 12.2 correspond to the n values for the GDS column and provide the fastest write-off allowed. Note that any asset not in a stated class is automatically assigned a 7-year recovery period under GDS.

The far right column of Table 12.4 lists the *alternative depreciation system (ADS)* recovery period range. This alternative method, which uses *SL depreciation over a longer recovery period* than GDS, removes the early-life tax advantages of MACRS. Since it takes longer to depreciate the asset to zero, and since the SL model is required, ADS is usually not considered for an economic analysis.

It is worth noting that, in general, an economic comparison that includes depreciation may be performed more rapidly and usually without altering the final decision by applying the classical straight-line model in lieu of the MACRS method.

12.5 SWITCHING BETWEEN CLASSICAL METHODS; RELATION TO MACRS RATES

The logic of switching between two methods over the depreciable life of an asset is of interest to understand how the MACRS rates are derived. MACRS includes switching from the DB to the SL model as an implicit property of the method. Additionally, applying the logic is necessary if tax law allows switching. Many countries other than the United States allow switching in order to accelerate the depreciation in the first years of life. The goal is to switch in order to *maximize* the present worth of total depreciation using the equation

$$PW_D = \sum_{t=1}^{t=n} D_t(P/F,i,t) \qquad [12.13]$$

By maximizing PW_D, the PW of taxes are also minimized, though the total amount of taxes paid through the life of the asset are not reduced.

Switching from a DB model to the SL method offers the best advantage, especially if the DB model is the DDB. General rules of switching are as follows:

1. Switch when the depreciation for year t by the current method is less than that for a new method. The selected depreciation is the larger amount.
2. Only one switch can take place during the recovery period.
3. Regardless of the method, BV < S is not allowed.
4. If no switch is made in year t, the undepreciated amount, that is, BV_t, is used as the new basis to select the larger depreciation for the switching decision the next year, $t + 1$.

The specific procedure for step 1 (above) to switch from DDB to SL depreciation is as follows:

1. For each year t, compute the two depreciation charges.

   For DDB: $$D_{DDB} = d(BV_{t-1}) \qquad [12.14]$$

   For SL: $$D_{SL} = \frac{BV_{t-1}}{n - t + 1} \qquad [12.15]$$

2. The depreciation for each year is

   $$D_t = \max[D_{DDB}, D_{SL}] \qquad [12.16]$$

The VDB function in Excel determines annual depreciation for the DB-to-SL switch by applying this procedure. Refer to Section 12.7 for an example.

Hemisphere Bank purchased a $100,000 online document imaging system with a depreciation recovery period of 5 years. Determine the depreciation for (*a*) the SL method and (*b*) DDB-to-SL switching. (*c*) Use a rate of $i = 15\%$ per year to determine the PW_D values. (MACRS is not involved in this example.)

EXAMPLE 12.4

TABLE 12.5   Depreciation and Present Worth for DDB-to-SL Switching, Example 12.4

| Year | DDB Model | | SL Model | Larger | (P/F,15%,t) | Present Worth of |
t	D_{DDB}	BV_t	D_{SL}	D_t	Factor	D_t
0	—	$100,000				
1	$40,000	60,000	$20,000	$ 40,000	0.8696	$34,784
2	24,000	36,000	15,000	24,000	0.7561	18,146
3	14,400	21,600	12,000	14,400	0.6575	9,468
4*	8,640	12,960	10,800	10,800	0.5718	6,175
5	5,184	7,776		10,800	0.4972	5,370
Totals	$92,224			$100,000		$73,943

*Indicates year of switch from DDB to SL depreciation.

Solution

a. Equation [12.1] determines the annual SL depreciation, which is the same each year.

$$D_t = \frac{100,000 - 0}{5} = \$20,000$$

b. Use the DDB-to-SL switching procedure.

 1. The DDB depreciation rate is $d = 2/5 = 0.40$. D_t amounts for DDB (Table 12.5) are compared with the D_{SL} values from Equation [12.15]. The D_{SL} values change each year because BV_{t-1} is different. Only in year 1 is $D_{SL} = \$20,000$, the same as computed in part (a). For illustration, compute D_{SL} values for years 2 and 4. For $t = 2$, $BV_1 = \$60,000$ by the DDB method and

$$D_{SL} = \frac{60,000 - 0}{5 - 2 + 1} = \$15,000$$

 For $t = 4$, $BV_3 = \$21,600$ by the DDB method and

$$D_{SL} = \frac{21,600 - 0}{5 - 4 + 1} = \$10,800$$

 2. The column "Larger D_t" indicates a switch in year 4 with $D_4 = \$10,800$. Total depreciation with switching is $100,000 compared to the DDB amount of $92,224.

c. The present worth values for depreciation are calculated using Equation [12.13]. For SL, $D_t = \$20,000$ per year, and the P/A factor replaces P/F.

$$PW_D = 20,000(P/A,15\%,5) = 20,000(3.3522) = \$67,044$$

With switching, $PW_D = \$73,943$ as detailed in Table 12.5. This is an increase of $6899 over the SL PW_D.

12.6 DEPLETION METHODS

Up to this point, we have discussed depreciation for assets that can be replaced. Depletion is another method to write off investment that is applicable only to *natural resources*. When the resources are removed, they cannot be replaced or repurchased in the same manner as can a machine, computer, or structure. Depletion is applicable to mines, wells, quarries, geothermal deposits, forests, and the like. There are two methods of depletion—*percentage* and *cost depletion*. (Details for U.S. taxes are provided in IRS Publication 535, *Business Expenses*.)

Percentage depletion is a special consideration given for natural resources. A constant, stated percentage of the resource's gross income may be depleted each year *provided it does not exceed 50% of the company's taxable income*. The annual depletion amount is

$$\text{Percentage depletion amount} = \text{percentage} \times \text{gross income from property} \qquad [12.17]$$

Using percentage depletion, total depletion charges may exceed first cost with no limitation. The U.S. government does not generally allow percentage depletion to be applied to timber or oil and gas wells (except small independent producers). The annual percentage depletion for some common natural deposits is listed below. These percentages may change from time to time.

Deposit	Percentage
Sulfur, uranium, lead, nickel, zinc, and some other ores and minerals	22%
Gold, silver, copper, iron ore, and some oil shale	15
Oil and natural gas wells (varies)	15–22
Coal, lignite, sodium chloride	10
Gravel, sand, some stones	5
Most other minerals, metallic ores	14

EXAMPLE 12.5

A gold mine was purchased for $10 million. It has an anticipated gross income of $8.0 million per year for years 1 to 5 and $5.0 million per year after year 5. Assume that depletion charges do not exceed 50% of taxable income. Compute the annual depletion amount and determine how long it will take to recover the initial investment.

Solution

A 15% depletion applies for gold.

Years 1 to 5:   0.15(8.0 million) = $1.2 million
Years thereafter: 0.15(5.0 million) = $750,000

A total of $6 million is written off in 5 years, and the remaining $4 million is written off at $750,000 per year. Total recovery is attained in

$$5 + \frac{\$4 \text{ million}}{\$750,000} = 5 + 5.3 = 10.3 \text{ years}$$

Cost depletion, also called factor depletion, is based on the level of activity or usage. It may be applied to most types of natural resources. The annual cost depletion factor p_t is the ratio of the first cost to the estimated number of units recoverable.

$$p_t = \frac{\textbf{first cost}}{\textbf{resource capacity}} \qquad [12.18]$$

The annual depletion charge is p_t times the year's usage. *The total cost depletion cannot exceed the first cost of the resource.* If the capacity of the property is reestimated as higher or lower in the future, a new p_t is determined based upon the new undepleted amount.

EXAMPLE 12.6   Temple-Inland Corporation has negotiated the rights to cut timber on privately held forest acreage for $700,000. An estimated 350 million board feet of lumber are harvestable. Determine the depletion amount for the first 2 years if 15 million and 22 million board feet are removed.

Solution

Use Equation [12.18] for p_t in dollars per million board feet.

$$p_t = \frac{\$700,000}{350} = \$2000 \text{ per million board feet}$$

Multiply p_t by the annual harvest to obtain depletion of $30,000 in year 1 and $44,000 in year 2. Continue using p_t until a total of $700,000 is written off or the remaining timber requires a reestimate of total board feet harvestable.

When allowed by law, the depletion each year may be determined using either the cost method or the percentage method. Usually, the percentage depletion amount is chosen because of the possibility of writing off more than the original investment. However, the law does require that the cost depletion amount be chosen if the percentage depletion is smaller in any year.

12.7 USING SPREADSHEETS FOR DEPRECIATION COMPUTATIONS

Excel spreadsheet functions calculate the annual depreciation for methods discussed and for switching between declining balance and straight line. All methods except MACRS have specific functions, but the VDB (variable declining balance) function may be slightly adjusted to obtain the correct MACRS rates, including the initial year and extra year depreciation adjustments. The next example illustrates the functions in the same order in which the methods are discussed in previous sections. The final part illustrates the use of an Excel chart of book values to compare methods.

EXAMPLE 12.7

BA Aerospace purchased new aircraft engine diagnostics equipment for its maintenance support facility in France. Installed cost was $500,000 with a depreciable life of 5 years and an estimated 1% salvage. Use a spreadsheet to determine the following:

a. Straight line (SL) depreciation and book value schedule.
b. Double declining balance (DDB) depreciation and book value schedule.
c. Declining balance (DB) at 150% of the SL rate depreciation and book value schedule.
d. MACRS depreciation and book value schedule.
e. Depreciation and book value schedule allowing a switch from DDB to SL depreciation.
f. An Excel *xy* scatter chart to plot the book value curves for all five schedules.

Solution

Assume that any method can be used to depreciate this equipment. All five schedules and the chart may be developed using a single worksheet; however, several are developed here to show the progression through different methods. The use of the Excel functions applied here—SLN, DDB, and VDB—is detailed in Appendix A and summarized inside the front cover of this text.

a. Figure 12.4*a*. The SLN function, which determines SL depreciation in column B, has the same entry each year. The entire amount $B - S = \$495,000$ is depreciated. Yearly book value for this and all other methods is determined by subtraction of annual depreciation D_t from BV_{t-1}.
b. Figure 12.4*b*. Column B indicates double declining balance depreciation using the DDB function = DDB(500000,5000,5,t,2) in which the period value t (second from right) changes each year. The last entry is optional; if omitted, 2 for DDB is assumed. $BV_5 = \$38,880$ does not reach the estimated salvage of $5000.

	A	B	C	D	E	F
1	First cost	$500,000		Life	5 years	
2	Salvage	$5,000		Method	SL	
3						
4			Straight line			
5	Year	Depreciation	Book value			
6	0		$ 500,000			
7	1	$ 99,000	401,000		=C6-B7	
8	2	99,000	302,000			
9	3	99,000	203,000			
10	4	99,000	104,000			
11	5	99,000	5,000			
12	Total	$ 495,000			=SLN(500000,5000,5)	
13						
14	**Straight line method**					
15						

FIGURE 12.4   (a) Straight line depreciation and book value schedule, Example 12.7a.

	A	B	C	D	E	F	G
1	First cost	$500,000		Life	5 years		
2	Salvage	$5,000		Method	DDB and DB at 150% SL		
3							
4		Double declining balance (DDB)		Declining balance (DB) at 150% SL			
5	Year	Depreciation	Book value	Depreciation	Book value		
6	0		$ 500,000		$ 500,000		
7	1	$ 200,000	300,000	$ 150,000	350,000		
8	2	120,000	180,000	105,000	245,000		
9	3	72,000	108,000	73,500	171,500		
10	4	43,200	64,800	51,450	120,050	= E9-D10	
11	5	25,920	38,880	36,015	84,035		
12	Total	$ 461,120		$ 415,965			
13							
14		= DDB(500000,5000,5,5,2)			= DDB(500000,5000,5,5,1.5)		
15							
16		**Declining balance method**					
17							

(b) DDB and DB at 150% SL rate depreciation and book value schedules, Example 12.7b and c.

c. Figure 12.4b. Declining balance depreciation at 150% of the SL rate (column D) is best determined using the DDB function with the last field entered as 1.5. (Refer to the DB function description in Appendix A for more details.) The ending book value (column E) is now higher at $84,035 since the depreciation rate is $1.5/5 = 0.3$, compared to the DDB rate of 0.4.

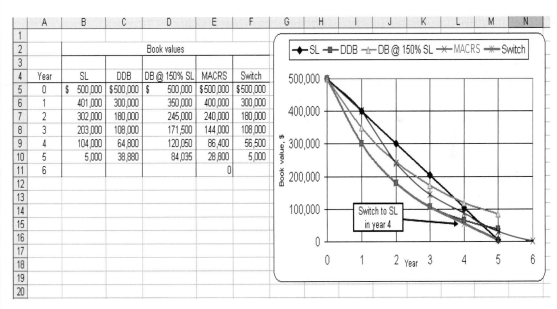

	A	B	C	D	E	F	G	H
1	First cost	$500,000		Life	5 years			
2	Salvage	$5,000		Method	MACRS and DDB-to-SL switch			
3								
4			MACRS		DDB-to-SL switch			
5	Year	Depreciation	Book value	Depreciation	Book value			
6	0		$ 500,000		$ 500,000		=VDB(500000,5000,5,A6,A7)	
7	1	$ 100,000	400,000	$ 200,000	300,000			
8	2	160,000	240,000	120,000	180,000			
9	3	96,000	144,000	72,000	108,000			
10	4	57,600	86,400	51,500	56,500			
11	5	57,600	28,800	51,500	5,000		=VDB(500000,5000,5,A10,A11)	
12	6	28,800	0					
13	Total	$ 500,000		$ 495,000				
14								
15		=VDB(500000,0,5,MAX(0,A7-1.5),MIN(5,A7-0.5),2)						
16								
17		=VDB(500000,0,5,MAX(0,A12-1.5),MIN(5,A12-0.5),2)						
18								
19								
20	MACRS and switching							
21								

(c) MACRS and DDB-to-SL switch depreciation using the VDB function and book value schedules, Example 12.7d and e.

	A	B	C	D	E	F
2				Book values		
4	Year	SL	DDB	DB @ 150% SL	MACRS	Switch
5	0	$ 500,000	$500,000	$ 500,000	$500,000	$500,000
6	1	401,000	300,000	350,000	400,000	300,000
7	2	302,000	180,000	245,000	240,000	180,000
8	3	203,000	108,000	171,500	144,000	108,000
9	4	104,000	64,800	120,050	86,400	56,500
10	5	5,000	38,880	84,035	28,800	5,000
11	6					0

(d) Book value curves using an Excel xy scatter chart, Example 12.7f.

d. Figure 12.4c. Column B displays the MACRS depreciation. The VDB function calculates depreciation when switching from the declining balance to the straight line method. To determine the correct MACRS rates it is necessary to embed the MAX and MIN Excel functions into the VDB function so that only one-half of the DDB depreciation is allowed in year 1 and an additional one-half of the final year depreciation is carried over to year $n + 1$. The VDB format for each year t is

$$= \text{VDB}(\text{cost}, 0, \text{life}, \text{MAX}(0, t - 1.5), \text{MIN}(\text{life}, t - 0.5), \text{factor})$$

The salvage entry is 0 since MACRS assumes $S = 0$. The factor entry is 2 for DDB (MACRS lives of 3, 5, 7, or 10) or 1.5 (lives of 15 or 20). In this case, for $t = 1$, for example, the function is VDB(500000,0,5,MAX(0,1 − 1.5), MIN(5,1 − 0.5),2) as shown in cell B7.

e. Figure 12.4c. Column D shows the larger of DDB or SL depreciation using the VDB function in its normal format. Now the estimated salvage of $5000 is entered. The factor entry (last field) of 2 is optional. When the SL amount of $51,500 is larger than the DDB amount of $43,200 (see cell B10 in Figure 12.4b) for the first time in year 4, the switch occurs.

f. Figure 12.4d. Book values are copied from previous spreadsheets for construction of the *xy* scatter chart. Some conclusions are: DDB and switching are the most accelerated compared to SL; the switch to SL from DDB in year 4 is indicated by the separation of the BV curves, DDB and switching do not come close to the estimated salvage; and, MACRS ignores salvage completely.

SUMMARY

This chapter discussed depreciation and depletion. Depreciation, which itself is not actual cash flow, writes off the investment in assets. For the purpose of income tax reduction, the MACRS method is applied (in the U.S.A. only). For book value estimation, the classical methods of straight line and declining balance are available.

Information about each method follows, and formulas are summarized in Table 12.6.

Modified Accelerated Cost Recovery System (MACRS)

- It is the only approved tax depreciation system in the United States.
- The rates automatically switch from DDB or DB to SL.
- It always depreciates to zero; that is, it assumes $S = 0$.

TABLE 12.6 Summary of Depreciation Method Relations

Model	MACRS	SL	DDB
Fixed depreciation rate d	Varies	$\dfrac{1}{n}$	$\dfrac{2}{n}$
Annual depreciation rate d_t	Table 12.2	$\dfrac{1}{n}$	$d(1 - d)^{t-1}$
Annual depreciation D_t	$d_t B$	$\dfrac{B - S}{n}$	$d(\text{BV}_{t-1})$
Book value BV_t	$\text{BV}_{t-1} - D_t$	$B - tD_t$	$B(1 - d)^t$

- Recovery periods are specified by property classes.
- The actual recovery period is 1 year longer due to the imposed half-year convention.

Straight Line (SL)

- It writes off capital investment linearly over n years.
- The estimated salvage value is always considered.
- This is the classical, nonaccelerated depreciation model.

Declining Balance (DB)

- The model accelerates depreciation compared to straight line.
- The book value is reduced each year by a fixed percentage.

- The most used rate is twice the SL rate, which is called double declining balance (DDB).
- It has an implied salvage that may be different than the estimated salvage.
- Not approved for tax depreciation in the United States, but it is frequently used for book depreciation.

Depletion writes off investment in natural resources. Percentage depletion, which can recover more than the initial investment, reduces the value of the resource by a constant percentage of gross income each year. Alternatively, cost depletion is applied to the amount of resource removed. No more than the initial investment can be recovered with cost depletion.

PROBLEMS

Depreciation Terms and Computations

12.1 Puritan Cement Products placed a new sand sifter into production 3 years ago. It had an installed cost of $100,000, a life of 5 years, and an anticipated salvage of $20,000. Book depreciation charges for the 3 years are $40,000, $24,000, and $14,000, respectively.
  a. Determine the book value after 3 years.
  b. If the sifter's market value today is $20,000, determine the difference between current book value, market value, and the expected salvage were it retained for 2 more years.
  c. Determine the yearly depreciation rate d_t ($t =$ 1, 2, 3) and total percentage of installed cost written off thus far.

12.2 Exactly 3 years ago, Sports Injury Analysis, Inc. paid $495,000 for a used 1998 GE 1.5T MRI for analyzing spinal and joint injuries caused by sports-related accidents. Depreciation is charged using the government-approved rates for tax purposes over a 4-year period. Annual rates are 33.33%, 44.45%, 14.81%, and 7.41%.
  a. Develop the book value schedule for the 4 years.
  b. If the MRI is sold now at the expected salvage of $150,000, compare this amount with the current book value.

12.3 Quantum Electronic Services paid $P = $40,000 for its networked computer system. Both tax and book depreciation accounts are maintained. The annual tax recovery rate is based on the previous year's book value (BV), while the book depreciation rate is based on the original first cost (P). Use the rates listed to plot (a) annual depreciation and (b) book values for each method.

Year of Ownership	1	2	3	4
Tax rate, % of BV	40	40	40	40
Book rate, % of P	25	25	25	25

12.4 There are 3 different life (recovery period) values associated with a depreciable asset. Identify each by name and explain how it is correctly used.

Straight Line Depreciation

12.5 Butler Buildings purchased semiautomated assembly and riveting robotics equipment for constructing its modular warehouse buildings. The first cost was $475,000 and installation costs were $75,000; life is estimated at 10 years with a salvage of 15% of first cost. Use the SL method to determine (*a*) annual recovery rate, (*b*) annual depreciation, (*c*) book value after 5 years, and (*d*) book value after 10 years.

12.6 Columbia Construction purchased new equipment for its project to transform an existing, vacant facility into a milk and butter processing plant. For the equipment, $B = \$350,000$, and $S = \$50,000$. Book depreciation will use the SL method with $n = 5$ years. Use calculator or spreadsheet-based computations (or both, as directed) to plot annual depreciation, accumulated depreciation, and book value on one graph.

12.7 Disna's supervisor asked her to comment on the appropriateness of the recovery period and depreciation schedule for the 3-D CAD system she uses for automotive electronic component design at Howard Designs. Tax and book depreciation data are listed. Plot the two book value curves for Disna using the SL method.

Book depreciation: $B = \$150,000$; $S = \$25,000$;
$n = 9.5$ years
Tax depreciation: $B = \$150,000$; $S = 0$;
$n = 5$ years

12.8 Carl is curious about the original cost of the digital imaging equipment he uses at the First National Bank. Accounting cannot tell him the cost, but they know the annual depreciation over an 8-year period is $18,900 per year. If all items are straight-line depreciated and the salvage is always 25% of the first cost, estimate the original cost.

Declining Balance Depreciation

The following information is used in problems 12.9 through 12.12.

Halcrow Yolles purchased equipment for new highway construction in Manitoba, Canada, costing $500,000 Canadian. Estimated salvage at the end of the expected life of 5 years is $50,000. Various acceptable depreciation methods are being studied currently.

12.9 Determine the depreciation for year 3 using the DDB, 150% DB and SL methods.

12.10 For the DDB and 150% DB methods, determine the implied salvage after 5 years. Compare these with the $S = \$50,000$ estimate.

12.11 Calculate the depreciation rate d_t for each year t for the DDB method.

12.12 Plot the book value curves for DDB and SL depreciation.

12.13 When declining balance (DB) depreciation is applied, there can be three different depreciation rates involved—d, $d_{\max}$, and d_t. Explain the differences between these rates.

12.14 An engineer with Accenture Middle East BV in Dubai was asked by her client to help understand the difference between 150% DB and DDB depreciation. Answer the questions if $B = \$180,000$, $n = 12$ years, and $S = \$30,000$.

    a. What are the book values after 12 years for both methods?

    b. How do the estimated salvage and the two book values after 12 years compare in value?

    c. Which of the two methods, when calculated correctly considering $S = \$30,000$, writes off more of the first cost over the 12 years?

12.15 Exactly 10 years ago, Boyditch Professional Associates purchased $100,000 in depreciable assets with an estimated salvage of $10,000. For tax depreciation the SL method with $n = 10$ years was used, but for book depreciation, Boyditch applied the DDB method with $n = 7$ years and neglected the salvage estimate. The company sold the assets today for $12,500. (*a*) Compare this amount with the book values using the SL and DDB methods. (*b*) If the salvage of $12,500 had been estimated exactly 10 years ago, determine the depreciation for each method in year 10.

12.16 Shirley is studying depreciation in her engineering management course. The instructor asked her to graphically compare the total percent of first cost depreciated for an asset costing B dollars over a life of $n = 5$ years for DDB and 125% DB depreciation. Help her by developing the plots of percent of B depreciated versus years. Use a spreadsheet unless instructed otherwise.

MACRS Depreciation

12.17 Develop the depreciation and book value schedules using the GDS MACRS method for oil and gas drilling equipment that cost $1.2 million. A salvage value of $300,000 is estimated.

12.18 Explain the difference between an accelerated depreciation method and one that is not accelerated. Give an example of each.

12.19 A 120 metric ton telescoping crane that cost $320,000 is owned by Upper State Power. Salvage is estimated at $75,000. (*a*) Compare book values for MACRS and standard SL depreciation over a

7-year recovery period. (*b*) Explain how the estimated salvage is treated using MACRS.

The following information is used in problems 12.20 through 12.22.

A new real-time vision/software system automatically tracks a sports ball (football, soccer, baseball, golf, rugby, etc.) while in flight to determine details of in-bounds, out-of-bounds, foul, and other questionable plays. Installed cost is $250,000 with a 5-year tax-deductible life and $S = \$50,000$.

12.20 Plot the tax depreciation schedule for MACRS.

12.21 Plot the alternative SL depreciation schedule (on the same graph) for $n = 8$ years with the half-year convention included (only 50% of the SL depreciation allowed in years 1 and 9).

12.22 Compare the accumulated percentage of first cost depreciated after 4 years for the two methods.

12.23 Youngblood Shipbuilding Yard just purchased $800,000 in capital equipment for ship repairing functions on dry-docked ships. Estimated salvage is $150,000 for any year after 5 years of use. Compare the depreciation and book value for year 3 for each of the following depreciation methods:

    a. GDS MACRS with a recovery period of 10 years.

    b. Double declining balance with a recovery period of 15 years.

    c. ADS straight line as an alternative to MACRS, with a recovery period of 15 years and the half year convention enforced.

12.24 Fairfield Properties owns real property that is MACRS depreciated with $n = 39$ years. They paid $3.4 million for the apartment complex and hope to sell it after 10 years of ownership for 50% more than the book value at that time. Compare the hoped-for selling price with the amount that Fairfield paid for the property.

12.25 Blackwater Spring and Metal utilizes the same computerized spring forming machinery in its U.S. and Malaysian plants. Purchased in 2007, the first cost was $750,000 with $S = \$150,000$ after 10 years. MACRS depreciation with $n = 5$ years is applied in the United States and standard SL depreciation with $n = 10$ years is used by the Malaysian facility. (*a*) Develop and graph the book value curves for both plants. (*b*) If the equipment is sold after 10 years for $100,000, calculate the over or under depreciated amounts for each method.

12.26 Aaron Pipeline has the service contract for part of the Black Mesa slurry coal pipeline in Arizona. The company placed $300,000 worth of depreciable capital equipment into operation. Use the VDB spreadsheet function to calculate the MACRS depreciation and book value schedules for a 5-year recovery.

Switching Depreciation Methods

12.27 MACRS rates incorporate a switch from one method of depreciation to another to obtain the rates detailed in Table 12.2. (*a*) Identify the MACRS recovery periods for which the DDB method is initially applied. (*b*) Describe the switching approach used to obtain the MACRS rates.

12.28 Use the Web to compare international depreciation methods with SL, DB, and the U.S.-imposed MACRS you have learned here. (Hint: Performing a search using "Depreciation methods in (country name)" is a good start.)

12.29 Henry has an assignment from his boss at Czech Glass and Wood Sculpting to evaluate depreciation methods for writing off the $200,000 first cost of a newly acquired Trotec CO_2 laser system for engraving and cutting. Productive life is 8 years and salvage is estimated at $10,000. Compare the PW of depreciation at 10% per year for DDB-to-SL switching with that for MACRS with $n = 7$ years. Which is the preferred method?

12.30 ConocoPhillips alkylation processes are licensed to produce high-octane, low-sulphur blendstocks domestically and internationally. Halliburton Industries has new licensed alkylation equipment costing $1 million per system at its Moscow, Houston, and Abu Dhabi refinery service operations. Russia requires a 10-year, straight line recovery with a 10% salvage value. The United States allows a 7-year MACRS recovery with no salvage considered. The United Arab Emirates allows a 7-year recovery with switching from DDB to SL method and no salvage considered. Which of the country's methods has the largest PW of depreciation at $i = 15\%$ per year?

12.31 To determine the MACRS rates in Table 12.2, the switching procedure in Section 12.5 must be altered slightly to accommodate the half-year convention imposed by MACRS. The first difference is in year 1, where the DDB rate is only one-half of the DDB rate. The second difference is in the denominator for the SL depreciation. The term is $(n - t + 1.5)$, instead of the $(n - t + 1)$ shown. The third and final change is that one-half of the SL depreciation allowed in year n is taken in the year $n + 1$. Use this procedure and three changes to verify the MACRS rates for $n = 5$.

Depletion Methods

12.32 When is it correct and necessary to use depletion in lieu of depreciation?

12.33 State the monetary limits placed on the amount of an investment (first cost) that can be recovered annually and over the expected life for both depletion methods.

The following information is used in problems 12.34 through 12.35.

Four years ago, International Uranium Mines paid $350 million for rights to remove uranium ore (pitchblend) for refining into U_3O_8 (triuranium octaoxide) for use in nuclear reactor fuels. Of the estimated 6.5 million pounds of ore available, the yearly amounts removed in pounds are 275,000; 250,000; 320,000; and 425,000. Uranium prices per pound over the 4 years have averaged $70, $69, $73, and $75 annually.

12.34 Use percentage depletion to determine the annual allowance. Determine the percentage of the original investment removed thus far.

12.35 Last month a yield reevaluation indicated that about 4 million pounds U_3O_8 ore remains in the mine. If the expected volume this year is 625,000 pounds sold at an average of $78 per pound, determine whether percentage or cost depletion will

offer the larger amount of write-off. (Assume company tax lawyers determined that either method could now be used.)

12.36 Carrolton Oil and Gas, an independent oil and gas producer, is approved to use a 20% of gross income depletion allowance. The write-off last year was $500,000 on its horizontal directional drill wells. Determine the estimated total reserves in barrels, if the volume pumped last year amounted to 1% of the total and the delivered-product price averaged $55 per barrel.

12.37 Ederly Quarry sells a wide variety of cut limestone for residential and commercial building construction. A recent quarry expansion cost $2.9 million and added an estimated 100,000 tons of reserves. Estimate the cost depletion allowance for the next 5 years, using the projections made by John Ederly, owner.

Year	1	2	3	4	5
Volume, 1000 tons	10	12	15	15	18
Price, $ per ton	85	90	90	95	95

12.38 For the last 10 years, Am-Mex Coal has used the cost depletion factor of $2500 per 100 tons to write off the investment of $35 million in its Pennsylvania anthracite coal mine. Depletion thus far totals $24.8 million. A new study to appraise mine reserves indicates that no more than 800,000 tons of salable coal remains. Determine next year's percentage and cost depletion amounts, if estimated gross income is expected to be between $6.125 and $8.50 million on a production level of 72,000 tons.

PROBLEMS FOR TEST REVIEW AND FE EXAM PRACTICE

12.39 Standard straight line depreciation of a $100,000 asset takes place over a 7-year recovery period. If the salvage value is 20% of first cost, the book value at the end of 3 years is closest to
  a. $57,140.
  b. $65,715.
  c. $11,430.
  d. $80,000.

12.40 If straight line and double declining balance depreciation rates for $n = 5$ years are calculated and compared, the rates (in percent) for the second year are
  a. 20% and 40%.
  b. 40% and 24%.

  c. 20% and 24%.
  d. 20% and 16%.

12.41 The following are correct for declining balance depreciation:

1—accelerates depreciation compared to straight line.

2—book value in the last year of recovery always equals zero.

3—not an approved depreciation method in the United States.

4—book value is reduced by a fixed percentage each year.

a. 1 and 2
b. All four
c. 1, 3, and 4
d. 2 and 3

12.42 Gisele is performing a make/buy study involving the retention or disposal of a 4-year-old machine that was to be in production for 8 years. It cost $500,000 originally. Without a market value estimate, she decided to use the current book value plus 20%. If DDB depreciation is applied, the market value estimate is closest to
a. $158,200.
b. $253,125.
c. $217,900.
d. $189,840.

12.43 Of the following, *incorrect* statement(s) about MACRS depreciation is (are):

1—has a lower present worth of depreciation than SL depreciation.
2—cannot be used to depreciate land.
3—has a recovery period of 7 years for any asset not specifically classified by MACRS.
4—must be used for book and tax depreciation purposes.
a. 1 and 4
b. 1 and 2
c. 2, 3, and 4
d. 4

12.44 MACRS depreciation switches from declining balance to straight line in the year when
a. more than 50% of the first cost is written off.
b. the depreciation by declining balance is less than that for straight line.
c. the straight line depreciation rate for year 1 exceeds the declining balance fixed rate of $2/n$.
d. all three of the above occur.

12.45 The correct relation used to determine book value after t years using MACRS depreciation is
a. book value in year t + depreciation in year t.
b. book value in year t − book value in year $(t - 1)$.
c. first cost − depreciation in years 1 through $(t - 1)$.
d. first cost − sum of all depreciation for t years.

12.46 The recovery rate most often used for comparison with other methods' rates is
a. $1/n$.
b. DDB rate.
c. $2/n$.
d. MACRS rate.

12.47 When the present worth of depreciation is maximized,
a. total taxes paid over the recovery period are smaller.
b. the present worth of taxes is minimized.
c. the future worth of depreciation is minimized.
d. the depreciation method applied had to be MACRS.

12.48 South African Gold Mines, Inc. is writing off its $210 million investment using the cost depletion method. An estimated 700,000 ounces of gold are available in its developed mines. This year 35,000 ounces were produced at an average price of $400 per ounce. The depletion for the year is closest to
a. $2.1 million, which is 15% of the gross income of $14.0 million.
b. $2.4 million.
c. $10.5 million.
d. $42 million, which is 15% of the gross income of $280 million.

12.49 The following are *incorrect* statement(s) about percentage depletion:

1—total depletion charges are limited to the first cost.
2—a fixed percentage of book value in year $(t - 1)$ is removed each year.
3—The depletion amount must not exceed 50% of gross income.
a. 1
b. 2
c. 1 and 3
d. All three statements

12.50 A stone and gravel quarry in Texas can use a percentage depletion rate of 5% of gross income, or a cost depletion rate of $1.28 per ton.

Quarry first cost = $3.2 million	Estimated total tonnage = 2.5 million tons
Tonnage this year = 65,000	Gross income = $40 per ton

Of the two depletion charges, the method and larger amount are
a. percentage at $83,200.
b. percentage at $130,000.
c. cost at $80,000.
d. cost at $130,000.

13 Chapter

After-Tax Economic Analysis

Royalty-Free/CORBIS

This chapter provides an overview of tax terminology and equations pertinent to an after-tax economic analysis. The transfer from estimating cash flow before taxes (CFBT) to cash flow after taxes (CFAT) involves a consideration of significant tax effects that may alter the final decision, as well as estimate the magnitude of the tax effect on cash flow over the life of the alternative.

Mutually exclusive alternative comparisons using after-tax PW, AW, and ROR methods are explained with major tax implications considered. Replacement studies are discussed with tax effects that occur at the time that a defender is replaced. Additionally, the cost of debt and equity investment capital and its relation to the MARR is examined.

Additional information on U.S. federal taxes—tax law, and annually updated tax rates—is available through Internal Revenue Service publications and, more readily, on the IRS website www.irs.gov. Publications 542, *Corporations,* and 544, *Sales and Other Dispositions of Assets,* are especially applicable.

Objectives

Purpose: Perform an economic evaluation of alternatives considering the effect of income taxes and pertinent tax regulations.

| Terminology and rates | 1. Correctly use the basic terminology and income tax rates. |

| CFAT analysis | 2. Calculate after-tax cash flow and evaluate alternatives. |

| Taxes and depreciation | 3. Demonstrate the tax impact of depreciation recapture, accelerated depreciation, and a shortened recovery period. |

| After-tax replacement | 4. Evaluate a defender and challenger in an after-tax replacement study. |

| Cost of capital | 5. Determine the weighted average cost of capital (WACC) and its relation to MARR. |

| Spreadsheets | 6. Use a spreadsheet to perform an after-tax PW, AW, or ROR analysis. |

| Value-added analysis | 7. Evaluate alternatives using after-tax economic value-added analysis (EVA). |

| Taxes outside the United States | 8. Understand the tax impact of selected tax laws in countries outside the United States. |

13.1 INCOME TAX TERMINOLOGY AND RELATIONS

Some basic tax terms and relationships are explained here.

Gross income *GI* is the total income realized from all revenue-producing sources of the corporation, plus any income from other sources such as sale of assets, royalties, and license fees.

Income tax is the amount of taxes based on some form of income or profit levied by the federal (or lower-level) government. A large percentage of U.S. tax revenue is based upon taxation of corporate and personal income. Taxes are actual cash flows.

Operating expenses *E* include all corporate costs incurred in the transaction of business. These expenses are tax deductible for corporations. For engineering economy alternatives, these are the AOC (annual operating cost) and M&O (maintenance and operating) costs.

Taxable income *TI* is the amount upon which income taxes are based. For corporations, depreciation *D* and operating expenses *E* are tax-deductible.

$$\text{TI} = \text{gross income} - \text{expenses} - \text{depreciation}$$
$$= \text{GI} - E - D \qquad \text{[13.1]}$$

Tax rate *T* is a percentage, or decimal equivalent, of TI owed in taxes. The tax rate is graduated; that is, higher rates apply as TI increases.

$$\text{Taxes} = (\text{taxable income}) \times (\text{applicable tax rate})$$
$$= (\text{TI})(T) \qquad \text{[13.2]}$$

Effective tax rate T_e is a single-figure rate used in an economy study to estimate the effects of federal, state, and local taxes. The T_e rate, which usually ranges from 25% to 50% of TI, includes an allowance for state (and possibly local) taxes that are deductible when determining federal taxes.

$$T_e = \text{state and local rate} + (1 - \text{state and local rate})(\text{federal rate}) \qquad \text{[13.3]}$$

$$\text{Taxes} = \text{TI}(T_e) \qquad \text{[13.4]}$$

Different bases (and taxes) are used. The most common is a basis of income (and income taxes). Others are total sales (sales tax); appraised value of property (property tax); value-added tax (VAT); net capital investment (asset tax); winnings from gambling (part of income tax); and retail value of items imported (import tax).

The annual U.S. federal tax rate *T* is based upon the principle of *graduated tax rates,* which means that higher rates go with larger taxable incomes. Table 13.1 presents recent *T* values for corporations. Each year the IRS alters the TI limits to account for inflation and other factors. This action is called *indexing.* The portion of each new dollar of TI is taxed at what is called the *marginal tax rate.* As an illustration, scan the tax rates in Table 13.1. A business with an annual TI of $50,000 has a marginal rate of 15%. However, a business with TI = $100,000

TABLE 13.1 **U.S. Corporate Federal Income Tax Rate Schedule (2007)**
(mil = million $)

TI Limits (1)	TI Range (2)	Tax Rate T (3)	Maximum Tax for TI Range (4) = (2)T	Maximum Tax Incurred (5) = Sum of (4)
$1–$50,000	$ 50,000	0.15	$ 7,500	$ 7,500
$50,001–75,000	25,000	0.25	6,250	13,750
$75,001–100,000	25,000	0.34	8,500	22,250
$100,001–335,000	235,000	0.39	91,650	113,900
$335,001–10 mil	9.665 mil	0.34	3.2861 mil	3.4 mil
Over $10–15 mil	5 mil	0.35	1.75 mil	5.15 mil
Over $15–18.33 mil	3.33 mil	0.38	1.267 mil	6.417 mil
Over $18.33 mil	Unlimited	0.35	Unlimited	Unlimited

pays 15% for the first $50,000, 25% on the next $25,000, and 34% on the remainder.

$$\text{Taxes} = 0.15(50,000) + 0.25(75,000 - 50,000) + 0.34(100,000 - 75,000)$$
$$= \$22,250$$

To simplify tax computations, an average federal tax rate may be used in Equation [13.3].

Corporate Federal

EXAMPLE 13.1

If the video division of Marvel Comics has an annual gross income of $2,750,000 with expenses and depreciation totaling $1,950,000, (*a*) compute the company's exact federal income taxes. (*b*) Estimate total federal and state taxes if a state tax rate is 8% and a 34% federal average tax rate applies.

Solution

a. Compute the TI by Equation [13.1] and the income taxes using Table 13.1 rates.

$$\text{TI} = 2,750,000 - 1,950,000 = \$800,000$$
$$\text{Taxes} = 50,000(0.15) + 25,000(0.25) + 25,000(0.34)$$
$$+ 235,000(0.39) + (800,000 - 335,000)(0.34)$$
$$= 7500 + 6250 + 8500 + 91,650 + 158,100$$
$$= \$272,000$$

A faster approach uses the amount in column 5 of Table 13.1 that is closest to the total TI and adds the tax for the next TI range.

$$\text{Taxes} = 113,900 + (800,000 - 335,000)(0.34) = \$272,000$$

b. Equations [13.3] and [13.4] determine the effective tax rate and taxes.

$$T_e = 0.08 + (1 - 0.08)(0.34) = 0.3928$$
$$\text{Taxes} = (800,000)(0.3928) = \$314,240$$

These two tax amounts are not comparable, because the tax in part (a) does not include state taxes.

It is important to understand how corporate tax and individual tax computations compare. Gross income for an individual taxpayer is, for the most part, the total of salaries and wages. However, for an individual's taxable income, most of the expenses for living and working are not tax-deductible to the same degree as corporate operating expenses. For individual taxpayers,

$$\text{GI} = \text{salaries} + \text{wages} + \text{interest and dividends} + \text{other income}$$
$$\text{Taxable income} = \text{GI} - \text{personal exemption} - \text{deductions}$$
$$\text{Taxes} = (\text{taxable income})(\text{applicable tax rate}) = (\text{TI})(T)$$

For individuals, corporate operating expenses are replaced by exemptions and (standard or itemized) deductions. Exemptions are yourself, spouse, children, and other dependents. Each exemption reduces TI by approximately \$3000 per year, depending upon current exemption allowances. Like the corporate tax structure, rates for individuals are graduated. As shown in Table 13.2, they range from 10% to 35% of TI. These rates change more often than corporate rates.

TABLE 13.2   2007 U.S. Federal Income Tax Rates for Individuals Who File "Single" or "Married and Jointly"

	Amount of Taxable Income, TI	
Tax Rate, T	**Filling Single**	**Filing Married and Jointly**
0.10	$0–7,825	$0–15,650
0.15	$7,826–31,850	$15,651–63,700
0.25	$31,851–77,100	$63,701–128,500
0.28	$77,101–160,850	$128,501–195,850
0.33	$160,851–349,700	$195,851–349,700
0.35	Over $349,700	Over $349,700

Individual   Federal

EXAMPLE 13.2   Josh and Allison submit a married-filing-jointly return. During the year, their two jobs provided them with a combined income of $82,000. They had their second child during the year, and they plan to use the standard deduction of $9500 applicable for the year. Dividends, interest, and capital gains amounted to $6050. Personal exemptions are $3100 currently. (*a*) Compute their exact federal tax liability. (*b*) What percent of their gross income is consumed by federal taxes?

Solution

a. Josh and Allison have four personal exemptions and the standard deduction of $9500.

$$\text{Gross income} = \text{salaries} + \text{interest and dividends and capital gains}$$
$$= 82{,}000 + 3550 + 2500 = \$88{,}050$$
$$\text{Taxable income} = \text{gross income} - \text{exemptions} - \text{deductions}$$
$$= 88{,}050 - 4(3100) - 9500 = \$66{,}150$$

Table 13.2 indicates the rates for federal taxes.

$$\text{Taxes} = 15{,}650(0.10) + (63{,}700 - 15{,}650)(0.15)$$
$$+ (66{,}150 - 63{,}700)(0.25)$$
$$= \$9385$$

b. Of the total $88,050, the percent paid in federal taxes is 9385/88,050 = 10.7%.

13.2 BEFORE-TAX AND AFTER-TAX ALTERNATIVE EVALUATION

Early in the text, the term *net cash flow* (*NCF*) was identified as the best estimate of actual cash flow. The NCF, calculated as annual cash inflows minus cash outflows, is used to perform alternative evaluations via the PW, AW, ROR, and B/C methods. Now that the impact of taxes will be considered, it is time to expand our terminology. NCF is replaced by the terms *cash flow before taxes* (*CFBT*) and *cash flow after taxes* (*CFAT*). Both are actual cash flows and are related as

$$\text{CFAT} = \text{CFBT} - \text{taxes} \qquad \text{[13.5]}$$

The CFBT should include the initial investment P or the salvage value estimate S in the year it occurs. *Depreciation D is included in TI, but not directly in the CFAT estimate since depreciation is not an actual cash flow.* This is very important since the engineering economy study must be based on actual cash flow. Using the effective tax rate, equations are:

$$\text{CFBT} = \text{gross income} - \text{expenses} - \text{initial investment} + \text{salvage value}$$
$$= \text{GI} - E - P + S \qquad \text{[13.6]}$$

$$\text{CFAT} = \text{CFBT} - \text{TI}(T_e)$$
$$= \text{GI} - E - P + S - (\text{GI} - E - D)(T_e) \qquad \text{[13.7]}$$

Suggested table column headings for CFBT and CFAT calculations are shown in Table 13.3. Numerical relations are shown in column numbers, with the effective

TABLE 13.3   Table Column Headings for Calculation of (a) CFBT and (b) CFAT

(a) CFBT table headings

Year	Gross Income GI	Operating Expenses E	Investment P and Salvage S	$1-2-4$ CFBT (4) =
	(1)	(2)	(3)	(1) + (2) + (3)

(b) CFAT table headings

Year	Gross Income GI	Operating Expenses E	Investment P and Salvage S	Depreciation D	Taxable Income TI	Taxes $(TI)(T_e)$	CFAT (7) =
	(1)	(2)	(3)	(4)	(5) = (1) + (2) − (4)	(6)	(1) + (2) + (3) − (6)

tax rate T_e used for income taxes. Expenses E and initial investment P are negative. It is possible that in some years TI may be negative due to a depreciation amount that is larger than $(GI - E)$. In this case, *the associated negative income tax is considered a tax savings for the year.* The assumption is that the negative tax is an advantage to the alternative and that it will offset taxes for the same year in other income-producing areas of the corporation.

Once the CFAT series is developed, apply an evaluation method—PW, AW, ROR, or B/C—and use exactly the same guidelines as in Chapters 4 through 7 to justify a single project or to select one mutually exclusive alternative. Example 13.3 illustrates CFAT computation and after-tax analysis.

If estimating the after-tax ROR is important, but after-tax detailed numbers are not of interest, or are too complicated to tackle, the before-tax ROR can be used to approximate the effects of taxation by using T_e and the relation

$$\text{After-tax ROR} = \text{Before-tax ROR}(1 - T_e) \qquad [13.8]$$

Applying the same logic, the required after-tax MARR to use in a PW- or AW-based study is approximated as

$$\text{After-tax MARR} = \text{Before-tax MARR}(1 - T_e) \qquad [13.9]$$

If an alternative's PW or AW value is close to zero, a more detailed analysis of the impact of taxes should be undertaken.

EXAMPLE 13.3   AMRO Engineering is evaluating a very large flood control program for several southern U.S. cities. One component is a 4-year project for a special-purpose transport ship-crane for use in building permanent storm surge protection against hurricanes on the New Orleans coastline. The estimates are $P =$

$300,000, $S = 0$, $n = 3$ years. MACRS depreciation with a 3-year recovery is indicated. Gross income and expenses are estimated at $200,000 and $80,000, respectively, for each of 4 years. (*a*) Perform before-tax and after-tax ROR analyses. AMRO uses $T_e = 35\%$ and a before-tax MARR of 15% per year. (*b*) Approximate the after-tax ROR with Equation [13.8] and comment on its accuracy.

Solution

a. Table 13.4 uses the format of Table 13.3 to determine the CFBT and CFAT series.

Before taxes: By Equation [13.6], CFBT = 200,000 − 80,000 = $120,000 for each year 1 to 4. The PW relation to estimate before-tax ROR per year is

$$PW = -300,000 + 120,000(P/A,i,4)$$

This is easily solved using factor tables or the Excel IRR function to obtain $i^* = 21.86\%$. Comparing i^* with the before-tax MARR of 15%, the project is justified.

After taxes: Table 13.4 (bottom half) details MACRS depreciation (rates from Table 12.2), TI using Equation [13.1], and CFAT by Equation [13.7].

TABLE 13.4   **CFBT, CFAT, and ROR Calculations Using MACRS and $T_e = 35\%$, Example 13.3**

Year	GI	E	P and S	CFBT			
			Before-Tax ROR Analysis				
0			$−300,000	$−300,000			
1	$200,000	$−80,000		120,000			
2	200,000	−80,000		120,000			
3	200,000	−80,000	0	120,000			
4	200,000	−80,000		120,000			

Year	GI	E	P and S	Depr	TI	Taxes	CFAT
			After-Tax ROR Analysis				
0			$−300,000				$−300,000
1	$200,000	$−80,000		$ 99,990	$ 20,010	$ 7,003	112,997
2	200,000	−80,000		133,350	−13,350	−4,673	124,673
3	200,000	−80,000	0	44,430	75,570	26,450	93,551
4	200,000	−80,000		22,230	97,770	34,220	85,781

[13.7]

After tax ROR MARR

Because depreciation exceeds (GI − E) in year 2, the tax savings of $4673 will increase after-tax ROR. The PW relation and after-tax ROR are

$$PW = -300,000 + 112,997(P/F,i\%,1) + \ldots + 85,781(P/F,i\%,4)$$

After-tax <u>ROR</u> = $i^* = 15.54\%$

By Equation [13.9], the after-tax <u>MARR</u> is $15(1 - 0.35) = 9.75\%$. The project is economically justified since $i^* > 9.75\%$.

b. Equation [13.8] yields

After-tax ROR = $21.86(1 - 0.35) = 14.21\%$

This is an underestimate of the calculated amount 15.54%, in part due to the neglect of the tax savings in year 2, which is explicitly considered in the after-tax analysis.

After-tax evaluation of 2 or more alternatives follows the same guidelines detailed in Chapters 4 through 7. A couple of important points to remember follow.

PW: For different-life alternatives, the LCM of lives must be used to ensure an equal service comparison.

ROR: Incremental analysis of CFAT must be performed, since overall i^* values do not guarantee a correct selection (Section 6.5 provides details.)

Similar to the before-tax situation, the use of AW relations is preferable since neither of these complications is present. Additionally, spreadsheet usage is recommended for an after-tax analysis. Section 13.6 demonstrates several Excel functions useful in PW, AW, and ROR analyses.

13.3 HOW DEPRECIATION CAN AFFECT AN AFTER-TAX STUDY

Several things may significantly alter the amount and timing of taxes and CFAT: depreciation method, length of recovery period, time at which the asset is disposed of, and relevant tax laws. A few of the tax implications should be considered in the economic evaluation (such as depreciation recovery). Others, such as capital gain or loss, can be neglected since they commonly occur toward the end of the asset's useful life and, consequently, are not reliably predictable at evaluation time. As discussed below, the key to estimating the potential tax effect is having a good sense of the selling price (salvage value) relative to the book value at disposal time.

Depreciation recapture (DR) occurs when a depreciable asset is sold for more than the current book value. As shown in Figure 13.1 (shaded area),

Depreciation recapture = selling price − book value

$$DR = SP - BV \qquad\qquad [13.10]$$

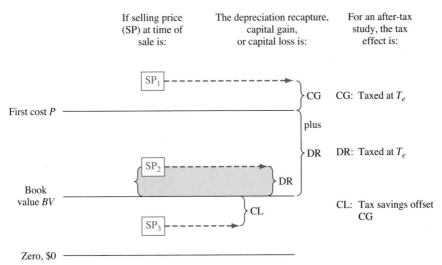

FIGURE 13.1
Summary of calculations and tax treatment for depreciation recapture and capital gains (losses).

Depreciation recapture is often present in the after-tax study. In the United States, an amount equal to the estimated salvage value can always be anticipated as DR when the asset is disposed of at the end of or after the recovery period. This occurs because MACRS depreciates to zero in $n + 1$ years. The amount DR is treated as ordinary taxable income.

When the selling price is expected to exceed first cost, a capital gain (discussed next) is also incurred and the TI due to the sale is the gain *plus* the depreciation recapture, as shown in Figure 13.1 for a selling price of SP_1. In this case, DR is the total depreciation taken thus far.

Capital gain (CG) is an amount incurred when the selling price exceeds its first cost. See Figure 13.1. At the time of asset disposal,

$$\text{Capital gain} = \text{selling price} - \text{first cost}$$
$$\text{CG} = \text{SP} - P \qquad \qquad \text{[13.11]}$$

Since future capital gains are difficult to predict, they are usually not detailed in an after-tax study. An exception is for assets that historically increase in value, such as buildings and land. *If included, the gain is taxed as ordinary TI.*

Capital loss (CL) occurs when a depreciable asset is disposed of for less than its current book value.

$$\text{Capital loss} = \text{book value} - \text{selling price}$$
$$\text{CL} = \text{BV} - \text{SP} \qquad \qquad \text{[13.12]}$$

An economic analysis does not commonly account for capital loss, simply because it is not estimable for a specific alternative. However, an after-tax replacement study should account for any capital loss if the defender must be traded at a "sacrifice" price. For tax purposes, CL offsets CG from other activities.

Tax law can present pertinent rules that are special, time-limited incentives offered by government agencies to boost capital, and possibly foreign investment, through allowances of increased depreciation and reduced taxes. These benefits, which come and go depending on the "health of the economy," can alter CFAT. IRS publications 334 and 544 may be consulted at www.irs.gov.

Now the expression for TI in Equation [13.1] can be expanded to include the additional cash flow estimates for asset disposal.

TI = **gross income − expenses − depreciation + depreciation recapture**
      + capital gain − capital loss

$$= GI - E - D + DR + CG - CL \qquad\qquad [13.13]$$

EXAMPLE 13.4   Biotech, a medical imaging and modeling company, must purchase a bone cell analysis system for use by a team of bioengineers and mechanical engineers studying bone density in athletes. This particular part of a 3-year contract with the NBA will provide additional gross income of $100,000 per year. The effective tax rate is 35%.

	Analyzer 1	Analyzer 2
First cost, $	150,000	225,000
Operating expenses, $ per year	30,000	10,000
MACRS recovery, years	5	5

a. The Biotech president, who is very tax conscious, wishes to use a criterion of minimizing total taxes incurred over the 3 years of the contract. Which analyzer should be purchased?

b. Assume that after 3 years the company will sell the selected analyzer. Using the same total tax criterion, does either analyzer have an advantage? The selling prices are expected to be $130,000 for analyzer 1 and $225,000 for analyzer 2.

Solution

a. Table 13.5 details the tax computations. First, the yearly MACRS depreciation is determined. Equation [13.1], TI = GI − E − D, is used to calculate TI, after which the 35% tax rate is applied. Taxes for the 3-year period are summed.

        Analyzer 1 tax total: $36,120   Analyzer 2 tax total: $38,430

The two analyzers are very close, but analyzer 1 wins with $2310 less in total taxes.

TABLE 13.5 Comparison of Total Taxes for Two Alternatives, Example 13.4a

Year	Gross Income GI	Operating Expenses E	First Cost P	MACRS Depreciation D	Book Value BV	Taxable Income TI	Taxes at 0.35TI
				Analyzer 1 P, 296			
0			$150,000	↓	$150,000	—	
1	$100,000	$30,000		$30,000 .2	120,000	$40,000	$14,000
2	100,000	30,000		48,000 .32	72,000	22,000	7,700
3	100,000	30,000		28,800	43,200	41,200	14,420
							$36,120
				Analyzer 2			
0			$225,000		$225,000		
1	$100,000	$10,000		$45,000	180,000	$45,000	$15,750
2	100,000	10,000		72,000	108,000	18,000	6,300
3	100,000	10,000		43,200	64,800	46,800	16,380
							$38,430

b. When the analyzer is sold after 3 years of service, there is a depreciation recapture (DR) that is taxed at the 35% rate. This tax is in addition to the third-year tax in Table 13.5. For each analyzer, account for the DR by Equation [13.10]; then determine the TI, using Equation [13.13], TI = GI − E − D + DR. Again, find the total taxes for 3 years, and select the analyzer with the smaller total.

Analyzer 1:   DR = SP − BV_3 = 130,000 − 43,200 = $86,800

     Year 3 TI = 100,000 − 30,000 − 28,800 + 86,800 = $128,000

    Year 3 taxes = 128,000(0.35) = $44,800

    Total taxes = 14,000 + 7700 + 44,800 = $66,500

Analyzer 2:   DR = 225,000 − 64,800 = $160,200

     Year 3 TI = 100,000 − 10,000 − 43,200 + 160,200 = $207,000

    Year 3 taxes = 207,000(0.35) = $72,450

    Total taxes = 15,750 + 6300 + 72,450 = $94,500

Now, analyzer 1 has a considerable advantage in total taxes.

It may be important to understand why accelerated depreciation rates and shortened recovery periods included in MACRS and DB methods give the corporation a tax advantage relative to that offered by the straight line method. Larger depreciation rates in earlier years require less in taxes due to the larger reductions in taxable income. The key is to choose the depreciation rates and n value that result in the *minimum present worth of taxes*.

$$PW_{tax} = \sum_{t=1}^{t=n} (\text{taxes in year } t)(P/F,i,t) \qquad [13.14]$$

To compare depreciation methods or recovery periods, assume the following: (1) There is a constant single-value tax rate, (2) CFBT exceeds the annual depreciation amount, and (3) the method reduces book value to the same salvage value. The following are then correct:

- Total taxes paid are *equal* for all depreciation methods or recovery periods.
- PW_{tax} is *less* for accelerated depreciation methods with the same n value.
- PW_{tax} is less for smaller n values with the same depreciation method.

MACRS is the prescribed tax depreciation model in the United States, and the only alternative is MACRS straight line depreciation with an extended recovery period. The accelerated write-off of MACRS always provides a smaller PW_{tax} compared to less accelerated models. If the DDB model were still allowed directly, it would not fare as well as MACRS, because DDB does not reduce the book value to zero.

EXAMPLE 13.5

Pw Analysis

An after-tax analysis for a new $50,000 machine proposed for a fiber optics manufacturing line is in process. The CFBT for the machine is estimated at $20,000. If a recovery period of 5 years applies, use the present-worth-of-taxes criterion, an effective tax rate of 35%, and a return of 8% per year to compare the following: straight line, DDB, and MACRS depreciation. Use a 6-year period to accommodate the MACRS half-year convention.

Solution

Table 13.6 presents a summary of depreciation, taxable income, and taxes for each model. For straight line depreciation with $n = 5$, $D = \$10,000$ for 5 years and $D_6 = 0$ (column 3). The CFBT of $20,000 is fully taxed at 35% in year 6.

The DDB percentage of $d = 2/n = 0.40$ is applied for 5 years. The implied salvage value is $\$50,000 - 46,112 = \$3888$, so not all $50,000 is tax deductible. The taxes using DDB are $3888 (0.35) = \$1361$ larger than for the SL model.

MACRS writes off the $50,000 in 6 years using the rates of Table 12.2. Total taxes are $24,500, the same as for SL depreciation.

The annual taxes (columns 5, 8, and 11) are accumulated year by year in Figure 13.2. Note the pattern of the curves, especially the lower total taxes relative to the SL model after year 1 for MACRS and in years 1 through 4 for DDB. These higher tax values for SL cause PW_{tax} for SL depreciation to be larger. The PW_{tax} values are at the bottom of Table 13.6. The MACRS PW_{tax} value is the smallest at $18,162.

TABLE 13.6 Comparison of Taxes and Present Worth of Taxes for Different Depreciation Methods

(1) Year	(2) CFBT	Straight Line			Double Declining Balance			MACRS		
		(3) D	(4) TI	(5) Taxes	(6) D	(7) TI	(8) Taxes	(9) D	(10) TI	(11) Taxes
1	+20,000	$10,000	$10,000	$ 3,500	$20,000	$ 0	$ 0	$10,000	$10,000	$ 3,500
2	+20,000	10,000	10,000	3,500	12,000	8,000	2,800	16,000	4,000	1,400
3	+20,000	10,000	10,000	3,500	7,200	12,800	4,480	9,600	10,400	3,640
4	+20,000	10,000	10,000	3,500	4,320	15,680	5,488	5,760	14,240	4,984
5	+20,000	10,000	10,000	3,500	2,592	17,408	6,093	5,760	14,240	4,984
6	+20,000	0	20,000	7,000	0	20,000	7,000	2,880	17,120	5,992
Totals		$50,000		$24,500	$46,112		$25,861*	$50,000		$24,500
PW$_{tax}$				$18,386			$18,549			$18,162

*Larger than other values since there is an implied salvage value of $3888 not recovered.

(handwritten annotations)

$\dfrac{2}{n} = .4$

$D_n = (1 - 2/n)\,D_{n-1}$

$TI = CFBT - D$

$.2$
$.32$
$.1192$
$.1152$
$.1152$
$.2$

20 k

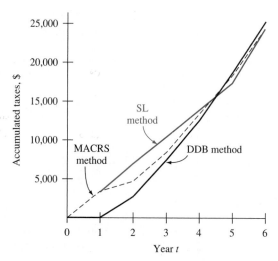

FIGURE 13.2 Taxes incurred by different depreciation rates for a 6-year comparison period, Example 13.5.

Comment: If a similar example is developed that applies only one depreciation method for different n values, the PW_{tax} value will be less for the smaller n.

13.4 AFTER-TAX REPLACEMENT STUDY

When a currently installed asset (the defender) is challenged with possible replacement, the effect of taxes can have an impact upon the decision of the replacement study. The final decision may not be reversed by taxes, but the difference between before-tax PW or AW values may be significantly different from the after-tax difference. There may be tax considerations in the year of the possible replacement due to *depreciation recapture* or *capital gain,* or there may be tax savings due to a sizable *capital loss,* if the trade of the defender is expected to occur at a sacrifice price. Additionally, the after-tax replacement study considers tax-deductible *depreciation* and *operating expenses* not accounted for in a before-tax analysis. The same procedure is applied for CFAT estimates as for the before-tax replacement study. A review of Sections 9.3 and 9.4 is recommended.

EXAMPLE 13.6   Savannah Power purchased railroad transport equipment for coal 3 years ago for $600,000. Management has discovered that it is technologically outdated now. New equipment has been identified. If the market value of $400,000 is offered as the trade-in for the current equipment, perform a replacement study using a 7% per year after-tax MARR. Assume an effective tax rate of 34% and straight line depreciation with $S = 0$ for both alternatives.

	Defender	Challenger
Market value, $	400,000	
First cost, $		$-1,000,000$
Annual cost, $/year	$-100,000$	$-15,000$
Recovery period, years	8 (originally)	5

Solution

For the defender *after-tax replacement study,* there are no tax effects other than income tax. The annual SL depreciation is $600,000/8 = \$75,000$, determined when the equipment was purchased 3 years ago. Now $P_D = \$-400,000$, which is the current market value.

Table 13.7 shows the TI and taxes at 34%, which are the same each year. The taxes are actually tax savings of $59,500 per year, as indicated by the minus sign. Since only costs are estimated, the annual CFAT is negative, but the $59,500 tax savings has reduced it. The CFAT and AW at 7% per year are

$$\text{CFAT} = \text{CFBT} - \text{taxes} = -100,000 - (-59,500) = \$-40,500$$
$$\text{AW}_D = -400,000(A/P,7\%,5) - 40,500 = \$-138,056$$

For the challenger, depreciation recapture on the defender occurs when it is replaced, because the trade-in amount of $400,000 is larger than the current

TABLE 13.7 After-Tax Replacement Analyses, Example 13.6

Defender Age	Year	**Before Taxes** Expenses E	P	CFBT	**After Taxes** Depreciation D	Taxable Income TI	Taxes* at 0.34TI	CFAT
				DEFENDER				
3	0		$-400,000	$-400,000				$-400,000
4–8	1–5	$-100,000		-100,000	$75,000	$-175,000	$-59,500	-40,500
AW at 7%								$-138,056
				CHALLENGER				
	0		$-1,000,000	$-1,000,000		$ +25,000†	$ 8,500	$-1,008,500
	1–5	$-15,000		-15,000	$200,000	-215,000	-73,100	+58,100
AW at 7%								$-187,863

*Minus sign indicates a tax savings for the year.
†DR_3, Depreciation recapture on defender trade-in.

book value. In year 0 for the challenger, Table 13.7 includes the following computations to arrive at a tax of $8500:

Defender book value, year 3: $BV_3 = 600,000 - 3(75,000) = \$375,000$

Depreciation recapture: $DR_3 = TI = 400,000 - 375,000 = \$25,000$

Taxes on the trade-in, year 0: $\text{Taxes} = 0.34(25,000) = \$8500$

The SL depreciation is $\$1,000,000/5 = \$200,000$ per year. This results in tax saving and CFAT as follows:

$$\text{Taxes} = (-15,000 - 200,000)(0.34) = \$-73,100$$
$$\text{CFAT} = \text{CFBT} - \text{taxes} = -15,000 - (-73,100) = \$+58,100$$

In year 5, it is assumed the challenger is sold for $0; there is no depreciation recapture. The AW for the challenger at the 7% after-tax MARR is

$$AW_C = -1,008,500(A/P,7\%,5) + 58,100 = \$-187,863$$

The defender is selected.

13.5 CAPITAL FUNDS AND THE COST OF CAPITAL

The funds that finance engineering projects are called *capital,* and the interest rate paid on these funds is called the *cost of capital.* The specific MARR used in an economic analysis is established such that MARR exceeds the cost of capital, as discussed initially in Section 1.3 of this text. To understand how a MARR is set, it is necessary to know the two types of capital.

Debt capital—These are funds borrowed from sources outside the corporation and its owners/stockholders. Debt financing includes loans, notes, mortgages, and bonds. The debt must be repaid at some stated interest rate using a specific time schedule, for example, over 15 years at 10% per year simple interest based on the declining loan balance. A corporation indicates outstanding debt financing in the liability section of its balance sheet. (See Appendix B, Table B.1.)

Equity capital—These are retained earnings previously kept within the corporation for future capital investment, and owners' funds obtained from stock sales (public corporations) or individual owner's funds (private corporations). Equity funds are indicated in the net worth section of the balance sheet. (Again, see Table B.1.)

Suppose a $55 million (U.S. dollars) project to expand the capacity for generated electricity by Mexico City's Luz y Fuerza del Centro is financed by $33 million from funds set aside over the last 5 years as retained earnings and

$22 million in municipal bond sales. The project's *capital pool* is developed as follows:

Equity capital: $33 million or $33/55 = 60\%$ equity funds

Debt capital: $22 million or $22/55 = 40\%$ debt funds

The *debt-to-equity (D-E) mix* is the ratio of debt to equity capital for a corporation or a project. In this example, the D-E mix is 40-60 with 40% debt from bonds and 60% equity from retained earnings.

Once the D-E mix is known, the *weighted average cost of capital (WACC)* can be calculated. The MARR is then established to exceed (or at least equal in the case of a not-for-profit project) this value. WACC is an estimate of the interest rate in percentage for all funds in the capital pool used to finance corporate projects. The relation is

$$\textbf{WACC} = \textbf{(equity fraction)(cost of equity capital)} \qquad \textbf{[13.15]}$$
$$\textbf{+ (debt fraction)(cost of debt capital)}$$

The cost terms, expressed as percentages, are the rates associated with each type of capital. If debt capital is acquired via loans that carry a 7% per year interest rate, the cost of debt capital in Equation [13.15] is 7%. Similarly, if invested retained earnings are making 12.5% per year, this is the cost of equity capital. The D-E mix and WACC are illustrated in the next example.

Figure 13.3 indicates the usual shape of cost of capital curves. If 100% of the capital is derived from equity or 100% is from debt sources, the WACC equals the cost of capital for that source of funds. There is virtually always a mixture of capital

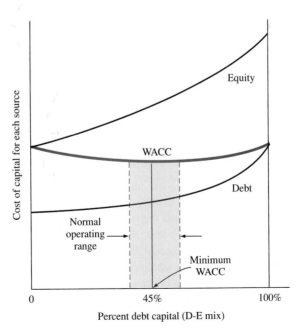

FIGURE 13.3
General shape of different cost of capital curves.

sources involved for any capitalization program. *As an illustration only,* Figure 13.3 indicates a minimum WACC at about 45% debt capital. Most firms operate over a range of D-E mixes for different projects.

EXAMPLE 13.7   A new program in genetics engineering at Gentex will require $10 million in capital. The chief financial officer (CFO) has estimated the following amounts of capital at the indicated interest rates per year.

Stock sales	$5 million at 13.7%
Use of retained earnings	$2 million at 8.9%
Debt financing through bonds	$3 million at 7.5%

Historically, Gentex has financed projects using a D-E mix of 40% from debt sources costing 7.5% per year and 60% from equity sources, which return 10% per year. (*a*) Compare the historical WACC value with that for this current genetics program. (*b*) Determine the MARR if a return of 5% per year is required by Gentex.

Solution

a. Equation [13.15] estimates the *historical* WACC.

$$\text{WACC} = 0.6(10) + 0.4(7.5) = 9.0\%$$

For the current program, the equity financing is comprised of two forms of equity capital—50% stock ($5 million out of $10 million) and 20% retained earnings. The remaining 30% is from debt sources. To calculate the *current* program's WACC, rewrite Equation [13.15] for equity funds.

$$\text{WACC} = \text{stock portion} + \text{retained earnings portion} + \text{debt portion}$$
$$= 0.5(13.7) + 0.2(8.9) + 0.3(7.5) = 10.88\%$$

This is higher than the 9% historical average.

b. The program should be evaluated using a MARR of 10.88 + 5.0 = 15.88% per year.

The use of debt capital has some real advantages, but high D-E mixes, such as 50–50 and larger, are usually unhealthy. *The leverage offered by larger debt capital percentages increases the riskiness of projects undertaken by the company.* When large debts are already present, additional financing using debt (or equity) sources gets more difficult to justify, and the corporation can be placed in a situation where it owns a smaller and smaller portion of itself. This is sometimes referred to as a *highly leveraged* corporation. Inability to obtain operating and investment capital means increased difficulty for the company and its projects. Thus, a reasonable balance between debt and equity financing is important for the financial health of a corporation.

Three automobile subcontract companies have the following debt and equity cap-   **EXAMPLE 13.8**
ital amounts and D-E mixes. Assume all equity capital is in the form of stock.

	Amount of Capital		
Company	Debt ($ in millions)	Equity ($ in millions)	D-E Mix (% – %)
A	10	40	20–80
B	20	20	50–50
C	40	10	80–20

Assume the annual revenue is $15 million for each corporation and that, after
interest on debt is considered, the net incomes are $14.4, $13.4, and $10.0 mil-
lion, respectively. Compute the return on stock for each company, and comment
on the return relative to the D-E mixes.

Solution

Divide the net income by the equity amount to compute the stock return. In million dollars,

A:
$$\text{Return} = \frac{14.4}{40} = 0.36 \quad (36\%)$$

B:
$$\text{Return} = \frac{13.4}{20} = 0.67 \quad (67\%)$$

C:
$$\text{Return} = \frac{10.0}{10} = 1.00 \quad (100\%)$$

As expected, the return is by far the largest for highly leveraged C, where only
20% of the company is in the hands of the ownership. The return on owner's
equity is excellent, but the risk associated with this firm is high compared to A,
where the D-E mix is only 20% debt.

The same principles discussed previously for corporations are applicable to
individuals. The person who is highly leveraged has large debts in terms of credit
card balances, personal loans, and house mortgages. As an example, assume two
engineers each have a take-home amount of $60,000 after all income tax, social
security, and insurance premiums are deducted from their annual salaries. Further,
assume that the cost of their debt (money borrowed via credit cards and loans)
averages 15% per year and that the current debt principal is being repaid in equal
amounts over 20 years. If Jamal has a total debt of $25,000 and Barry owes
$150,000, the remaining amounts of the annual take-home pay vary considerably,
as shown on the next page. Jamal has $55,000, or 91.7% of his $60,000 take-home
available, while Barry has only 50% available.

Person	Total Debt, $	Annual Cost of Debt at 15%, $	Annual Repayment $	Amount Remaining from $60,000, $
Jamal	25,000	3,750	1,250	55,000
Barry	150,000	22,500	7,500	30,000

The previous computations are correct for a before-tax analysis. If an after-tax analysis is performed, the tax advantage of debt capital should be considered. In the United States, and many other countries, interest paid on all forms of debt capital (loans, bonds, and mortgages) is considered a corporate expense, and is therefore, tax deductible. Equity capital does not have this advantage; dividends on stocks, for example, are not considered tax deductible. Though a detailed analysis can be performed, the easiest way to estimate an after-tax WACC via Equation [13.15] is to approximate the cost of debt capital using the effective tax rate T_e.

$$\textbf{After-tax cost of debt = (before-tax cost)}(1 - T_e) \qquad \textbf{[13.16]}$$

The cost of debt capital will be less after taxes are considered, but the cost of equity will remain the same in the after-tax WACC computation. Once determined, the after-tax MARR can be set. (Alternatively, the after-tax MARR can be approximated using the logic of Section 13.2.)

EXAMPLE 13.9 If Gentex's effective tax rate is 38%, determine the after-tax MARR to be applied when evaluating the program described in Example 13.7.

Solution

The after-tax WACC will *decrease* from the previous result of WACC = 10.88% based on the tax-advantaged cost of debt capital approximated by Equation [13.16].

$$\text{After-tax cost of debt} = (7.5\%)(1 - 0.38) = 4.65\%$$
$$\text{After-tax WACC} = 0.5(13.7) + 0.2(8.9) + 0.3(4.65) = 10.03\%$$

Assuming that the expected 5% per year return is after taxes, the after-tax MARR is 15.03% per year.

13.6 USING SPREADSHEETS FOR AFTER-TAX EVALUATION

The next example illustrates spreadsheet use for several after-tax evaluation techniques, including CFAT computation (using Table 13.3 format), PW and incremental ROR evaluations, and finally graphical breakeven ROR analysis. The solution stages build on each other in order to allow study of selected areas.

Two units offer similar features to perform bone density analysis for a skeletal diagnostic clinic for each of 10 NBA teams during a 3-year contract period. (The situation here is the same as that in Example 13.4.) Use MACRS 5-year recovery, $T_e = 35\%$, and an after-tax MARR of 10% per year to perform the following analyses for the 3-year contract:

EXAMPLE 13.10

a. Present worth and annual worth
b. Incremental ROR
c. Graphical determination of breakeven ROR to compare with MARR $= 10\%$.

	Analyzer 1	Analyzer 2
First cost, $	150,000	225,000
Gross income, $/year	100,000	100,000
AOC, $/year	30,000	10,000
MACRS recovery, years	5	5
Estimated selling price after 3 years, $	130,000	225,000

Solution

Depreciation, income taxes, depreciation recapture, and salvage (selling price) are all included in this evaluation. The combination of Equations [13.7] and [13.13] define the CFAT relation for the spreadsheets. (No capital gains or losses are anticipated.)

$$\text{CFAT} = \text{CFBT} - \text{taxes}$$
$$= \text{GI} - E - P + S - (\text{GI} - E - D + \text{DR})(T_e)$$

The spreadsheet and cell tags in Figure 13.4 detail CFAT calculation. At the end of the contract, year 3 computations of CFAT use the relation above. In $1000 units, they are

Analyzer 1: $\text{CFAT}_3 = 100 - 30 + 130 - [100 - 30 - 28.8 + (130 - 43.2)](0.35)$
$$= 200 - 128(0.35)$$
$$= \$155.20$$

Analyzer 2: $\text{CFAT}_3 = 100 - 10 + 225 - [100 - 10 - 43.2 + (225 - 64.8)](0.35)$
$$= 315 - 207(0.35)$$
$$= \$242.55$$

In practice, only one of the following analyses is necessary. For illustration purposes only, all four are presented here.

a. Figure 13.4: Since the 3-year contract period is the same for both alternatives, the NPV functions at MARR $= 10\%$ are excellent for determining PW values for the CFAT series (column I). Likewise, the PMT function for 3 years is used to obtain AW from PW. (The minus sign on PMT ensures

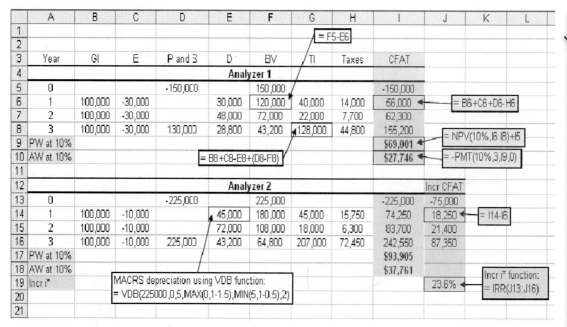

FIGURE 13.4   After-tax PW, AW, and incremental ROR analyses including depreciation recapture, Example 13.10a and b.

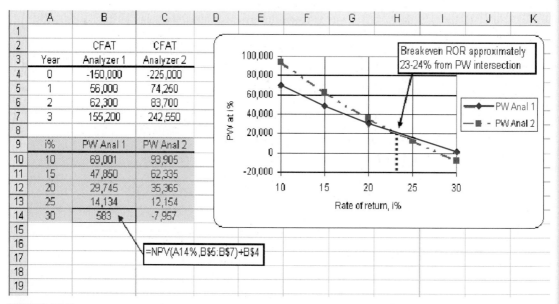

FIGURE 13.5   Graphical determination of breakeven ROR using PW, Example 13.10c.

that AW maintains the same sign as PW.) Analyzer 2 is selected with the larger PW.

Calculating AW is not necessary since the conclusion will always be the same as PW analysis for equal-life alternatives. Were lives unequal, the LCM years would be used in the NPV and PMT functions to ensure equal service comparison.

b. Figure 13.4: Column J presents the incremental CFAT of the larger investment in analyzer 2 over analyzer 1. The incremental $i^* = 23.6\%$ obtained using the IRR function (cell J19) shows that analyzer 2 significantly exceeds the 10% after-tax requirement.

c. Figure 13.5: The upper table repeats the two CFAT series. In the lower table, the NPV functions determine the PW values at different i values. An xy scatter chart shows the PW curves to cross in the range of 23 or 24% per year. (The exact value was determined previously to be 23.6%.) Since this breakeven value exceeds 10%, the larger investment in analyzer 2 is justified.

13.7 AFTER-TAX VALUE-ADDED ANALYSIS

Value added indicates that a product or service has added economic worth from the perspective of the owner, an investor, or a consumer. It is possible to highly leverage the value-added activities of a process. For example, onions are sold at the farm level for cents per pound. They may be purchased in a store for $1 to $2 per pound. When onions are cut and coated with a special batter, they may be fried in hot oil and sold as onion rings for dollars per pound. From the perspective of the consumer's willingness to pay, there has been great value added by the processing.

Value-added analysis of an alternative is performed in a slightly different way than CFAT analysis. But the decision about an alternative will be the same, because the AW of economic value added (EVA) is the same as the AW of CFAT.*

Value added analysis starts with the net profit after taxes (NPAT). This is the amount remaining each year when income taxes are subtracted from taxable income.

$$\text{NPAT} = \text{taxable income} - \text{taxes} = \text{TI} - (\text{TI})(T_e)$$
$$= (\text{TI})(1 - T_e) \qquad\qquad [13.17]$$

Also called net income, NPAT implicitly includes depreciation accumulated thus far when TI is calculated. This is different from CFAT, where the depreciation is specifically removed so that only *actual* cash flows are used.

The annual EVA is the amount of NPAT remaining after removing the *cost of invested capital* during the year. That is, EVA indicates the project's contribution to the net worth of the corporation after taxes. The cost of invested capital is the after-tax MARR multiplied by the book value of the asset during the year. This is

---

*W. G. Sullivan and K. L. Needy, "Determination of Economic Value Added for a Proposed Investment in New Manufacturing." *The Engineering Economist*, vol. 45, no. 2 (2000), pp. 166–181.

the interest incurred by the current level of capital invested in the asset. Computationally, including Equation [13.17],

$$\begin{aligned}
\text{EVA} &= \textbf{NPAT} - \textbf{cost of invested capital} \\
&= \textbf{NPAT} - \textbf{(after-tax interest rate)(book value in year } t - 1) \\
&= \text{TI}(1 - T_e) - (i)(\text{BV}_{t-1}) \qquad\qquad\qquad\qquad\qquad \textbf{[13.18]}
\end{aligned}$$

Since both TI and BV consider depreciation, EVA is a measure of worth that mingles actual cash flows with noncash flows to determine financial worth. This financial worth is the amount used in public documents of the corporation (balance sheet, income statement, stock reports, etc.). Because corporations want to present the largest value possible to stockholders, the EVA method is often more appealing than the AW method.

The result of an EVA analysis is a series of annual EVA estimates. Calculate the AW of EVA estimates and select the alternative with the larger AW value. If only one project is evaluated, AW > 0 means the after-tax MARR is exceeded, thus making the project value adding. Since the final AW of EVA and the AW of CFAT values are equal, either method can be used. The annual EVA estimates indicate added worth to the corporation generated by the alternative, while the annual CFAT estimates describe how cash will flow. This comparison is made in the next (spreadsheet) example.

EXAMPLE 13.11   Electrical engineers and medical doctors at First Hope Health Center have developed a project for investing in new capital equipment with the expectation of increased revenue from its medical diagnostic services to cancer patients. The estimates are summarized below. (*a*) Use classical straight line depreciation, an after-tax MARR of 12%, and an effective tax rate of 40% to perform two annual worth after-tax analyses: EVA and CFAT. (*b*) Explain the fundamental difference between the results of the two analyses.

Initial investment	$500,000
Gross income − expenses	$170,000 per year
Estimated life	4 years
Salvage value	None

Solution

a. *EVA evaluation:* All the necessary information for EVA estimation is in columns B through G of Figure 13.6. The NPAT (column H) is calculated by Equation [13.17]. The book values (column E) are used to determine the cost of invested capital in column I, using the second term in Equation [13.18], that is, $i(\text{BV}_{t-1})$, where i is 12%. This represents the amount of interest for the currently invested capital. The annual EVA is the sum of columns H and I. *Notice there is no EVA estimate for year 0, since NPAT and the cost of* ▼

	A	B	C	D	E	F	G	H	I	J	K	L
1												
2									EVA analysis		CFAT analysis	
3				SL					Cost of			
4	Year	GI - E	P	Depreciation	BV	TI	Taxes	NPAT	invested capital	EVA	CFAT	
5	0		-500,000		500,000						-500,000	
6	1	170,000		125,000	375,000	45,000	18,000	27,000	-60,000	-33,000	152,000	
7	2	170,000		125,000	250,000	45,000	18,000	27,000	-45,000	-18,000	152,000	
8	3	170,000		125,000	125,000	45,000	18,000	27,000	-30,000	-3,000	152,000	
9	4	170,000		125,000	0	45,000	18,000	27,000	-15,000	12,000	152,000	
10	AW									-$12,617	-$12,617	
11					= B9 - D9		= F9 - G9		= - 0.12*E8			
12											= B6 +C6 -G6	
13							= PMT(12%,4,-(NPV(12%,J6:J9)))					
14												

FIGURE 13.6   Project evaluation using EVA and CFAT approaches, Example 13.11.

invested capital are estimated for years 1 through n. The negative AW of EVA (J10) indicates that plan A does not make the 12% return.

CFAT evaluation: The CFAT (column K) is calculated as $GI - E - P -$ taxes. The AW of CFAT again concludes that plan A does not return the 12%.

b. The series in columns J and K are clearly equivalent since the AW values are numerically the same. To explain the difference, consider, the constant CFAT of $152,000 per year. To obtain the AW of EVA estimate of $-12,617$ for years 1 through 4, the initial investment of $500,000 is distributed over the 4-year life using the A/P factor at 12%. That is, an equivalent amount of $500,000(A/P,12\%,4) = \$164,617$ is "charged" against the cash inflows in each of years 1 through 4. In effect, the yearly CFAT is reduced by this charge.

$$CFAT - (\text{initial investment})(A/P,12\%,4) = \$152,000 - 500,000(A/P,12\%,4)$$
$$152,000 - 164,617 = \$-12,617$$

This is the AW value for both series, demonstrating that the two methods are economically equivalent. The EVA method indicates yearly contribution to the *value of the corporation,* which is negative for the first three years. The CFAT method estimates the *actual cash flows* to the corporation.

13.8 TAX CONSIDERATIONS FOR INTERNATIONAL PROJECTS

Primary questions to be answered prior to performing a corporate-based after-tax analysis for international settings revolve around tax-deductible allowances—depreciation, business expenses, capital investment—and the effective tax rate T_e. As discussed in Chapter 12, most countries use the SL and DB methods of depreciation with some variations. Expense deductions vary widely. Some examples are summarized here.

Canada

Depreciation: This is deductible and is normally based on DB calculations, although SL may be used. An equivalent of the half-year convention is applied in the first year of ownership. The annual tax-deductible allowance is termed *capital cost allowance (CCA)*. As in the U.S. system, recovery rates are standardized.

Class and CCA rate: Asset classes are defined and annual depreciation rates are specified by class. No specific recovery period is identified, in part because assets of a particular class are grouped together and the annual CCA is determined for the entire class, not individual assets. There are some 44 classes, and CCA rates vary from 4% per year (the equivalent of a 25-year-life asset) for buildings (class 1) to 100% (1-year life) for applications software, chinaware, dies, etc. (class 12). Most rates are in the range of 10% to 30% per year.

Expenses: Business expenses are deductible in calculating TI. Expenses related to capital investments are not deductible, since they are accommodated through the CCA.

Internet: Further details are available on the Revenue Canada website at www.ccra-adrc.gc.ca in the Forms and Publications section.

Mexico

Depreciation: This is fully deductible for calculating TI. The SL method is applied with an index for inflation considered each year. For some asset types, an immediate deduction of a percentage of the first cost is allowed.

Class and rates: Major classes are identified, and annual recovery rates vary from 5% for buildings (the equivalent of a 20-year life) to 100% for environmental machinery. Most rates range from 10% to 30% per year.

Profit tax: The income tax is levied on profits on income earned from carrying on business in Mexico. Most business expenses are deductible. Corporate income is taxed only once, at the federal level; no state-level taxes are imposed.

Tax on Net Assets (TNA): A tax of 1.8% of the average value of assets located in Mexico is paid annually in addition to income taxes, but income taxes paid are credited toward TNA.

Internet: The best information is via websites for corporations that assist international corporations located in Mexico. One example is PriceWaterhouseCoopers at www.pwcglobal.com/mx.

Japan

Depreciation: This is fully tax-deductible and is based on classical SL or DB methods. A total of 95% of the first cost may be recovered through depreciation, but an asset's salvage value is assumed to be 10% of its acquisition cost. The capital investment is recovered for each asset or groups of similarly classified assets.

TABLE 13.8 **Summary of International Corporate Tax Rates (2006)***

Tax Rate Levied on Taxable Income, %	For These Countries
≥ 40	USA, Japan
36 to < 40	Canada, Germany, South Africa
32 to < 36	China, France, India, New Zealand, Spain
28 to < 32	Australia, Indonesia, United Kingdom, Mexico
24 to < 28	Russia, Taiwan, Republic of Korea
20 to < 24	Singapore, Albania, United Arab Emirates (non–oil related)
< 20	Hong Kong, Iceland, Ireland, Hungary, Poland, Chile, Oman

Sources: Extracted from KPMG's Corporate Tax Rates Survey, January 2006, available at www.kpmg.com/
Services/tax/IntCorp/CTR and from specific country websites on taxation for corporations.

Class and life: A statutory useful life ranging from 4 to 24 years, with a
50-year life for reinforced concrete buildings, is specified.

Expenses: Business expenses are deductible in calculating TI.

Internet: Further details are available on the Japanese Ministry of Finance
at www.mof.go.jp.

Average corporate tax rates vary considerably between countries. Some overall
rates are summarized in Table 13.8. Though these average rates will vary from year
to year, especially as tax reform is enacted, most corporations face average rates of
about 20% to 40% of taxable income. Corporate tax rate reductions are an effective
way to encourage foreign and domestic corporations to expand their capital investments within a country. As noted in the KPMG report identified below Table 13.8,
worldwide overall corporate tax rates have decreased considerably over the last 10 to
15 years, especially in Europe. European Union (EU) country rates decreased from
33.9% (2000) to 25.8% (2006), led primarily by Eastern European countries relatively
new to the EU. This is in contrast to the United States, which has maintained an average rate of 40% over the same time frame (except for 34% in 2003 and 2004).

SUMMARY

After-tax analysis does not usually change the decision to select one alternative over another; however,
it does offer a much clearer estimate of the monetary
impact of taxes. After-tax PW, AW, and ROR evaluations are performed on the CFAT series using exactly
the same procedures as in previous chapters.

U.S. income tax rates are graduated—higher
taxable incomes pay higher income taxes. A single-
value, effective tax rate T_e is usually applied in
an after-tax economic analysis. Taxes are reduced

because of tax-deductible items, such as depreciation and operating expenses.

If an alternative's estimated contribution to
corporate financial worth is the economic measure,
the economic value added (EVA) should be determined. Unlike CFAT, the EVA includes the effect
of depreciation. The AW of CFAT and EVA are the
same numerically; they interpret the annual cost of
the capital investment in different, but equivalent
manners.

PROBLEMS

Tax Terms and Computations

13.1 Explain the following three tax terms: graduated tax rates, marginal tax rate, and indexing.

13.2 Divisions of Doubleday Computers and Merritt-Douglas Computing make competing products for the commercial IT market worldwide. Use the data for each company to compare the following amounts in a tabular format:

a. Effective total tax rate T_e.

b. Estimate of federal income taxes using the average tax rate.

c. Federal taxes using the graduated tax rates and the percent of total revenue paid in federal income taxes.

	Doubleday	M-D
Sales, $	2.8 million	4.7 million
Other revenue, $	900,000	250,000
Expenses, $	1.4 million	3.1 million
Depreciation, $	850,000	970,000
Total of state and local tax rate, %	9.2	7.5
Average federal tax rate, %	34	34

13.3 Carl read the annual report of Harrison Engineering's 3-D Imaging Division. From it he deduced that GI = $4.9 million, E = $2.1 million, and D = $1.4 million. If the average federal tax rate is 31% and state/local tax rates total 9.8%, estimate (*a*) federal income taxes, and (*b*) the percent of GI that the federal government takes in income taxes.

13.4 Last year, Marylynn opened Baron's Appliance Sales and Service. Her tax accountant provided the year's results.

Gross income = $320,000
Business expenses = $149,000
MACRS depreciation = $95,000
Average federal tax rate = 18.5%
Average state tax rate = 6%
City and country flat tax rates combined = 4.5%

Marylynn wants to know the following. Determine them for her.

a. Taxable income.

b. Exact amount of federal income taxes.

c. Percent of the gross income taken by income taxes—federal, state, and local.

13.5 Write the taxable income (TI) relations for a corporation and for an individual. Identify at least three fundamental differences between the two relations as the TI is calculated.

13.6 Last year Carolyn was single; this year she is married to Vijay. Salaries, deductions, and other information are listed for the 2 years. Determine the federal income taxes for each when they were single and now as married and filing jointly using Table 13.2 rates. Is there a financial benefit or penalty for being married and filing as such? What is the size of the benefit or penalty?

	Carolyn, Last Year	Vijay, Last Year	Carolyn/Vijay, This Year
Salary, $	75,000	75,000	150,000
Other income, $	5,000	5,000	10,000
Exemptions and deductions, $	8,450	8,450	16,900
Filing status	Single	Single	Married and jointly

13.7 CJ operates Ads++, a legally registered Internet ad business. Assume he can file his federal income tax return as a small corporation or as a single individual. The GI of $173,000 is for Ads++ as a corporation, while the $95,000 salary is the amount CJ paid himself. (The excess of $20,000 between GI − E = $115,000 and CJ's salary is called retained earnings for Ads++. Appendix B describes accounting reports.)

	Ads++	Filing Single
Gross income, $	173,000	
Expenses, $	58,000	
Depreciation, $	10,000	
Salary, $		95,000
Exemptions, $		3,300
Deductions, $		5,150

CJ estimates an effective federal tax rate of $T_e = 25\%$ regardless of the method used to calculate taxes. For both methods of filing, determine (*a*) the estimated taxes using T_e and (*b*) taxes using the tax rate tables. Which method results in lower taxes for CJ?

Before-Tax and After-Tax Cash Flows

13.8 Danielle performed an after-tax AW analysis based on CFBT estimates and the effective corporate tax rate of 40% that includes all federal, state, and local taxes. The asset's first cost is $4 million with a life of 10 years. If the resulting AW is $−35.52 per year, should she recommend rejecting the project? If yes, why? If no, explain the next step she should take and explain why.

The following information is used in problems 13.9 through 13.12.

After 4 years of use, Procter and Gamble has decided to replace capital equipment used on its Zest bath soap line. Cash flow data is tabulated in $1000 units, and MACRS 3-year depreciation was used. After-tax MARR is 10% per year, and T_e is 35% in the United States.

Year	0	1	2	3	4
Purchase, $	−1900				
Gross income, $		800	950	600	300
Expenses, $		−100	−150	−200	−250
Salvage, $					700

13.9 Utilize the CFBT and AW value to determine if the cash flows over 4 years exceeded MARR.

13.10 Calculate MACRS depreciation and estimate the CFAT series over the 4 years. Neglect any tax impact caused by the $700 salvage received in year 4.

13.11 Utilize the CFAT and AW value to determine if the investment's cash flows over 4 years exceeded MARR.

13.12 Compare the after-tax ROR values using both methods—CFAT series and approximated from the CFBT values using the before-tax ROR and T_e.

13.13 Advanced Anatomists, Inc., researchers in medical science, is contemplating a commercial venture concentrating on proteins based on the new X-ray technology of free-electron lasers. To recover the huge investment needed, an annual $2.5 million CFAT is needed. A favored average federal tax rate of 20% is expected; however, state taxing authorities will levy an 8% tax on TI. Over a 3-year period, the deductible expenses and depreciation are estimated to total $1.3 million the first year, increasing by $500,000 per year thereafter. Of this, 50% is expenses and 50% is depreciation. What is the required gross income for each of the 3 years?

13.14 Two years ago, on the recommendation of its construction engineers, United Homebuilders purchased dumpsters, front-end loader, and truck equipment to carry off construction debris, rather than subcontracting the service. Information is listed below for the project with depreciation determined using 5-year MACRS.

	0	1	2	3	4	5	6
First cost, $	−350,000						
Savings, $		150,000	150,000				
Expenses, $		−25,000	−25,000				
Depreciation, $		70,000	112,000	67,200	40,320	40,320	20,160

a. The United president does not want to continue ownership, as he prefers subcontracting such services. What must the realizable market value of the owned capital equipment be to equal the current book value?

b. If savings are the equivalent of gross income, determine the CFBT series and rate of return over the 2-year period. Assume the market value from (*a*) is paid by an exporter of used heavy equipment.

c. Determine CFAT and after-tax ROR for the 2 years of ownership. Use the information from previous parts. Let $T_e = 40\%$.

d. Determine the minimum market value required to make the after-tax ROR exactly zero over the 2-year ownership period.

13.15 The information shown on the next page is for a marginally successful project. (*a*) Determine the CFBT and CFAT series. The effective tax rate is 32%. (*b*) Use manual computations or the spreadsheet function to obtain before-tax and after-tax rates of return to determine the change that depreciation and taxes make. Assume that tax savings are used to offset taxes in other parts of the corporation.

Year	Gross Income, $	Operating Expenses, $	Depreciation, $	First Cost and Salvage, $
0				−150,000
1	60,000	−55,000	35,000	
2	75,000	−50,000	35,000	
3	90,000	−45,000	35,000	
4	105,000	−40,000	35,000	10,000

13.16 Elias wants to perform an after-tax evaluation of equivalent methods to electrostatically remove airborne particulate matter from clean rooms used to package liquid pharmaceutical products. Using the information shown, MACRS depreciation with $n = 3$ years, a 5-year study period, after-tax MARR = 7% per year, and $T_e = 34\%$ and Excel, he obtained the results $AW_A = \$-2176$ and $AW_B = \$3545$. Any tax effects when the equipment is salvaged were neglected. Method B is the better method. Use classical SL depreciation with $n = 5$ years to select the better method. Is the decision different from that reached using MACRS?

	Method A	Method B
First cost, $	−100,000	−150,000
Salvage value, $	10,000	20,000
Savings, $ per year	35,000	45,000
AOC, $ per year	−15,000	−6,000
Expected life, years	5	5

Depreciation Effects on Taxes

13.17 Last month, a company specializing in wind power plant design and engineering made a large capital investment of $400,000 in physical simulation equipment that will be used for at least 5 years, then sold for approximately 25% of the first cost. By law, the assets are MACRS depreciated using a 3-year recovery period.

a. Explain why there is a predictable tax implication when the assets are sold.

b. By how much will the sale cause TI and taxes to change in year 5?

13.18 Though capital gains and losses can make significant differences in CFAT estimates in the year that a depreciable asset is salvaged, as a matter of practice gains and losses are generally neglected when the evaluation is performed. Why is this? Identify at least one recommended exception to this practice.

The following information is used in problems 13.19 through 13.22.

Open Access, Inc. is an international provider of computer network communications gear. Different depreciation, recovery period, and tax law practices in the three countries where depreciable assets are located are summarized in the table below. Also, information about assets purchased 5 years ago at each location and sold this year is provided. After-tax MARR = 9% per year and $T_e = 30\%$ can be used for all countries.

Practice or Estimate	Country 1	Country 2	Country 3
Depreciation method	SL with $n = 5$	MACRS with $n = 3$	DDB with $n = 5$
Depreciation recapture	Not taxed	Taxed as TI	Taxed as TI
First cost, $	−100,000	−100,000	−100,000
Gross income-expenses, $/year	25,000	25,000	25,000
Estimated salvage, $	0 in year 5	0 in year 5	20,000 in year 5
Life, years	5	5	5
Actual selling price, $	20,000 in year 5	20,000 in year 5	20,000 in year 5

13.19 For Country 1, SL depreciation is $20,000 per year. Determine the (a) CFAT series and (b) PW of depreciation, taxes, and CFAT series.

13.20 For Country 2, MACRS depreciation for the 4 years is $33,333, $44,444, $14,815, and $7,407, respectively. Determine the (a) CFAT series and (b) PW of depreciation, taxes, and CFAT series.

13.21 For Country 3, DDB depreciation for the 5 years is $40,000, $24,000, $14,400, $1,600, and 0, respectively. Determine the (a) CFAT series and (b) PW of depreciation, taxes, and CFAT series.

13.22 If you worked the previous three problems, develop a table that summarizes, for each country, the total taxes paid and the PW values of the depreciation, taxes, and CFAT series. For each criterion, select the country that provides the best PW value. Explain why the same country is not selected for all three criteria. (Hint: The PW should be minimized for some and maximized for other criteria. Review first to be sure you choose correctly.)

13.23 Cheryl, a EE student who is working on a business minor, is studying depreciation and finance in her engineering management course. The assignment is to demonstrate that shorter recovery periods result in the same total taxes, but they offer a time value of taxes advantage for depreciable assets. Help her using asset estimates made for a 6-year study period: $P = \$65,000$, $S = \$5000$, GI = $32,000 per year, AOC is $10,000 per year, SL depreciation, $i = 12\%$ per year, $T_e = 31\%$. The recovery period is either 3 or 6 years.

13.24 A bioengineer is evaluating methods used to apply the adhesive onto microporous paper tape that is commonly used after surgery. The machinery costs $200,000, has no salvage value, and the CFBT estimate is $75,000 per year for up to 10 years. The $T_e = 38\%$ and $i = 8\%$ per year. The two depreciation methods to consider are:

> MACRS with $n = 5$ years
>
> SL with $n = 8$ years (neglect the half-year convention effect)

For a study period of 8 years, (a) determine which depreciation method and recovery period offers the better tax advantage, and (b) demonstrate that the same total taxes are paid for MACRS and SL depreciation.

13.25 Thomas completed an after-tax study of a $1 million 3-year-old DNA analysis and modeling system that DynaScope Enterprises wants to keep for 1 more year or dispose of now. His table (in $1000 units) details the analysis, including an anticipated $100,000 selling price (SP) next year, SL depreciation, taxes at the all-inclusive rate of $T_e = 52\%$, and PW at the after-tax MARR of 5% per year. Thomas recommends retention since PW > 0. Critique the analysis to determine if he made the correct recommendation.

Year	CFBT	SP	Depr	TI	Taxes	CFAT
0	$-1000					$-1000
1	275		$250	$25	$13	262
2	275		250	25	13	262
3	275		250	25	13	262
4	275	$100	250	25	13	362
PW @ 5%						$ 11.3

After-Tax Replacement Analysis

13.26 Though an after-tax replacement study may not change the decision to retain the defender or purchase the challenger, the difference between after-tax and before-tax PW, AW, or ROR values can change significantly. List at least five things considered in an after-tax evaluation that can cause these changes.

13.27 Capital gains and losses are usually neglected in an after-tax study. However, it may be important to consider a possible large capital loss in an after-tax replacement study. Why?

13.28 Justyne needs assistance with a replacement study for the information shown on the next page. Perform the PW study using after-tax MARR = 12% per year, $T_e = 35\%$, and a study period of 4 years. All monetary values are in $1000 units. (Notes: Assume that either asset is salvaged in the future at its original salvage estimate. Since no revenues are estimated, all taxes are negative and considered "savings" to the alternative. Neglect possible capital gains or losses.)

	Defender	Challenger
First cost, $	−45	−24
Estimated S at purchase, $	5	0
Market value now, $	35	—
AOC, $ per year	−7	−8
Depreciation method	SL	MACRS
Recovery period, years	8	3
Useful life, years	8	5
. Years owned	3	—

13.29 After 8 years of use, the heavy truck engine overhaul equipment at Pete's Truck Repair was evaluated for replacement. Pete's accountant used an after-tax MARR of 8% per year, $T_e = 30\%$, and a current market value of $25,000 to determine AW = $2100. The new equipment costs $75,000, uses SL depreciation over a 10-year recovery period, and has a $15,000 salvage estimate. Estimated CFBT is $15,000 per year. Pete asked his engineer son Ramon to determine if the new equipment should replace what is owned currently. From the accountant, Ramon learned the current equipment cost $20,000 when purchased and reached a zero book value several years ago. Help Ramon answer his father's question.

Debit/Equity Capital and WACC

13.30 Two public corporations, First Engineering and Midwest Development, each show capitalization of $175 million in their annual reports. The balance sheet for First indicates total debt of $87 million, and that of Midwest indicates net worth of $62 million. Determine the D-E mix for each company.

13.31 Two women with new jobs purchase the same amounts of clothes and shoes to wear at work. Amanda maintained a $2000 balance without any repayment on her credit card for exactly 1 year at an interest rate of 18% per year compounded monthly. There is no tax advantage for credit card interest. Charlotte took the $2000 from her savings account and did not earn the effective after-tax rate of 8% per year for exactly 1 year on this amount. (*a*) What type of financing did each woman use? (*b*) What is the total amount needed for Amanda to pay off the card in full at the end of the year, and for Charlotte to replenish her savings completely at the end of the year?

13.32 Determine if each of the following involves debt or equity financing.
   a. A privately owned oil and gas company purchased $5 million of offshore drilling equipment outright using funds from a corporate bank account.
   b. The Whitestones obtained a $20,000 home equity loan to upgrade bathrooms.
   c. MA Handley Corporation buys back $16 million of its own stock using internal funds.
   d. Garden City offers $58 million in 4.5% municipal bonds to build a new police headquarters.
   e. Nicole is forced to withdraw $55,800 from her 403(b) retirement plan to pay for medical expenses not covered by her husband's health insurance plan.

13.33 Business and engineering seniors are comparing methods of financing their college education during their senior year. The business student has $30,000 in student loans that come due at graduation. Interest is an effective 4% per year. The engineering senior owes $50,000; 50% from his parents with no interest due, and 50% from a credit union loan. This latter amount is also due at graduation with an effective rate of 7% per year.
   a. What is the D-E mix for each student?
   b. If their grandparents pay the loans in full at graduation, what are the amounts on the checks they write for each graduate?
   c. When grandparents pay the full amount at graduation, what percent of the loan's principal does the interest represent?

13.34 Fruit Transgenics Engineering (FTE) is contemplating the purchase of its rival. One of FTE's genetics engineers is interested in the financing strategy of the buyout. He learned of two plans. Plan A requires 50% equity funds from FTE retained earnings that currently earn 9% per year, with the balance borrowed externally at 6%, based on the company's excellent stock rating. Plan B requires only 20% equity funds with the balance borrowed at a higher rate of 8% per year.
   a. Which plan has the lower average cost of capital?
   b. If the current corporate WACC of 8.2% will not be exceeded, what is the maximum cost of debt capital allowed for each plan? Are these rates higher or lower than the current estimates?

13.35 Dougherty International has worked on eight major housing projects in the United States, Mexico, and Canada during the last year. The D-E mixes and rates are shown. Plot the WACC curve and identify the D-E mix that had the lowest WACC.

	Debt		Equity	
Project	Percent	Rate	Percent	Rate
203	100	10.9	0	0
206	50	7.0	50	8.5
306	65	11.6	35	7.5
367	15	8.2	85	6.0
456	0	0	100	8.9
913	10	5.5	90	7.2
914	73	11.4	27	8.4
987	80	10.5	20	8.1

13.36 Deavyanne Johnston, the engineering manager at TZO Chemicals, wants to complete an alternative evaluation study. She asked the finance manager for the corporate MARR. The finance manager gave her some data on the project and stated that all projects must clear their average (pooled) cost by at least 4%.

Funds Source	Amount, $	Average Cost, %
Retained earnings	4 million	7.4
Stock sales	6 million	4.8
Long-term loans	5 million	9.8
Budgeted funds for project	15 million	

    a. Use the data to determine the minimum MARR.

    b. The study is after-taxes and part (*a*) provided the before-tax MARR. Determine the correct MARR to use if T_e was 32% last year and the finance manager meant that the 4% above the cost is for after-tax evaluations.

13.37 Justin and Greg have bank cards that can be used as either credit or debit cards. Credit purchases are charged interest on the balance each month; debit charges carry no interest, but the amount is deducted immediately from the linked bank account.

    a. Normally, what is the type of financing when the credit card function is used? when the debit card function is applied?

    b. Over the last year, purchases have totaled to the amounts shown. Based on payment timing, assume both have paid an effective interest rate of 10.5% for the year on their credit purchases. Determine the D-E mix for each man.

	Credit, $	Debit, $
Justin	6590	2300
Greg	2300	6590

13.38 Bow Chemical will invest $14 million this year to upgrade its ethylene glycol processes. This chemical is used to produce polyester resins to manufacture products varying from construction materials to aircraft, and from luggage to home appliances. Equity capital costs 14.5% per year and will supply 65% of the capital funds. Debt capital costs 10% per year before taxes. The effective tax rate is 36%.

    a. Determine the amount of annual revenue after taxes that is consumed in covering the interest on the project's initial cost.

    b. If the corporation does not want to use 65% of its own funds, the financing plan may include 75% debt capital. Determine the amount of annual revenue needed to cover the interest with this plan, and explain the effect it may have on the corporation's ability to borrow in the future.

Economic Value Analysis

13.39 While an engineering manager may prefer to use CFAT estimates to evaluate the AW of a project, a financial manager may select AW of EVA estimates. Why are these preferences predictable?

13.40 Use the information in Example 13.3 (Table 13.4) to calculate the AW values of the CFAT and EVA series. They should have the same value. The after-tax MARR is 9.75%.

13.41 Triple Play Innovators Corporation (TPIC) plans to offer IPTV (Internet Protocol TV) service to North American customers starting soon. Perform an AW analysis of the EVA series for the two alternative suppliers available for the hardware and software. Let $T_e = 30\%$ and after-tax MARR = 8%; use SL depreciation (neglect half-year convention and MACRS, for simplicity) and a study period of 8 years.

Vendor	Hong Kong	Japan
First cost, $	4.2 million	3.6 million
Recovery period, years	8	5
Salvage value, $	0	0
GI − E, $ per year	1,500,000 in year 1; increasing by 300,000 per year up to 8 years	

International Corporate Taxes

13.42 Answer the following questions concerning international corporate tax rates and tax-deductible items.

a. How is the United States government unique in the world in its allowed methods for depreciating assets?

b. For what purpose may a country reduce domestic and foreign corporate tax rates?

c. What general statement can be made about the expected range of corporate tax rates as a percent of taxable income?

d. How do most countries define the allowable recovery period for depreciable assets?

13.43 Go to a website in your country that identifies tax rates. Compare them by TI range (or equivalent) and percentage with those (*a*) for corporations listed in Table 13.1, and (*b*) for individuals listed in Table 13.2. How do they compare?

PROBLEMS FOR TEST REVIEW AND FE EXAM PRACTICE

13.44 A graduated income tax system means

a. only taxable incomes above a certain level pay any taxes.

b. a higher flat rate goes with all of the taxable income.

c. higher tax rates go with higher taxable incomes.

d. rates are indexed each year to keep up with inflation.

13.45 Nicole, who files her tax return as a single individual, has an effective tax rate of 25%. This year she has the following: gross income of $55,000, other income of $4000, expenses of $15,000, and personal deductions and exemptions of $12,000. The income tax due is closest to

a. $11,750.

b. $8,750.

c. $10,750.

d. $13,750.

13.46

Year	Investment	GI − E	D	TI	Taxes	CFAT
0	−60,000					−60,000
1		30,000				26,000
2		35,000	15,000	6000		29,000

The after-tax analysis for a $60,000 investment with associated gross income minus expenses (GI − E) is shown above for the first 2 years only. If the effective tax rate is 40%, the values for depreciation (*D*), taxable income (TI), and taxes for year 1 are closest to

a. Depr = $5,000, TI = $25,000, Taxes = $10,000.

b. Depr = $30,000, TI = $30,000, Taxes = $4000.

c. Depr = $20,000, TI = $50,000, Taxes = $20,000.

d. Depr = $20,000, TI = $10,000, Taxes = $4000.

13.47 If the after-tax rate of return for a cash flow series is 11.2% and the corporate effective tax rate is 39%, the approximated before-tax rate of return is closest to

a. 6.8%.

b. 5.4%.

c. 18.4%.

d. 28.7%.

13.48 An asset purchased for $100,000 with S = $20,000 after 5 years was depreciated using the 5-year MACRS rates. Expenses average $18,000 per year and the effective tax rate is 30%. The asset is actually sold after 5 years of service for

$22,000. MACRS rates in years 5 and 6 are 11.53% and 5.76%, respectively. The after-tax cash flow from the sale is closest to

a. $27,760.

b. $17,130.

c. $26,870.

d. $20,585.

13.49 When accelerated depreciation methods or shortened recovery periods are applied, there are impacts on the income taxes due. Of the following, the statement(s) that are commonly *incorrect* are

1. Total taxes paid are the same for all depreciation methods.

2. Present worth of taxes is lower for smaller recovery periods.

3. Accelerated depreciation imposes more taxes in the later years of the recovery period.

4. Present worth of taxes is higher for smaller recovery periods.

a. 1, 2, and 3.

b. 1 and 4.

c. 2.

d. 4.

13.50 The Wilkins Company has maintained a 50-50 D-E mix for capital investments. Equity capital has cost 11%; however, debt capital that historically cost 9% has now *increased by* 20% per year. If Wilkins does not want to exceed its historical weighted average cost of capital (WACC), and it is forced to go to a D-E mix of 75-25, the maximum cost of equity capital that Wilkins can accept is closest to

a. 9.8%.

b. 10.9%.

c. 7.6%.

d. 9.2%.

A

Appendix

Using Spreadsheets and Microsoft Excel©

This appendix explains the layout of a spreadsheet and the use of Microsoft Excel (hereafter called Excel) functions in engineering economy. Refer to the Excel help system for your particular computer and version of Excel for additional detail.

A.1 INTRODUCTION TO USING EXCEL

Enter a Formula or Use an Excel Function

The = sign is necessary to perform any formula or function computation in a cell. The formulas and functions on the worksheet can be displayed by pressing Ctrl and `. The symbol ` is usually in the upper left of the keyboard with the ~ (tilde) symbol. Pressing Ctrl+` a second time hides the formulas and functions. Some examples of entries follow.

1. Run Excel.
2. Move to cell C3. (Move the mouse pointer to C3 and left-click.)
3. Type = PV(5%,12,10) and <Enter>. This function will calculate the present value of 12 payments of $10 at a 5% per year interest rate.

Another example: To calculate the future value of 12 payments of $10 at 6% per year interest, do the following:

1. Move to cell B3, and type INTEREST.
2. Move to cell C3, and type 6% or = 6/100.
3. Move to cell B4, and type PAYMENT.
4. Move to cell C4, and type 10 (to represent the size of each payment).
5. Move to cell B5, and type NUMBER OF PAYMENTS.
6. Move to cell C5, and type 12 (to represent the number of payments).
7. Move to cell B7, and type FUTURE VALUE.
8. Move to cell C7, and type = FV(C3,C5,C4) and hit <Enter>. The answer will appear in cell C7.

348

To edit the values in cells,

1. Move to cell C3 and type 5% or $= 5/100$ (the previous value will be replaced).
2. The value in cell C7 will update.

Cell References in Formulas and Functions

If a cell reference is used in lieu of a specific number, it is possible to change the number once and perform sensitivity analysis on any variable (entry) that is referenced by the cell number, such as C5. This approach defines the referenced cell as a *global variable* for the worksheet. There are two types of cell references—relative and absolute.

Relative References

If a cell reference is entered, for example, A1, into a formula or function that is copied or dragged into another cell, the reference is changed relative to the movement of the original cell. If the formula in C5 is $= A1$, and it is copied into cell C6, the formula is changed to $= A2$. This feature is used when dragging a function through several cells, and the source entries must change with the column or row.

Absolute References

If adjusting cell references is not desired, place a $ sign in front of the part of the cell reference that is not to be adjusted—the column, row, or both. For example, $= \$A\$1$ will retain the formula when it is moved anywhere on the worksheet. Similarly, $= \$A1$ will retain the column A, but the relative reference on 1 will adjust the row number upon movement around the worksheet.

Absolute references are used in engineering economy for sensitivity analysis of parameters such as MARR, first cost, and annual cash flows. In these cases, a change in the absolute-reference cell entry can help determine the sensitivity of a measure such as PW or AW.

Print the Spreadsheet

First define the portion (or all) of the spreadsheet to be printed.

1. Move the mouse pointer to the top left corner of your spreadsheet.
2. Hold down the left click button. (Do not release the left click button.)
3. Drag the mouse to the lower right corner of your spreadsheet or to wherever you want to stop printing.
4. Release the left click button. (It is ready to print.)
5. Left-click the File top bar menu.
6. Move the mouse down to select Print and left-click.
7. In the Print dialog box, left-click the option Selection in the Print What box.
8. Left-click the OK button to start printing.

Create a Column Chart

1. Run Excel.
2. Move to cell A1 and type 1. Move down to cell A2 and type 2. Type 3 in cell A3, 4 in cell A4, and 5 in cell A5.
3. Move to cell B1 and type 4. Type 3.5 in cell B2; 5 in cell B3; 7 in cell B4; and 12 in cell B5.
4. Move the mouse pointer to cell A1, left-click and hold, while dragging the mouse to cell B5. (All the cells with numbers should be highlighted.)
5. Left-click on the Chart Wizard button on the toolbar.
6. Select the Column option in step 1 of 4 and choose the first subtype of column chart.
7. Left-click and hold the Press and Hold to View Sample button to determine you have selected the type and style of chart desired. Click Next.
8. Since the data were highlighted previously, step 2 can be passed. Left-click Next.
9. For step 3 of 4, click the Titles tab and the Chart Title box. Type Sample 1.
10. Left-click Category (X) axis box and type Year, then left-click Value (Y) axis box and type Rate of return. There are other options (gridlines, legend, etc.) on additional tabs. When finished, left-click Next.
11. For step 4 of 4, left-click As Object In; Sheet1 is highlighted.
12. Left-click Finish, and the chart appears on the spreadsheet.
13. To adjust the size of the chart window, left-click anywhere inside the chart to display small dots on the sides and corners. The words Chart Area will appear immediately below the arrow. Move the mouse to a dot, left-click and hold, then drag the dot to change the size of the chart.
14. To move the chart, left-click and hold within the chart frame, but outside of the graphic itself. A small crosshairs indicator will appear as soon as any movement in the mouse takes place. Changing the position of the mouse moves the entire chart to any location on the worksheet.
15. To adjust the size of the plot area (the graphic itself) within the chart frame, left-click within the graphic. The words Plot Area will appear. Left-click and hold any corner or side dot, and move the mouse to change the size of the graphic up to the size of the chart frame.

Other features are available to change the specific characteristics of the chart. Left-click within the chart frame and click the Chart button on the toolbar at the top of the screen. Options are to alter Chart Type, Source Data, and Chart Options. To obtain detailed help on these, see the help function, or experiment with the sample Column Chart.

Create an xy (Scatter) Chart

This chart is one of the most commonly used in scientific analysis, including engineering economy. It plots pairs of data and can place multiple series of entries on the Y axis. The *xy* scatter chart is especially useful for results such as the PW vs.

i graph, where *i* is the X axis and the Y axis displays the results of the NPV function for several alternatives.

1. Run Excel.
2. Enter the following numbers in columns A, B, and C, respectively.
   Column A, cell A1 through A6: Rate *i%*, 4, 6, 8, 9, 10
   Column B, cell B1 through B6: $ for A, 40, 55, 60, 45, 10
   Column C, cell C1 through C6: $ for B, 100, 70, 65, 50, 30.
3. Move the mouse to A1, left-click, and hold while dragging to cell C6. All cells will be highlighted, including the title cell for each column.
4. If all the columns for the chart are not adjacent to one another, first press and hold the Control key on the keyboard during the entirety of step 3. After dragging over one column of data, momentarily release the left click, then move to the top of the next (nonadjacent) column for the chart. Do not release the Control key until all columns to be plotted have been highlighted.
5. Left-click on the Chart Wizard button on the toolbar.
6. Select the *xy* (scatter) option in step 1 of 4, and choose a subtype of scatter chart.

The rest of the steps (7 and higher) are the same as detailed earlier for the Column chart. The Legend tab in step 3 of 4 of the Chart Wizard process displays the series labels from the highlighted columns. (Only the bottom row of the title can be highlighted.) If titles are not highlighted, the data sets are generically identified as Series 1, Series 2, etc. on the legend.

A.2 ORGANIZATION (LAYOUT) OF THE SPREADSHEET

A spreadsheet can be used in several ways to obtain answers to numerical questions. The first is as a rapid solution tool, often with the entry of only a few numbers or one predefined function. For example, to find the future worth in a single-call operation, move the mouse to cell B4 and type = FV(8%,5,−2500). The number $14,666.50 is displayed as the 8% per year future worth at the end of the fifth year of five payments of $2500 each. A second use is more formal; it may present data, answers, graphs and tables that identify what problem(s) the spreadsheet solves. Some fundamental guidelines useful in setting up the spreadsheet follow. A very simple layout is presented in Figure A.1. As the solutions become more complex, an orderly arrangement of information makes the spreadsheet easier to read and use by you and others.

Cluster the data and the answers. It is advisable to organize the given or estimated data in the top left of the spreadsheet. A very brief label should be used to identify the data, for example, MARR = in cell A1 and the value, 12%, in cell B1. Then B1 can be the referenced cell for all entries requiring the MARR. Additionally, it may be worthwhile to cluster the answers into one area and frame it using the Outside Border button on the toolbar. Often, the answers are best placed at the bottom or top of the column of entries used in the formula or predefined function.

Enter titles for columns and rows. Each column or row should be labeled so its entries are clear to the reader. It is very easy to select from the wrong column or row when no brief title is present.

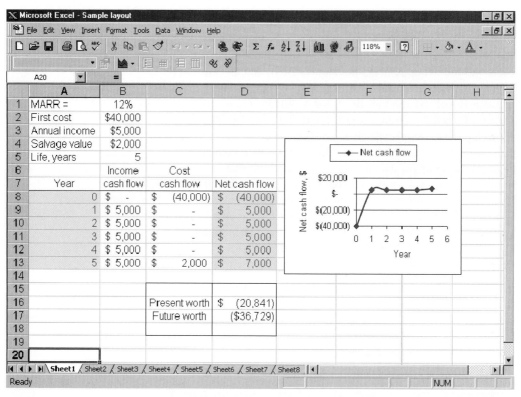

FIGURE A.1   Sample spreadsheet layout with estimates, results of formulas and functions, and an *xy* scatter chart.

Enter income and cost cash flows separately. When there are both income and cost cash flows involved, it is strongly recommended that the cash flow estimates for revenue (usually positive) and first cost, salvage value, and annual costs (usually negative, with salvage a positive number) be entered into two adjacent columns. Then a formula combining them in a third column displays the net cash flow. There are two immediate advantages to this practice: this reduces errors made when performing the summation and subtraction mentally, and changes for sensitivity analysis are more easily made.

Use cell references. The use of absolute and relative cell references is a must when any changes in entries are expected. For example, suppose the MARR is entered in cell B1, and three separate references are made to the MARR in functions on the spreadsheet. The absolute cell reference entry B1 in the three functions allows the MARR to be changed one time, not three.

Obtain a final answer through summing and embedding. When the formulas and functions are kept relatively simple, the final answer can be obtained using the SUM function. For example, if the present worth values (PW) of two columns of cash flows are determined separately, then the total PW is the SUM of the subtotals. This practice is especially useful when the cash flow series are complex.

Prepare for a chart. If a chart (graph) will be developed, plan ahead by leaving sufficient room on the right of the data and answers. Charts can be placed on the same worksheet or on a separate worksheet when the Chart Wizard is used, as discussed in Section A.1 on creating charts. Placement on the same worksheet is recommended, especially when the results of sensitivity analysis are plotted.

A.3 EXCEL FUNCTIONS IMPORTANT TO ENGINEERING ECONOMY (alphabetical order)

DB (Declining Balance)

Calculates the depreciation amount for an asset for a specified period n using the declining balance method. The depreciation rate d used in the computation is determined from asset values S (salvage value) and B (basis or first cost) as $d = 1 - (S/B)^{1/n}$. Three-decimal-place accuracy is used for d.

= DB(**cost, salvage, life, period, month**)

cost	First cost or basis of the asset.
salvage	Salvage value.
life	Recovery period.
period	The period, year, for which the depreciation is to be calculated.
month	(optional entry) If this entry is omitted, a full year is assumed for the first year.

Example   A new machine costs $100,000 and is expected to last 10 years. At the end of 10 years, the salvage value of the machine is $50,000. What is the depreciation of the machine in the first year and the fifth year?

Depreciation for the first year: = DB(100000,50000,10,1)
Depreciation for the fifth year: = DB(100000,50000,10,5)

DDB (Double Declining Balance)

Calculates the depreciation of an asset for a specified period n using the double declining balance method. A factor can also be entered for some other declining balance depreciation method by specifying a factor in the function.

= DDB(**cost, salvage, life, period, factor**)

cost	First cost or basis of the asset.
salvage	Salvage value of the asset.
life	Recovery period.
period	The period, year, for which the depreciation is to be calculated.
factor	(optional entry) If this entry is omitted, the function will use a double declining method with 2 times the straight line rate. If, for example, the entry is 1.5, the 150% declining balance method will be used.

Example   A new machine costs $200,000 and is expected to last 10 years. The salvage value is $10,000. Calculate the depreciation of the machine for the first and the eighth years. Finally, calculate the depreciation for the fifth year using the 175% declining balance method.

Depreciation for the first year: = DDB(200000,10000,10,1)

Depreciation for the eighth year: = DDB(200000,10000,10,8)

Depreciation for the fifth year using 175% DB: = DDB(200000,10000,10,5,1.75)

Because of the manner in which the DB function determines the fixed percentage d and the accuracy of the computations, it is recommended that the DDB function be used for all declining balance depreciation rates. Simply use the optional factor entry for rates other than $d = 2/n$.

Effect (Effective Interest Rate)

Calculates the effective annual interest rate for a stated nominal annual rate and a given number of compounding periods per year. Excel uses Equation [3.2] to calculate the effective rate.

= **EFFECT(nominal, npery)**

nominal   Nominal interest rate for the year.

npery   Number of times that interest is compounded per year.

Example   Claude has applied for a $10,000 loan. The bank office told him the interest rate is 8% per year and that interest is compounded monthly to conveniently match his monthly payments. What effective annual rate will Claude pay?

Effective annual rate: = EFFECT(8%,12)

EFFECT can also be used to find effective rates other than annually. Enter the nominal rate for the time period of the required effective rate; npery is the number of times compounding occurs during the time period of the effective rate.

Example   Interest is stated as 3.5% per quarter with quarterly compounding. Find the effective semiannual rate.

The 6-month nominal rate is 7% and compounding is 2 times per 6 months.

Effective semiannual rate: = EFFECT(7%,2)

FV (Future Value)

Calculates the future value (worth) based on periodic payments at a specific interest rate.

= **FV(rate, nper, pmt, pv, type)**

rate   Interest rate per compounding period.

nper   Number of compounding periods.

pmt   Constant payment amount.

pv   The present value amount. If pv is not specified, the function will assume it to be 0.

type (optional entry) Either 0 or 1. A 0 represents payments made at the end of the period, and 1 represents payments at the beginning of the period. If omitted, 0 is assumed.

Example   Jack wants to start a savings account that can be increased as desired. He will deposit \$12,000 to start the account and plans to add \$500 to the account at the beginning of each month for the next 24 months. The bank pays 0.25% per month. How much will be in Jack's account at the end of 24 months?

Future value in 24 months: $= \text{FV}(0.25\%, 24, 500, 12000, 1)$

IPMT (Interest Payment)

Calculates the interest due for a specific period based on constant periodic payments and interest rate.

$= \textbf{IPMT(rate, per, nper, pv, fv, type)}$

rate Interest rate per compounding period.

per   Period for which interest is to be calculated.

nper Number of compounding periods.

pv   Present value. If pv is not specified, the function will assume it to be 0.

fv   Future value. If fv is omitted, the function will assume it to be 0. The fv can also be considered a cash balance after the last payment is made.

type (optional entry) Either 0 or 1. A 0 represents payments made at the end of the period, and 1 represents payments made at the beginning of the period. If omitted, 0 is assumed.

Example   Calculate the interest due in the tenth month for a 48-month, \$20,000 loan. The interest rate is 0.25% per month.

Interest due: $= \text{IPMT}(0.25\%, 10, 48, 20000)$

IRR (Internal Rate of Return)

Calculates the internal rate of return between -100% and infinity for a series of cash flows at regular periods.

$= \textbf{IRR(values, guess)}$

values A set of numbers in a spreadsheet column (or row) for which the rate of return will be calculated. The set of numbers must consist of at least *one* positive and *one* negative number.

guess   (optional entry) To reduce the number of iterations, a *guessed rate of return* can be entered. In most cases, a guess is not required, and a 10% rate of return is initially assumed. If the #NUM!

error appears, try using different values for guess. Inputting different guess values makes it possible to determine the multiple roots for the rate of return equation of a nonconventional cash flow series.

Example John wants to start a printing business. He will need $25,000 in capital and anticipates that the business will generate the following incomes during the first 5 years. Calculate his rate of return.

Year 1	$5,000
Year 2	$7,500
Year 3	$8,000
Year 4	$10,000
Year 5	$15,000

Set up an array in the spreadsheet.

In cell A1, type −25000 (negative for payment).
In cell A2, type 5000 (positive for income).
In cell A3, type 7500.
In cell A4, type 8000.
In cell A5, type 10000.
In cell A6, type 15000.

Note that any years with a zero cash flow must have a zero entered to ensure that the year value is correctly maintained for computation purposes.

To calculate the internal rate of return after 5 years and specify a guess value of 5%, type = IRR(A1:A6,5%).

MIRR (Modified Internal Rate of Return)

Calculates the modified internal rate of return for a series of cash flows and reinvestment of income and interest at a stated rate.

= **MIRR(values, finance_rate, reinvest_rate)**

values	Refers to an array in the spreadsheet. The series must occur at regular periods and must contain at least *one* positive number and *one* negative number.
finance_rate	Interest rate of money.
reinvest_rate	Interest rate for reinvestment on positive cash flows. (This is not the same reinvestment rate on the net investments when the cash flow series is nonconventional. See Section 6.8 for comments.)

Example Jane opened a hobby store 4 years ago. When she started the business, Jane borrowed $50,000 from a bank at 12% per year interest. Since then, the business has yielded $10,000 the first year, $15,000 the second year, $18,000 the third year, and $21,000 the fourth year. Jane reinvests her profits, earning 8% per year. What is the modified rate of return after 4 years?

In cells A1 through A5, type -50000, 10000, 15000, 18000, and 21000.

Modified rate of return after 4 years: = MIRR(A1:A5,12%,8%).

Nominal (Nominal Interest Rate)

Calculates the nominal annual interest rate for a stated effective annual rate and a given number of compounding periods per year.

= NOMINAL(effective, npery)

effective   Effective interest rate for the year.

npery   Number of times interest is compounded per year.

Example Last year, a corporate stock earned an effective return of 12.55% per year. Calculate the nominal annual rate, if interest is compounded quarterly and if interest is compounded continuously.

Nominal annual rate, quarterly compounding: = NOMINAL(12.55%,4)

Nominal annual rate, continuous compounding: = NOMINAL(12.55%,100000)

NPER (Number of Periods)

Calculates the number of periods for the present worth of an investment to equal the future value specified, based on uniform regular payments and a stated interest rate.

= NPER(rate, pmt, pv, fv, type)

rate   Interest rate per compounding period.

pmt   Amount paid during each compounding period.

pv   Present value (lump-sum amount).

fv   (optional entry) Future value or cash balance after the last payment. If fv is omitted, the function will assume a value of 0.

type   (optional entry) Enter 0 if payments are due at the end of the compounding period, and 1 if payments are due at the beginning of the period. If omitted, 0 is assumed.

Example Sally plans to open a savings account which pays 0.25% per month. Her initial deposit is $3000, and she plans to deposit $250 at the beginning of every month. How many payments does she have to make to accumulate $15,000 to buy a used car?

Number of payments: = NPER(0.25%,$-250,-3000$,15000,1)

NPV (Net Present Value)

Calculates the net present value of a series of future cash flows at a stated interest rate.

= NPV(rate, series)

rate   Interest rate per compounding period.

series   Series of costs and incomes set up in a range of cells in the spreadsheet. Any cash flow in year 0 (now) is not included in the series entry, since it is already a present value.

Example   Mark is considering buying a sports franchise for $100,000 and expects to receive the following income during the next 6 years of business: $25,000, $40,000, $42,000, $44,000, $48,000, $50,000. The interest rate is 8% per year.

In cells A1 through A7, enter -100000, followed by the six estimated annual incomes.

Present value: $= \text{NPV}(8\%, \text{A2:A7}) + \text{A1}$

PMT (Payments)

Calculates equivalent periodic amounts based on present value and/or future value at a stated interest rate.

$= \textbf{PMT(rate, nper, pv, fv, type)}$

rate   Interest rate per compounding period.

nper   Total number of periods.

pv   Present value.

fv   Future value.

type   (optional entry) Enter 0 for payments due at the end of the compounding period, and 1 if payment is due at the start of the compounding period. If omitted, 0 is assumed.

Example   Jim plans to take a $15,000 loan to help him buy a new car. The interest rate is 7%. He wants to pay the loan off in 5 years (60 months). What are his monthly payments?

Monthly payments: $= \text{PMT}(7\%/12, 60, 15000)$

PV (Present Value)

Calculates the present value of a future series of equal cash flows and a single lump sum in the last period at a constant interest rate.

$= \textbf{PV(rate, nper, pmt, fv, type)}$

rate   Interest rate per compounding period.

nper   Total number of periods.

pmt   Cash flow at regular intervals. Negative numbers represent payments (cash outflows), and positive numbers represent income.

fv   Future value or cash balance at the end of the last period.

type   (optional entry) Enter 0 if payments are due at the end of the compounding period, and 1 if payments are due at the start of each compounding period. If omitted, 0 is assumed.

There are two primary differences between the PV function and the NPV function: PV allows for end or beginning of period cash flows, and PV requires that all amounts have the same value, whereas they may vary for the NPV function.

Example   Jose is considering leasing a car for $300 a month for 3 years (36 months). After the 36-month lease, he can purchase the car for $12,000. Using an interest rate of 8% per year, find the present value of this option.

$$\text{Present value: } = \text{PV}(8\%/12,36,-300,-12000)$$

Note the minus signs on the pmt and fv amounts.

RATE (Interest Rate)

Calculates the interest rate per compounding period for a series of equal cash flows.

$$= \textbf{RATE(nper, pmt, pv, fv, type, guess)}$$

nper   Total number of periods.

pmt   Payment amount made each compounding period.

pv   Present value.

fv   Future value (not including the pmt amount).

type   (optional entry) Enter 0 for payments due at the end of the
          compounding period, and 1 if payments are due at the start of
          each compounding period. If omitted, 0 is assumed.

guess   (optional entry) To minimize computing time, include a guessed
          interest rate. If a value of guess is not specified, the function
          will assume a rate of 10%. This function usually converges to
          a solution, if the rate is between 0% and 100%.

Example   Mary wants to start a savings account at a bank. She will make an initial deposit of $1000 to open the account and plans to deposit $100 at the beginning of each month. She plans to do this for the next 3 years (36 months). At the end of 3 years, she wants to have at least $5000. What is the minimum interest required to achieve this result?

$$\text{Interest rate: } = \text{RATE}(36,-100,-1000,5000,1)$$

SLN (Straight Line Depreciation)

Calculates the straight line depreciation of an asset for a given year.

$$= \textbf{SLN(cost, salvage, life)}$$

cost   First cost or basis of the asset.

salvage   Salvage value.

life   Recovery period.

Example   Marisco, Inc., purchased a printing machine for $100,000. The machine has a life of 8 years and an estimated salvage value of $15,000. What is the depreciation each year?

$$\text{Depreciation: } = \text{SLN}(100000,15000,8)$$

VDB (Variable Declining Balance)

Calculates the depreciation using the declining balance method with a switch to straight line depreciation in the year in which straight line has a larger depreciation amount. This function automatically implements the switch from DB to SL depreciation, unless specifically instructed to not switch.

= VDB (cost, salvage, life, start_period, end_period, factor, no_switch)

cost	First cost of the asset.
salvage	Salvage value.
life	Recovery period.
start_period	First period for depreciation to be calculated.
end_period	Last period for depreciation to be calculated.
factor	(optional entry) If omitted, the function will use the double declining rate of $2/n$.
no_switch	(optional entry) If omitted or entered as FALSE, the function will switch from declining balance to straight line depreciation when the latter is greater than DB depreciation. If entered as TRUE, the function will not switch to SL depreciation at any time.

Example   Newly purchased equipment with a first cost of $300,000 has a depreciable life of 10 years with no salvage value. Calculate the 175% declining balance depreciation for the first year and the ninth year if switching to SL depreciation is acceptable, and if switching is not permitted.

Depreciation for first year, with switching: = VDB(300000,0,10,0,1,1.75)

Depreciation for ninth year, with switching: = VDB(300000,0,10,8,9,1.75)

Depreciation for first year, no switching: = VDB(300000,0,10,0,1,1.75,TRUE)

Depreciation for ninth year, no switching: = VDB(300000,0,10,8,9,1.75,TRUE)

If the start_period and end_period are replaced with the MAX and MIN functions, respectively, the VDB function generates the MACRS depreciation. As above, the factor option should be entered if other than DDB rates start the MACRS depreciation. The VDB format is

= VDB(cost, 0, life, MAX(0,t−1.5), MIN(life,t−0.5), factor)

Example   Determine the MACRS depreciation for year 4 for a $350,000 asset that has a 20% salvage and a MACRS recovery period of 3 years. $D_4 = \$25,926$ is the display.

For year 4: = VDB(350000,0,3,MAX(0,4−1.5),MIN(3,4−.5),2)

As a second example, if the MACRS recovery period is $n = 15$ years, $D_{16} = \$10,334$.

For year 16: = VDB(350000,0,15,MAX(0,16−1.5),MIN(15,16−0.5),1.5)

The optional factor 1.5 is required here, since MACRS starts with 150% DB for $n = 15$ and 20 year recoveries.

A.4 GOAL SEEK—A SIMPLE TOOL FOR BREAKEVEN AND SENSITIVITY ANALYSES

GOAL SEEK, found on the Excel toolbar labeled Tools, changes the value in a specific cell based on a numerical value for another (changing) cell as input by the user. It is a good tool for sensitivity analysis as well as breakeven. The initial GOAL SEEK template is pictured in Figure A.2. One of the cells (set or changing cell) must contain an equation or Excel function that uses the other cell to determine a numeric value. Only a single cell can be identified as the changing cell; however, this limitation can be avoided by using equations rather than specific numerical inputs in any additional cells also to be changed.

Example A proposed asset will cost $25,000, generate an annual cash flow of $6000 over its 5-year life, and then have an estimated $500 salvage value. The rate of return using the IRR function is 6.94%. Determine the annual cash flow necessary to raise the return to 10% per year.

Figure A.3 (top left) shows the cash flows and return displayed using the function = IRR (B4:B9) prior to the use of GOAL SEEK. Note that the initial $6000 is input in cell B5, but other years' cash flows are input as equations that refer to B5. The $500 salvage is added for the last year. This format allows GOAL SEEK to change only cell B5 while forcing the other cash flows to the same value. The tool finds the required cash flow of $6506 to approximate the 10% per year return. The GOAL SEEK Status inset indicates that a solution is found. Clicking OK saves all changed cells; clicking Cancel returns to the original values.

More complicated analysis can be performed using the SOLVER tool. For example, SOLVER can handle multiple changing cells. Additionally, equality and inequality constraints can be developed. See the Excel Help system to utilize this more powerful tool. If SOLVER is not shown on the Tools toolbar, click Add-ins and install the SOLVER Add-In.

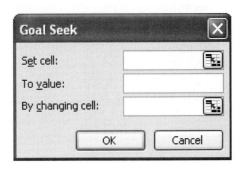

FIGURE A.2
GOAL SEEK template used to specify a cell and value and the changing cell.

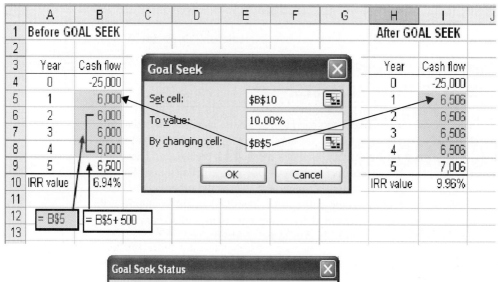

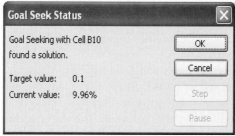

FIGURE A.3 Use of GOAL SEEK tool to determine an annual cash flow to increase the rate of return.

A.5 ERROR MESSAGES

If Excel is unable to complete a formula or function computation, an error message is displayed.

#DIV/0!	Requires division by zero.
#N/A	Refers to a value that is not available.
#NAME?	Uses a name that Excel doesn't recognize.
#NULL!	Specifies an invalid intersection of two areas.
#NUM!	Uses a number incorrectly.
#REF!	Refers to a cell that is not valid.
#VALUE!	Uses an invalid argument or operand.
#####	Produces a result, or includes a constant numeric value, that is too long to fit in the cell. (Widen the column.)

Accounting Reports and Business Ratios

This appendix provides a fundamental description of financial statements. The documents discussed here will assist in reviewing or understanding basic financial statements and in gathering information useful in an engineering economy study.

B.1 THE BALANCE SHEET

The fiscal year and the tax year are defined identically for a corporation or an individual—12 months in length. The fiscal year (FY) is commonly not the calendar year (CY) for a corporation. The U.S. government uses October through September as its FY. For example, October 2007 through September 2008 is FY2008. The fiscal or tax year is always the calendar year for an individual.

At the end of each fiscal year, a company publishes a *balance sheet*. A sample balance sheet for Delta Corporation is presented in Table B.1. This is a yearly presentation of the state of the firm at a particular time, for example, December 31, 2008; however, a balance sheet is also usually prepared quarterly and monthly. There are three main categories.

    Assets. This section is a summary of all resources owned by or owed to the company. *Current assets* represent shorter-lived working capital (cash, accounts receivable, etc.), which is more easily converted to cash, usually within 1 year. Longer-lived assets are referred to as *fixed assets* (land, equipment, etc.). Conversion of these holdings to cash in a short period of time would require a major corporate reorientation.

    Liabilities. This section is a summary of all financial obligations (debts, bonds, mortgages, loans, etc.) of a corporation.

    Net worth. Also called *owner's equity,* this section provides a summary of the financial value of ownership, including stocks issued and earnings retained by the corporation.

The balance sheet is constructed using the relation

$$\text{Assets} = \text{liabilities} + \text{net worth}$$

TABLE B.1 **Sample Balance Sheet**

DELTA CORPORATION **Balance Sheet** **December 31, 2008**			
Assets		**Liabilities**	
Current			
Cash	$10,500	Accounts payable	$19,700
Accounts receivable	18,700	Dividends payable	7,000
Interest accrued receivable	500	Long-term notes payable	16,000
Inventories	52,000	Bonds payable	20,000
Total current assets	$81,700	Total liabilities	$62,700
Fixed		**Net Worth**	
Land	$25,000	Common stock	$275,000
Building and equipment	438,000	Preferred stock	100,000
Less: Depreciation allowance $82,000	356,000	Retained earnings	25,000
Total fixed assets	381,000	Total net worth	400,000
Total assets	$462,700	Total liabilities and net worth	$462,700

In Table B.1 each major category is further divided into standard subcategories. For example, current assets is comprised of cash, accounts receivable, etc. Each subdivision has a specific interpretation; for example, accounts receivable, which represents all money owed to the company by its customers.

B.2 INCOME STATEMENT AND COST OF GOODS SOLD STATEMENT

A second important financial statement is the *income statement* (Table B.2), which summarizes the profits or losses for a stated period of time. Income statements always accompany balance sheets. The major categories of an income statement are

Revenues. This includes all sales and revenues that the company has received in the past accounting period.

Expenses. This is a summary of all expenses for the period. Some expense amounts are detailed in other statements, for example, cost of goods sold.

The income statement, published at the same time as the balance sheet, uses the basic equation

$$\text{Revenues} - \text{expenses} = \text{profit (or loss)}$$

The *cost of goods sold* is an important accounting term for a manufacturing company. It represents the net cost of producing the product marketed by the firm.

TABLE B.2 Sample Income Statement

DELTA CORPORATION Income Statement Year Ended December 31, 2008		
Revenues		
Sales	$505,000	
Interest revenue	3,500	
Total revenues		$508,500
Expenses		
Cost of goods sold (from Table B.3)	$290,000	
Selling	28,000	
Administrative	35,000	
Other	12,000	
Total expenses		365,000
Income before taxes		143,500
Taxes for year		64,575
Net profit for year		$ 78,925

TABLE B.3 Sample Cost of Goods Sold Statement

DELTA CORPORATION Statement of Cost of Goods Sold Year Ended December 31, 2008		
Materials		
Inventory, January 1, 2008	$ 54,000	
Purchases during year	174,500	
Total	$228,500	
Less: Inventory December 31, 2008	50,000	
Cost of materials		$178,500
Direct labor		110,000
Prime cost		288,500
Indirect costs		7,000
Factory cost		295,500
Less: Increase in finished goods inventory during year		5,500
Cost of goods sold (into Table B.2)		$290,000

Cost of goods sold may also be called *factory cost*. Note that the total of the cost of goods sold statement in Table B.3 is entered as an expense item on the income statement. This total is determined using the relations

$$\text{Cost of goods sold} = \text{prime cost} + \text{indirect cost}$$
$$\text{Prime cost} = \text{direct materials} + \text{direct labor}$$

[B.1]

Indirect costs include all indirect and overhead charges made to a product, process, service, or cost center. Indirect cost allocation is discussed in Chapter 11.

B.3 BUSINESS RATIOS

Accountants, financial analysts, and engineering economists frequently utilize business ratio analysis to evaluate the financial health of a company over time and in relation to industry norms. Because the engineering economist must continually communicate with others, she or he should have a basic understanding of several ratios. For comparison purposes, it is necessary to compute the ratios for several companies in the same industry. Industrywide median ratio values are published annually by firms such as Dun and Bradstreet in *Industry Norms and Key Business Ratios*. The ratios are classified according to their role in measuring the corporation.

> **Solvency ratios.** Assess ability to meet short-term and long-term financial obligations.
>
> **Efficiency ratios.** Measure management's ability to use and control assets.
>
> **Profitability ratios.** Evaluate the ability to earn a return for the owners of the corporation.

Numerical data for several important ratios are discussed here and are extracted from the Delta balance sheet and income statement, Tables B.1 and B.2.

Current Ratio

This ratio is utilized to analyze the company's working capital condition.

$$\text{Current ratio} = \frac{\text{current assets}}{\text{current liabilities}}$$

Current liabilities include all short-term debts, such as accounts and dividends payable. Note that only balance sheet data are utilized in the current ratio; that is, no association with revenues or expenses is made. For the balance sheet of Table B.1, current liabilities amount to $19,700 + $7000 = $26,700 and

$$\text{Current ratio} = \frac{81,700}{26,700} = 3.06$$

Since current liabilities are those debts payable in the next year, the current ratio value of 3.06 means that the current assets would cover short-term debts approximately 3 times. Current ratio values of 2 to 3 are common.

The current ratio assumes that the working capital invested in inventory can be converted to cash quite rapidly. Often, however, a better idea of a company's *immediate* financial position can be obtained by using the acid test ratio.

Acid Test Ratio (Quick Ratio)

$$\text{Acid test ratio} = \frac{\text{quick assets}}{\text{current liabilities}} = \frac{\text{current assets} - \text{inventories}}{\text{current liabilities}}$$

It is meaningful for the emergency situation when the firm must cover short-term debts using its readily convertible assets. For Delta Corporation,

$$\text{Acid test ratio} = \frac{81,700 - 52,000}{26,700} = 1.11$$

Comparison of this and the current ratio shows that approximately 2 times the current debts of the company are invested in inventories. However, an acid test ratio of approximately 1.0 is generally regarded as a strong current position, regardless of the amount of assets in inventories.

Debt Ratio

This ratio is a measure of financial strength.

$$\text{Debt ratio} = \frac{\text{total liabilities}}{\text{total assets}}$$

For Delta Corporation,

$$\text{Debt ratio} = \frac{62,700}{462,700} = 0.136$$

Delta is 13.6% creditor-owned and 86.4% stockholder-owned. A debt ratio in the range of 20% or less usually indicates a sound financial condition, with little fear of forced reorganization because of unpaid liabilities. However, a company with virtually no debts is inexperienced in dealing with short-term and long-term debt financing. The debt-equity (D-E) mix is another measure of financial strength.

Return on Sales Ratio

This often quoted ratio indicates the profit margin for the company.

$$\text{Return on sales} = \frac{\text{net profit}}{\text{net sales}}(100\%)$$

Net profit is the after-tax value from the income statement. This ratio measures profit earned per sales dollar and indicates how well the corporation can sustain adverse conditions over time, such as falling prices, rising costs, and declining sales. For Delta Corporation,

$$\text{Return on sales} = \frac{78,925}{505,000}(100\%) = 15.6\%$$

Corporations may point to small return on sales ratios, say, 2.5% to 4.0%, as indications of sagging economic conditions. In truth, for a relatively large-volume, high-turnover business, an income ratio of 3% is quite healthy.

Return on Assets Ratio

This is the key indicator of profitability since it evaluates the ability of the corporation to transfer assets into operating profit. The definition and value for Delta are

$$\text{Return on assets} = \frac{\text{net profit}}{\text{total assets}} (100\%)$$

$$= \frac{78{,}925}{462{,}700} (100\%) = 17.1\%$$

Efficient use of assets indicates that the company should earn a high return, while low returns usually accompany lower values of this ratio compared to the industry group ratios.

Inventory Turnover Ratio

This ratio indicates the number of times the average inventory value passes through the operations of the company.

$$\text{Net sales to inventory} = \frac{\text{net sales}}{\text{average inventory}}$$

where average inventory is the figure recorded in the balance sheet. For Delta Corporation this ratio is

$$\text{Net sales to inventory} = \frac{505{,}000}{52{,}000} = 9.71$$

This means that the average value of the inventory has been sold 9.71 times during the year. Values of this ratio vary greatly from one industry to another.

EXAMPLE B.1   Typical values for financial ratios or percentages for 4 industry sectors are presented below. Compare the corresponding Delta Corporation values with these norms.

Ratio or Percentage	Motor Vehicles and Parts Manufacturing 336105*	Air Transportation (Medium-Sized) 481000*	Industrial Machinery Manufacturing 333200*	Home Furnishings 442000*
Current ratio	2.4	0.4	1.7	2.6
Quick ratio	1.6	0.3	0.9	1.2
Debt ratio	59.3%	96.8%	61.5%	52.4%
Return on assets	40.9%	8.1%	6.4%	5.1%

*North American Industry Classification System (NAICS) code for this industry sector.
Source: L. Troy, *Almanac of Business and Industrial Financial Ratios,* 33d annual edition, Prentice-Hall, Paramus, NJ, 2002.

Solution

It is not correct to compare ratios for one company with indexes in different industries, that is, with indexes for different NAICS codes. So, the following comparison is for illustration purposes only. The corresponding values for Delta are

$$\text{Current ratio} = 3.06$$
$$\text{Quick ratio} = 1.11$$
$$\text{Debt ratio} = 13.5\%$$
$$\text{Return on assets} = 17.1\%$$

Delta has a current ratio larger than all four of these industries, since 3.06 indicates it can cover current liabilities 3 times compared with 2.6 and much less in the case of the "average" air transportation corporation. Delta has a significantly lower debt ratio than that of any of the sample industries, so it is likely more financially sound. Return on assets, which is a measure of ability to turn assets into profitability, is not as high at Delta as it is for motor vehicles, but Delta competes well with the other industry sectors.

C Appendix

Alternative Evaluation That Includes Multiple Attributes and Risk

This appendix introduces techniques that may tailor the evaluation to situations not included in a standard engineering economy evaluation. First, noneconomic attributes that may alter the straight economic decision are examined. Then the element of variation in parameter values is examined using simple probability and statistics. This allows the aspect of risk and uncertainty to be considered as the best alternative is selected.

C.1 MULTIPLE ATTRIBUTE ANALYSIS

In all evaluations thus far, only one attribute—the economic one—has been relied upon in selecting the best alternative by maximizing the PW, AW, ROR, or B/C value. However, noneconomic factors are indirectly considered in most alternative evaluations. These factors are mostly intangible and usually difficult to quantify in economic terms.

Public sector projects are excellent examples of multiple-attribute selection. For example, a project to construct a dam forming a lake usually has several purposes, such as flood control, industrial use, commercial development, drinking water, recreation, and nature conservation. Noneconomic attributes, evaluated in different ways by different stakeholders, make selection of the best dam alternative very complex.

In evaluations, key noneconomic attributes can be considered directly using several techniques. A technique popular in most engineering disciplines, the *weighted attribute method,* is described here.

Once the decision is made to consider multiple attributes in an evaluation, the following must be accomplished:

1. Identify the key attributes.
2. Determine each attribute's importance and weight.
3. Rate (value) each alternative by attribute.
4. Calculate the evaluation measure and select the best alternative.

1. Key Attribute Identification

Attributes are identified by several methods, depending upon the situation. Seeking input from other people is important because it helps focus on key attributes as determined by those with experience and those who will use the selected system. Some identification approaches are:

- Comparison with similar studies that include multiple attributes.
- Input from experts with relevant experience.
- Survey of stakeholders (customers, employees, managers)
- Small group discussions (brainstorming, focus groups)
- Delphi method, a formal procedure that reaches consensus from people with different perspectives.

As an example of identifying key attributes, consider the purchase of a car for the Kerry family of four versus a car purchase for Clare, a university student. Economics (possibly with different slants) will be a key attribute for both, but other attributes will be considered. Examples follow.

Kerry Family

Economics—first cost and operating cost (mileage and maintenance)

Safety—airbags; rollover and crash factors; traction

Inside design—seating space, cargo space, etc.

Reliability—warranty coverage; breakdown record

Clare

Economics—first cost and mileage cost

Style—exterior design, color, sleek and modern look

Inside design—cargo room; seating

Dependability—pick up and speed factors; required maintenance

2. Weights (Importance) for the Attributes

An importance score is set by a person or group experienced with each attribute compared to alternative attributes. If a group is involved, consensus is required to arrive at one score for each attribute. The resulting score is used to determine a weight W_i for each attribute i.

$$W_i = \frac{\text{importance score}_i}{\text{sum of all scores}} = \frac{\text{importance score}_i}{S} \qquad \text{[C.1]}$$

This normalizes the weights, making their sum equal to 1.0 over all attributes $i = 1, 2, \ldots, m$. Table C.1 is a tabular layout of attributes, weights, and alternatives used to implement the weighted attribute method. Attributes and weights from Equation [C.1] are entered on the left; value ratings for each alternative complete the table as discussed in the next step.

Three approaches to assigning weights are *equal, rank order,* and *weighted rank order. Equal weighting* means that all attributes are of the same importance,

TABLE C.1   Tabular Layout of Attributes and Alternatives Used for Multiple Attribute Evaluation

Attributes	Weights	Alternatives				
		1	**2**	**3**	**. . .**	**n**
1	W_1					
2	W_2					
3	W_3		Value ratings			
$\vdots$	$\vdots$					
m	W_m					

because there is no rationale or criterion to distinguish differences. This default approach sets all importance scores to 1, which makes each weight equal to $1/m$. To *rank order* attributes, place them in increasing importance, assigning a 1 to the least important and m to the most important attribute. This means that the difference between attribute importance is constant. Equation [C.1] results in weights of $1/S, 2/S, \ldots, m/S$.

A more practical and versatile approach is to assign importance using a *weighted rank order*. Place the attributes in decreasing order of importance first, assign a score of 100 to the most important one(s), and score other attributes relative to this using scores between 100 and 0. If s_i identifies the score for each attribute, Equation [C.1] is rewritten to determine attribute weights. This automatically normalizes them to sum to 1.0.

$$W_i = \frac{s_i}{S} \qquad [C.2]$$

This approach is commonly applied because one or more attributes can be heavily weighted, while attributes of minor importance can be included in the analysis. As an example, suppose the 4 key attributes developed for the Kerry family car purchase are ordered as safety, economics, inside design, and reliability. If the economics attribute is half as important as safety, while the remaining 2 are half as important as economics, the attribute list and weights for Table C.1 are as follows:

Attribute, i	Score, s_i	Weight, W_i
Safety	100	$100/200 = 0.50$
Economics	50	$50/200 = 0.25$
Inside design	25	$25/200 = 0.125$
Reliability	25	$25/200 = 0.125$
Sum	200	1.000

There are other weighting techniques, usually designed for group input where diverse opinions prevail. Some are utility functions, pairwise comparison, and the AHP (analytic hierarchy process), which is more complicated and comprehensive. These techniques provide a significant advantage compared to those described here; they guarantee consistency between ranks, scores, and individual scorers.

3. Value Rating for Each Alternative

A decision maker evaluates each alternative (j) based on each attribute (i) to determine a value rating V_{ij}. These are the right-side entries of Table C.1. The value rating scale uses some numeric basis, such as, 0 to 100, 1 to 10, -1 to $+1$, or -3 to $+3$, with larger scores indicating a higher value rating. The last two scales allow an individual to give negative input on an alternative. A commonly used technique is the *Likert* scale, which defines several gradations with a range of numbers for each one. For example, a scale from 0 to 10 may be described as follows:

If You Value the Alternative as	Give it a Value Rating between the Numbers
Very poor	0–2
Poor	3–5
Good	6–8
Very good	9–10

Each evaluator uses this scale to rate each alternative on each attribute, thus generating the V_{ij} values. Likert scales with an even number of choices, say, 4, are preferred so that the central tendency of "fair" is not overrated.

Continuing the car purchase example, assume that the father has value rated 3 car alternatives using a 0 to 10 scale against the 4 key attributes. The result may look like Table C.2 with all V_{ij} values entered. Mrs. Kelly and each child will have a table with (possibly) different V_{ij} entries.

TABLE C.2 **Value Ratings of Four Attributes and Three Alternatives for Multiple-Attribute Evaluation**

Attribute	Weight	Alternative 1	2	3
Safety	0.50	6	4	8
Econimics	0.25	9	3	1
Inside Design	0.125	5	6	6
Reliability	0.125	5	9	7

4. Evaluation Measure for Each Alternative

Alternative evaluation by the weighted attribute method results in a measure R_j for each alternative, which is the sum of each attribute's weight multiplied by a corresponding alternative's value rating.

$$R_j = \text{sum of weights} \times \text{value rating}$$

$$= \sum_{i=1}^{m} W_i \times V_{ij} \qquad \text{[C.3]}$$

The selection guideline is:

Select the alternative with the largest R_j value.

When several decision makers are involved, a different R_j can be calculated for each person and some resolution made if different alternatives are indicated to be the best. Alternatively, resolution to agree on one set of V_{ij} values can be reached before a single R_j is determined. As always, sensitivity analysis of scores, weights, and value ratings offers insight into the sensitivity of the final selection for different people and groups.

A final note on the evaluation measure R_j is in order. It is a single-dimension number that quite effectively combines the different dimensions addressed by the attributes, people's importance scores, and evaluators' value ratings. This type of aggregate measure, often called a *rank-and-rate method,* removes the complexity of balancing the different attributes, but it also eliminates much of the robust information captured in ranking attributes and rating alternatives against attributes.

EXAMPLE C.1   Two vendors offer chlorine gas distribution systems in 1-ton cylinders for use in industrial water cooling systems. Hartmix, Inc. engineers have completed a present worth analysis of both alternatives using a 5-year study period. The values are $PW_A = \$-432{,}500$ and $PW_B = \$-378{,}750$, leading to a recommendation of vendor B. This result and some noneconomic factors were considered by the regional manager, Herb, and district superintendent, Charlotte. They independently defined attributes and assigned importance scores (0 to 100) and value ratings (1 to 10) for the two vendor proposals. Higher scores and values are considered better. Use their results (Table C.3) to determine if vendor B is the better choice when these key attributes are considered.

Solution

Of the four steps in the weighted attribute method, the first (attribute definition) and third (value rating) are completed. Finish the other two steps using each evaluator's scores and ratings and compare the selections to determine if Herb, Charlotte, and the engineers all selected vendor B.

TABLE C.3 Importance Scores and Value Ratings Determined by Herb and Charlotte, Example C.1

Attribute	Importance Score		Herb's Values		Charlotte's Values	
	Herb	Charlotte	Vendor A	Vendor B	Vendor A	Vendor B
Safety	100	80	10	9	7	9
Economics	35	100	3	10	5	5
Flexibility	20	10	10	9	5	8
Maintainability	20	50	2	10	8	4
Total	175	240				

TABLE C.4 Evaluation Measure for Multiple Attribute Comparison, Example C.1

Attribute	Herb's Evaluation			Charlotte's Evaluation		
	Weights	A	B	Weights	A	B
Safety	0.5714	10	9	0.3333	7	9
Economics	0.2000	3	10	0.4167	5	5
Flexibility	0.1143	10	9	0.0417	5	8
Maintainability	0.1143	2	10	0.2083	8	4
Totals and R values	1.0000	7.69	9.31	1.0000	6.29	6.25

Equation [C.2] calculates the weights for each attribute for both evaluators. The totals are $S = 175$ for Herb and $S = 240$ for Charlotte.

Herb Safety: $100/175 = 0.5714$ Economics: $35/175 = 0.20$
 Flexibility: $20/175 = 0.1143$ Maintainability: $20/175 = 0.1143$

Charlotte Safety: $80/240 = 0.3333$ Economics: $100/240 = 0.4167$
 Flexibility: $10/240 = 0.0417$ Maintainability: $50/240 = 0.2083$

Table C.4 details the weights and value ratings as well as R_A and R_B values (last row). As an illustration, the measure for vendor B for Charlotte is calculated by Equation [C.3].

Charlotte: $R_B = 0.3333(9) + 0.4167(5) + 0.0417(8) + 0.2083(4) = 6.25$

Herb selects vendor B conclusively and Charlotte selects A by a very small margin. Since the engineers selected B on purely economic terms, some negotiated agreement is necessary; however, B is likely the better choice.

C.2 ECONOMIC EVALUATION WITH RISK CONSIDERED

All things in the world vary from one situation or environment to another and over time. We are guaranteed that variation will occur in engineering economy due to its emphasis on future estimates. All analyses thus far have used estimates (e.g., AOC = $-45,000 per year) and computations (PW, ROR, and other measures) that are *certain*, that is, no variation. We can observe and estimate outcomes with a high degree of certainty, but even this depends upon the precision and accuracy of our skills and tools.

When variation enters into estimation and evaluation, it is called *risk*. When there may be two or more observable values for a parameter *and* it is possible to estimate the chance that each value may occur, risk is considered. Decision making under risk is present, for example, if a cash flow estimate has a 50-50 chance of being $10,000 or $5,000. In reality, virtually all decisions are made under risk, but the risk may not be explicitly taken into account. This section introduces *decision making under risk*.

Decision making under certainty—Deterministic estimates are made and entered into engineering economy computations for PW, AW, FW, ROR, or B/C equivalency values. Estimates are the *most likely value,* also called *single-value estimates.* All previous examples and problems in this book are of this type. In fact, sensitivity analysis is simply another form of analysis with certainty, except that the computations are repeated with different values, *each estimated with certainty.*

Decision making under risk—The element of chance is formally taken into account; however, a clear-cut decision is harder to make because variation in estimates is allowed. One or more parameters (*P, A, S*, AOC, *i, n*, etc.) in an alternative can vary. The two fundamental ways to consider risk in the economic evaluation are expected value analysis and simulation analysis.

- *Expected value analysis* uses the possible values of an estimate and the chance of occurrence associated with each value to calculate simple statistical measures such as the expected value and standard deviation of a single estimate or an entire alternative's PW (or other measure). The alternative with the most favorable "statistics" is indicated as best.

- *Simulation analysis* uses the chance and parameter estimates to generate repeated computations of the PW (or other measure) by random sampling. Large enough samples are generated to conclude which alternative is best with a reasonable degree of confidence. Spreadsheets and software that is more advanced than treated in this text are generally used to conduct the sampling and plotting needed to reach a statistical conclusion.

The remainder of this section concentrates on decision making under risk using the expected value approach applied to the engineering economy methods we have learned in previous chapters. First, however, it is important to understand how to use several fundamentals of probability and statistics. (If these are already familiar to you, please skip forward.)

Random Variable (or Variable)   A random variable can take on any one of several values. It is classified as either *discrete* or *continuous* and is identified by a letter. Discrete variables have several isolated values, while continuous variables can assume any value between two stated limits. A project's life n that is estimated to have a value of, say, 4 or 6 or 10 years is a discrete variable, and is written $n = 4, 6, 10$. The rate of return i is a continuous variable that may range from -100% to ∞, that is, $-100\% \leq i < \infty$.

Probability   A probability is a number between 0 and 1.0 that expresses the chance in decimal form that a variable may take on any value identified as possible for it. Simply stated, it is the *chance that something specific will occur,* divided by 100. Probability for a variable X, for example, is identified by $P(X = 6)$ and is interpreted as the probability that X will equal the specific value 6. If the chance is 25%, the probability that $X = 6$ is written $P(X = 6) = 0.25$. In all probability statements, the $P(X)$ values over all possible X values must total to 1.0.

Probability Distribution   This is a graphical representation describing how probability is distributed over all possible values of a variable. Figure C.1a is a plot of the discrete variable estimated asset life n, which has possible values of 4, 6, and 10 years, all with equal probability of 1/3. The distribution for a continuous variable is described by a continuous curve over the range of the variable. Figure C.1b indicates an equal probability of 0.01 for all values of a uniform annual cost A between $400 and $500 per year. These are called *uniform distributions.*

Random Sample (or Sample)   A sample is a collection of N values drawn in a random fashion from all possible values of the variable using the probability distribution of the variable. Suppose a random sample of size $N = 10$ is taken from the salvage value variable S, where $S = \$500$ or

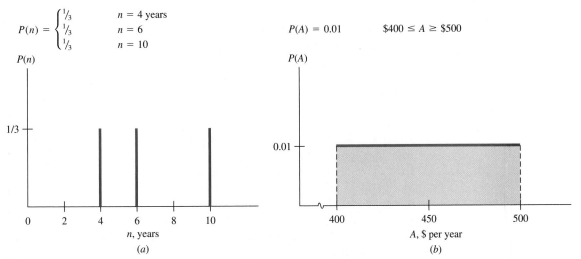

$$P(n) = \begin{cases} \frac{1}{3} & n = 4 \text{ years} \\ \frac{1}{3} & n = 6 \\ \frac{1}{3} & n = 10 \end{cases} \qquad\qquad P(A) = 0.01 \qquad \$400 \leq A \geq \$500$$

FIGURE C.1   Probability distribution of (a) discrete variable and (b) continuous variable.

$800, each with an equal probability of 0.5. If a coin is tossed with heads representing $500 and tails $800, the sample outcome may be HHTHTTTHTT, which translates to four $500s and six $800s. As larger samples are taken, the equal probability for each value of S will be clearly represented in the sample. Samples are used to "estimate" properties for the population of the variable. As discussed below, these estimates are used to make engineering economy calculations (PW, ROR, and other measures) necessary to select the best alternative. If the data is assumed to be uniformly distributed between any two numbers, the Excel functions RAND or RANDBETWEEN can be used to generate a random sample. Check the Excel software help function for more information.

EXAMPLE C.2   Carlos, a cost estimator for Deblack Chemicals, LLC, has estimated the annual net revenue R for a newly developed antifungal spray applied to post-harvest citrus fruit. He predicts the expected revenue to be $3.1 million per year for the next 5 years. This expected amount is based on his estimate that R could be $2.6, $2.8, $3.0, $3.2, $3.4, or $3.6 million per year, all with the same probability. Do the following for Carlos:

  a. Identify the variable R as discrete or continuous; describe it in engineering economy terms.
  b. Write the probability statements for the estimates identified.
  c. Plot the probability distribution of R.
  d. A sample of size 4 is developed from the distribution for R. The values in $ million are: 2.6; 3.0; 3.2; 3.0. If the interest rate is 12% per year, use the sample to calculate the PW values of R that could be included in an economic evaluation with risk considered.

Solution

  a. R is discrete since, as estimated by Carlos, it can take on only 6 specific values. In engineering economy terms, R is a uniform series amount, that is, an A value.
  b. Probability statements for 6 estimated values (in $ million) are all equal at $1/6$.

$$P(R = 2.6) = P(R = 2.8) = \ldots = P(R = 3.6) = 1/6 \text{ or } 0.16667$$

  c. Figure C.2 shows that probability is distributed as a uniform distribution with equal probability for each value of R.
  d. Let the symbol $PW(R_i)$ indicate the present worth of the ith sample value. Since R (in $ million) is a uniform series amount, the P/A factor is used.

$$PW(R_1) = R_1(P/A,12\%,5) = 2.6(3.6048) = \$9,372,480$$
$$PW(R_2) = 3.0(P/A,12\%,5) = \$10,814,400$$
$$PW(R_3) = 3.2(P/A,12\%,5) = \$11,535,360$$
$$PW(R_4) = 3.0(P/A,12\%,5) = \$10,814,400$$

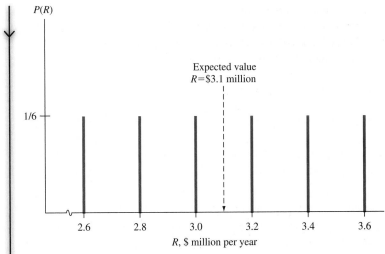

FIGURE C.2   Probability distribution of estimated annual revenue, Example C.2.

Discrete variables use summation operations and continuous variables use integration operations to calculate the properties discussed next. The development here is for discrete variables only.

The expected value of a variable X is the long-term average if a variable is sampled many times. For a complete project, the expected value of a measure, say, PW, is the long-run average of PW, were the project repeated many times. The expected value of the variable X is identified by $E(X)$. The *sample average* $\overline{X}$ estimates the expected value by summing all values in the sample and dividing by the sample size N.

E(X) estimate:
$$\overline{X} = \sum_{i=1}^{N} X_i/N \qquad\qquad [\text{C.4}]$$

In Example C.2, the variable is PW and $E(\text{PW})$ is estimated by $\overline{PW}$ from Equation [C.4].

$$\overline{PW} = (9{,}372{,}480 + \ldots + 10{,}814{,}400)/4 = \$10{,}634{,}160$$

The Excel function = AVERAGE() calculates the sample mean, which is the same as Equation [C.4], from a list of entries in spreadsheet cells or up to 30 individually entered values. Check the Excel software help function for details.

If the probability distribution is known or estimated, the expected value is calculated as

$$E(X) = \sum_{\text{all } i} X_i P(X_i) \qquad\qquad [\text{C.5}]$$

An expected value can be utilized in one of two ways in engineering economy. The first calculates $E(\text{parameter estimate})$, for example $E(P)$ or $E(\text{AOC})$, over all possible values of the variable. If an entire alternative is to be evaluated, once $E()$

is calculated, the evaluation is performed in the same way discussed in chapters of this text. The second way determines E(measure), for example E(PW) or E(ROR) of the entire alternative. Now the best alternative is selected from the best E(measure) value. The next two examples illustrate these two approaches in turn.

EXAMPLE C.3   Plumb Electric Cooperative is experiencing a difficult time obtaining natural gas for electricity generation. Fuels other than natural gas are purchased at an extra cost, which is transferred to the customer. Total monthly fuel expenses are now averaging $7,750,000. An engineer with this city-owned utility has calculated the average revenue for the past 24 months using three fuel-mix situations—gas plentiful, less than 30% other fuels purchased, and 30% or more other fuels. Table C.5 indicates the number of months that each fuel-mix situation occurred. Can the utility expect to meet future monthly expenses based on the 24 months of data, if a similar fuel-mix pattern continues?

TABLE C.5   Revenue and Fuel-Mix Data, Example C.3

Fuel-Mix Situation	Months in Past 24	Average Revenue, $ per Month
Gas plentiful	12	5,270,000
<30% other	6	7,850,000
≥30% other	6	12,130,000

Solution

Using the 24 months of data, estimate a probability for each fuel mix.

Fuel-Mix Situation	Probability of Occurrence
Gas plentiful	12/24 = 0.50
<30% other	6/24 = 0.25
≥30% other	6/24 = 0.25

Let R represent average monthly revenue. Use Equation [C.5] to determine expected revenue per month.

$$E(R) = 5,270,000(0.50) + 7,850,000(0.25) + 12,130,000(0.25)$$
$$= \$7,630,000$$

With expenses averaging $7,750,000, the average monthly revenue shortfall is $120,000. To break even, other sources of revenue must be generated, or the additional costs may be transferred to the customer.

Lite-Weight Wheelchair Company has a substantial investment in tubular steel bend- **EXAMPLE C.4**
ing equipment. A new piece of equipment costs $5000 and has a life of 3 years.
Estimated cash flows (Table C.6) depend on economic conditions classified as receding, stable, or expanding. A probability is estimated that each of the economic conditions will prevail during the 3-year period. Apply expected value and PW analysis
to determine if the equipment should be purchased. Use a MARR of 15% per year.

TABLE C.6 Equipment Cash Flow and Probabilities, Example C.4

	Economic Condition		
Year	Receding (Prob. = 0.2)	Stable (Prob. = 0.6)	Expanding (Prob. = 0.2)
	Annual Cash Flow Estimates, $ per year		
0	$-5000	$-5000	$-5000
1	+2500	+2000	+2000
2	+2000	+2000	+3000
3	+1000	+2000	+3500

Solution

First determine the PW of the cash flows in Table C.6 for each economic condition, and then calculate $E(PW)$, using Equation [C.5]. Define subscripts R for
receding economy, S for stable, and E for expanding. The PW values for the
three scenarios are

$$PW_R = -5000 + 2500(P/F,15\%,1) + 2000(P/F,15\%,2) + 1000(P/F,15\%,3)$$
$$= -5000 + 4344 = \$-656$$
$$PW_S = -5000 + 4566 = \$-434$$
$$PW_E = -5000 + 6309 = \$+1309$$

Only in an expanding economy will the cash flows return the 15% and justify
the investment. Using the probabilities estimated for each economic condition
(alternative) the expected present worth is

$$E(PW) = \sum_{j=R,S,E} PW_j[P(j)]$$
$$= -656(0.2) - 434(0.6) + 1309(0.2)$$
$$= \$-130$$

At 15%, $E(PW) < 0$; the equipment is not justified using an expected value analysis.

The *average* is a *measure of central tendency* of data. Another measure, the *standard deviation* identified by the small letter *s*, is a *measure of the spread* of the data
points around the average. By definition, *s* is the dispersion or spread about the

expected value $E(X)$ or sample average $\overline{X}$. The standard deviation for a sample is calculated by taking each X minus the sample average $\overline{X}$, squaring it, adding all terms, dividing by $N-1$, then extracting the square root of the result.

$$s = \left[\frac{\sum\limits_{i=1}^{N} (X_i - \overline{X})^2}{N - 1} \right]^{1/2} \qquad \text{[C.6]}$$

An equivalent, easier way to compute s eliminates the subtractions.

$$s = \left[\frac{\sum\limits_{i=1}^{N} X_i^2}{N - 1} - \frac{N}{N - 1}\overline{X}^2 \right]^{1/2} \qquad \text{[C.7]}$$

From the random sample, the central tendency, estimated by $\overline{X}$, and dispersion, estimated by s, measures are calculated and combined to determine the fraction or percentage of the values that are expected to be within ± 1, ± 2, and ± 3 standard deviations of the average.

$$\overline{X} \pm ts \quad \text{for } t = 1, 2, 3 \qquad \text{[C.8]}$$

In probability terms, this can be stated as

$$P((\overline{X} - ts \leq X \leq \overline{X} + ts) \qquad \text{[C.9]}$$

This is a very good measure of the clustering of the data. The tighter the clustering about the expected value, the more confidence that the decision maker can place on the selection of one alternative. This logic is illustrated in Example C.5, which considers risk.

The Excel function = STDEV() calculates the sample standard deviation using Equation [C.6] from a series of entries in spreadsheet cells or up to 30 individually entered values. Check the Excel software help function for details.

EXAMPLE C.5   Jerry, an electrical engineer with TGS Nationwide Utility Services, is analyzing electric bills for 1-bedroom apartments in Atlanta and Chicago. In both cities, TGS charges apartment complex owners a flat rate of $125 per month, with this cost passed on to the resident through the rent. This eliminates the expensive alternative of metering and billing each resident monthly. Initially, small random samples of monthly bills from both cities will assist in decision making under risk about the future of "blanket" versus "individual" billing. Help Jerry by answering several questions using the sample data.

   a. Estimate the average monthly bill for each city. Do they appear to be approximately equal or significantly different?

   b. Estimate the standard deviations. How does the clustering about the average compare for the two cities?

   c. Determine the number of bills within the limits $\overline{X} \pm 1s$ for each sample. Based on the sample results, what decision seems reasonable concerning the flat rate of $125 per month versus individual billing in the future?

Sample point	1	2	3	4	5	6	7
Atlanta sample, A, $	65	66	73	92	117	159	225
Chicago sample, C, $	84	90	104	140	157		

Solution

The following solution uses manual calculations. It is followed by a spreadsheet-based solution.

a. Equation [C.4] estimates the expected values, which are quite close in amount.

Atlanta: $N = 7$ $\overline{X}_A = 797/7 = \$113.86$ per month
Chicago: $N = 5$ $\overline{X}_C = 575/5 = \$115.00$ per month

b. Only for illustration purposes, calculate the standard deviation using both formats; Equation [C.6] for Atlanta and Equation [C.7] for Chicago. Tables C.7 and C.8 summarize the calculations, resulting in $s_A = \$59.48$ and $s_C = \$32.00$. In terms of the sample averages, these vary considerably for the two cities.

Atlanta: s_A is 52% of the average $\overline{X}_A = \$113.86$
Chicago: s_C is 28% of the average $\overline{X}_C = \$115.00$

The clustering about the average for the Chicago sample is tighter than for Atlanta.

c. One standard deviation from the mean is calculated using $t = 1$ in Equation [C.8]. The number of sample points within the $\pm 1s$ is then determined.

Atlanta: $\$113.86 \pm 59.48$ results in limits of $\$54.38$ and $\$173.34$.
One data point is outside these limits.

TABLE C.7 **Calculation of Standard Deviation using Equation [C.6] for Atlanta Sample, Example C.5**

X	$(X - \overline{X})$	$(X - \overline{X})^2$
65	−48.86	2,387.30
66	−47.86	2,290.58
73	−40.86	1,669.54
92	−21.86	477.86
117	3.14	9.86
159	45.14	2,037.62
225	111.14	12,352.10
Total		21,224.86

$\overline{X}_A = 113.86$
$s_A = [(21,224.86)/(7 - 1)]^{1/2} = 59.48$

TABLE C.8 **Calculation of Standard Deviation using Equation [C.7] for Chicago Sample, Example C.5**

X	X^2
84	7,056
90	8,100
104	10,816
140	19,600
157	24,649
Total	70,221

$$\overline{X}_C = 115$$
$$s_c = [70{,}221/4 - 5/4(115)^2]^{1/2} = 32$$

Chicago: $115.00 $\pm$ 32.00 results in limits of $83.00 and $147.00. Again, one data point is outside these limits.

Based on these quite small samples, the clustering is tighter for Chicago bills, but the same pattern of bills is experienced within the $\pm 1s$ range. Therefore, the average charge of $125 per month adequately covers the expected cost. However, larger samples may prove this conclusion incorrect.

Solution using a spreadsheet is presented in Figure C.3. The AVERAGE and STDEV functions are used to determine $\overline{X}$ and s. The results and conclusions about clustering are the same as discussed above.

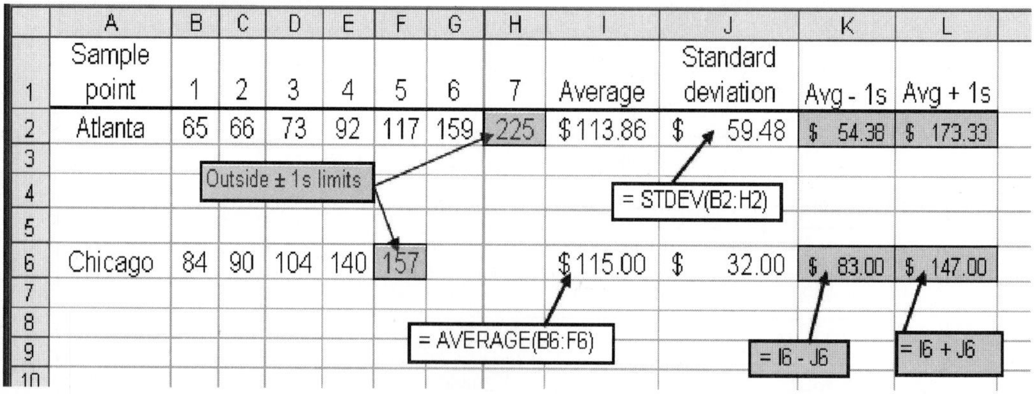

FIGURE C.3 Excel functions display sample average, standard deviation, and $\pm$ 1s limits, Example C.5.

Answers to Problems for Test Review and FE Exam Practice

Chapter 1—41 (c); 42 (c); 43 (b); 44 (d); 45 (c); 46 (d)

Chapter 2—75 (b); 76 (c); 77 (c); 78 (d); 79 (b); 80 (b); 81 (a); 82 (c); 83 (d); 84 (d); 85 (a); 86 (b)

Chapter 3—56 (d); 57 (c); 58 (d); 59 (b); 60 (c); 61 (d); 62 (a); 63 (c)

Chapter 4—45 (b); 46 (c); 47 (a); 48 (a); 49 (d); 50 (b); 51 (c); 52 (a); 53 (d); 54 (d); 55 (c); 56 (b); 57 (a); 58 (c)

Chapter 5—23 (d); 24 (b); 25 (b); 26 (c); 27 (d); 28 (a); 29 (d); 30 (c); 31 (c); 32 (b)

Chapter 6—54 (a); 55 (d); 56 (d); 57 (b); 58 (b); 59 (d); 60 (a); 61 (c); 62 (c)

Chapter 7—32 (d); 33 (b); 34 (b); 35 (b); 36 (c); 37 (a); 38 (c); 39 (d); 40 (b); 41 (a)

Chapter 8—54 (c); 55 (b); 56 (d); 57 (b); 58 (a); 59 (c); 60 (d); 61 (c); 62 (b); 63 (d); 64 (a)

Chapter 9—26 (b); 27 (b); 28 (b); 29 (c); 30 (a); 31 (b); 32 (a)

Chapter 10—39 (a); 40 (b); 41 (a); 42 (a); 43 (d); 44 (d); 45 (a); 46 (c); 47 (a); 48 (d)

Chapter 11—43 (b); 44 (b); 45 (c); 46 (a); 47 (c); 48 (d); 49 (b); 50 (c); 51 (a); 52 (b); 53 (a); 54 (c); 55 (d); 56 (c); 57 (a)

Chapter 12—39 (b); 40 (c); 41 (c); 42 (d); 43 (a); 44 (b); 45 (d); 46 (a); 47 (b); 48 (c); 49 (d); 50 (b)

Chapter 13—44 (c); 45 (a); 46 (d); 47 (c); 48 (b); 49 (d); 50 (c)

Reference Materials

TEXTBOOKS

Blank, L. T., A. Tarquin: *Engineering Economy,* 6th ed., McGraw-Hill, New York, 2005.

Blank, L. T., A. Tarquin, and S. Iverson: *Engineering Economy,* Canadian ed., McGraw-Hill Ryerson, Whitby, ON, 2008.

Bowman, M. S.: *Applied Economic Analysis for Technologists, Engineers, and Managers,* 2d ed., Pearson Prentice-Hall, Upper Saddle River, NJ, 2003.

Canada, J. R., W. G. Sullivan, D. Kulonda, and J. A. White: *Capital Investment Analysis for Engineering and Management*, 3d ed., Pearson Prentice-Hall, Upper Saddle River, NJ, 2005.

Collier, C. A., and C. R. Glagola: *Engineering and Economic Cost Analysis*, 3d ed., Pearson Prentice-Hall, Upper Saddle River, NJ, 1999.

Eschenbach, T. G.: *Engineering Economy: Applying Theory to Practice*, 2d ed., Oxford University Press, New York, 2003.

Fraser, N. M., E. M. Lewkes, I. Bernhardt, and M. Tajima: *Engineering Economics in Canada,* 3d ed., Pearson Prentice-Hall, Upper Saddle River, NJ, 2006.

Hartman, J. C.: *Engineering Economy and the Decision-Making Process,* Pearson Prentice-Hall, Upper Saddle River, NJ, 2007.

Levy, S. M.: *Build, Operate, Transfer: Paving the Way for Tomorrow's Infrastructure,* John Wiley & Sons, Hoboken, New Jersey, 1996.

Newnan, D. G., J. P. Lavelle, and T. G. Eschenbach: *Essentials of Engineering Economic Analysis,* 2d ed., Oxford University Press, New York, 2002.

Newnan, D. G., T. G. Eschenbach, and J. P. Lavelle: *Engineering Economic Analysis,* 9th ed., Oxford University Press, New York, 2004.

Ostwald, P. F.: *Construction Cost Analysis and Estimating,* Pearson Prentice-Hall, Upper Saddle River, NJ, 2001.

Ostwald, P. F., and T. S. McLaren: *Cost Analysis and Estimating for Engineering and Management*, Pearson Prentice-Hall, Upper Saddle River, NJ, 2004.

Park, C. S.: *Contemporary Engineering Economics,* 4th ed., Pearson Prentice-Hall, Upper Saddle River, NJ, 2007.

Park, C. S.: *Fundamentals of Engineering Economics,* Pearson Prentice-Hall, Upper Saddle River, NJ, 2004.

Peters, M. S., K. D. Timmerhaus, and R. E. West: *Plant Design and Economics for Chemical Engineers,* 5th ed., McGraw-Hill, New York, 2003.

Peurifoy, R. L., and G. D. Oberlender: *Estimating Construction Costs,* 5th ed., McGraw-Hill, New York, 2002.

Sullivan, W. G., E. Wicks, and J. Luxhoj: *Engineering Economy,* 13th ed., Pearson Prentice-Hall, Upper Saddle River, NJ, 2006.

Thuesen, G. J., and W. J. Fabrycky: *Engineering Economy,* 9th ed., Pearson Prentice-Hall, Upper Saddle River, NJ, 2001.

White, J. A., K. E. Case, D. B. Pratt, and M. H. Agee: *Principles of Engineering Economic Analysis,* 4th ed., John Wiley & Sons, Hoboken, New Jersey, 1998.

SELECTED JOURNALS AND PUBLICATIONS

Chemical Engineering, Access Intelligence, New York, monthly.

Corporations, Publication 542, Department of the Treasury, Internal Revenue Service, Government Printing Office, Washington, DC, annually.

Engineering News-Record, McGraw-Hill, New York, monthly.

How to Depreciate Property, Publication 946, U.S. Department of the Treasury, Internal Revenue Service, Government Printing Office, Washington, DC, annually.

Sales and Other Dispositions of Assets, Publication 542, Department of the Treasury, Internal Revenue Service, Government Printing Office, Washington, DC, annually.

The Engineering Economist, joint publication of the Engineering Economy Divisions of ASEE and IIE, published by Taylor and Francis, Philadelphia, PA, quarterly.

U.S. Master Tax Guide, Commerce Clearing House, Chicago, annually.

0.25%			TABLE 1	Discrete Cash Flow: Compound Interest Factors				0.25%
	Single Payments		Uniform Series Payments				Arithmetic Gradients	
	F/P Compound Amount	**P/F** Present Worth	**A/F** Sinking Fund	**F/A** Compound Amount	**A/P** Capital Recovery	**P/A** Present Worth	**P/G** Gradient Present Worth	**A/G** Gradient Uniform Series
n								
1	1.0025	0.9975	1.00000	1.0000	1.00250	0.9975		
2	1.0050	0.9950	0.49938	2.0025	0.50188	1.9925	0.9950	0.4994
3	1.0075	0.9925	0.33250	3.0075	0.33500	2.9851	2.9801	0.9983
4	1.0100	0.9901	0.24906	4.0150	0.25156	3.9751	5.9503	1.4969
5	1.0126	0.9876	0.19900	5.0251	0.20150	4.9627	9.9007	1.9950
6	1.0151	0.9851	0.16563	6.0376	0.16813	5.9478	14.8263	2.4927
7	1.0176	0.9827	0.14179	7.0527	0.14429	6.9305	20.7223	2.9900
8	1.0202	0.9802	0.12391	8.0704	0.12641	7.9107	27.5839	3.4869
9	1.0227	0.9778	0.11000	9.0905	0.11250	8.8885	35.4061	3.9834
10	1.0253	0.9753	0.09888	10.1133	0.10138	9.8639	44.1842	4.4794
11	1.0278	0.9729	0.08978	11.1385	0.09228	10.8368	53.9133	4.9750
12	1.0304	0.9705	0.08219	12.1664	0.08469	11.8073	64.5886	5.4702
13	1.0330	0.9681	0.07578	13.1968	0.07828	12.7753	76.2053	5.9650
14	1.0356	0.9656	0.07028	14.2298	0.07278	13.7410	88.7587	6.4594
15	1.0382	0.9632	0.06551	15.2654	0.06801	14.7042	102.2441	6.9534
16	1.0408	0.9608	0.06134	16.3035	0.06384	15.6650	116.6567	7.4469
17	1.0434	0.9584	0.05766	17.3443	0.06016	16.6235	131.9917	7.9401
18	1.0460	0.9561	0.05438	18.3876	0.05688	17.5795	148.2446	8.4328
19	1.0486	0.9537	0.05146	19.4336	0.05396	18.5332	165.4106	8.9251
20	1.0512	0.9513	0.04882	20.4822	0.05132	19.4845	183.4851	9.4170
21	1.0538	0.9489	0.04644	21.5334	0.04894	20.4334	202.4634	9.9085
22	1.0565	0.9466	0.04427	22.5872	0.04677	21.3800	222.3410	10.3995
23	1.0591	0.9442	0.04229	23.6437	0.04479	22.3241	243.1131	10.8901
24	1.0618	0.9418	0.04048	24.7028	0.04298	23.2660	264.7753	11.3804
25	1.0644	0.9395	0.03881	25.7646	0.04131	24.2055	287.3230	11.8702
26	1.0671	0.9371	0.03727	26.8290	0.03977	25.1426	310.7516	12.3596
27	1.0697	0.9348	0.03585	27.8961	0.03835	26.0774	335.0566	12.8485
28	1.0724	0.9325	0.03452	28.9658	0.03702	27.0099	360.2334	13.3371
29	1.0751	0.9301	0.03329	30.0382	0.03579	27.9400	386.2776	13.8252
30	1.0778	0.9278	0.03214	31.1133	0.03464	28.8679	413.1847	14.3130
36	1.0941	0.9140	0.02658	37.6206	0.02908	34.3865	592.4988	17.2306
40	1.1050	0.9050	0.02380	42.0132	0.02630	38.0199	728.7399	19.1673
48	1.1273	0.8871	0.01963	50.9312	0.02213	45.1787	1040.06	23.0209
50	1.1330	0.8826	0.01880	53.1887	0.02130	46.9462	1125.78	23.9802
52	1.1386	0.8782	0.01803	55.4575	0.02053	48.7048	1214.59	24.9377
55	1.1472	0.8717	0.01698	58.8819	0.01948	51.3264	1353.53	26.3710
60	1.1616	0.8609	0.01547	64.6467	0.01797	55.6524	1600.08	28.7514
72	1.1969	0.8355	0.01269	78.7794	0.01519	65.8169	2265.56	34.4221
75	1.2059	0.8292	0.01214	82.3792	0.01464	68.3108	2447.61	35.8305
84	1.2334	0.8108	0.01071	93.3419	0.01321	75.6813	3029.76	40.0331
90	1.2520	0.7987	0.00992	100.7885	0.01242	80.5038	3446.87	42.8162
96	1.2709	0.7869	0.00923	108.3474	0.01173	85.2546	3886.28	45.5844
100	1.2836	0.7790	0.00881	113.4500	0.01131	88.3825	4191.24	47.4216
108	1.3095	0.7636	0.00808	123.8093	0.01058	94.5453	4829.01	51.0762
120	1.3494	0.7411	0.00716	139.7414	0.00966	103.5618	5852.11	56.5084
132	1.3904	0.7192	0.00640	156.1582	0.00890	112.3121	6950.01	61.8813
144	1.4327	0.6980	0.00578	173.0743	0.00828	120.8041	8117.41	67.1949
240	1.8208	0.5492	0.00305	328.3020	0.00555	180.3109	19399	107.5863
360	2.4568	0.4070	0.00172	582.7369	0.00422	237.1894	36264	152.8902
480	3.3151	0.3016	0.00108	926.0595	0.00358	279.3418	53821	192.6699

0.5%			TABLE 2	Discrete Cash Flow: Compound Interest Factors			0.5%
	Single Payments		**Uniform Series Payments**				**Arithmetic Gradients**	
	F/P Compound Amount	*P/F* Present Worth	*A/F* Sinking Fund	*F/A* Compound Amount	*A/P* Capital Recovery	*P/A* Present Worth	*P/G* Gradient Present Worth	*A/G* Gradient Uniform Series
n								
1	1.0050	0.9950	1.00000	1.0000	1.00500	0.9950		
2	1.0100	0.9901	0.49875	2.0050	0.50375	1.9851	0.9901	0.4988
3	1.0151	0.9851	0.33167	3.0150	0.33667	2.9702	2.9604	0.9967
4	1.0202	0.9802	0.24813	4.0301	0.25313	3.9505	5.9011	1.4938
5	1.0253	0.9754	0.19801	5.0503	0.20301	4.9259	9.8026	1.9900
6	1.0304	0.9705	0.16460	6.0755	0.16960	5.8964	14.6552	2.4855
7	1.0355	0.9657	0.14073	7.1059	0.14573	6.8621	20.4493	2.9801
8	1.0407	0.9609	0.12283	8.1414	0.12783	7.8230	27.1755	3.4738
9	1.0459	0.9561	0.10891	9.1821	0.11391	8.7791	34.8244	3.9668
10	1.0511	0.9513	0.09777	10.2280	0.10277	9.7304	43.3865	4.4589
11	1.0564	0.9466	0.08866	11.2792	0.09366	10.6770	52.8526	4.9501
12	1.0617	0.9419	0.08107	12.3356	0.08607	11.6189	63.2136	5.4406
13	1.0670	0.9372	0.07464	13.3972	0.07964	12.5562	74.4602	5.9302
14	1.0723	0.9326	0.06914	14.4642	0.07414	13.4887	86.5835	6.4190
15	1.0777	0.9279	0.06436	15.5365	0.06936	14.4166	99.5743	6.9069
16	1.0831	0.9233	0.06019	16.6142	0.06519	15.3399	113.4238	7.3940
17	1.0885	0.9187	0.05651	17.6973	0.06151	16.2586	128.1231	7.8803
18	1.0939	0.9141	0.05323	18.7858	0.05823	17.1728	143.6634	8.3658
19	1.0994	0.9096	0.05030	19.8797	0.05530	18.0824	160.0360	8.8504
20	1.1049	0.9051	0.04767	20.9791	0.05267	18.9874	177.2322	9.3342
21	1.1104	0.9006	0.04528	22.0840	0.05028	19.8880	195.2434	9.8172
22	1.1160	0.8961	0.04311	23.1944	0.04811	20.7841	214.0611	10.2993
23	1.1216	0.8916	0.04113	24.3104	0.04613	21.6757	233.6768	10.7806
24	1.1272	0.8872	0.03932	25.4320	0.04432	22.5629	254.0820	11.2611
25	1.1328	0.8828	0.03765	26.5591	0.04265	23.4456	275.2686	11.7407
26	1.1385	0.8784	0.03611	27.6919	0.04111	24.3240	297.2281	12.2195
27	1.1442	0.8740	0.03469	28.8304	0.03969	25.1980	319.9523	12.6975
28	1.1499	0.8697	0.03336	29.9745	0.03836	26.0677	343.4332	13.1747
29	1.1556	0.8653	0.03213	31.1244	0.03713	26.9330	367.6625	13.6510
30	1.1614	0.8610	0.03098	32.2800	0.03598	27.7941	392.6324	14.1265
36	1.1967	0.8356	0.02542	39.3361	0.03042	32.8710	557.5598	16.9621
40	1.2208	0.8191	0.02265	44.1588	0.02765	36.1722	681.3347	18.8359
48	1.2705	0.7871	0.01849	54.0978	0.02349	42.5803	959.9188	22.5437
50	1.2832	0.7793	0.01765	56.6452	0.02265	44.1428	1035.70	23.4624
52	1.2961	0.7716	0.01689	59.2180	0.02189	45.6897	1113.82	24.3778
55	1.3156	0.7601	0.01584	63.1258	0.02084	47.9814	1235.27	25.7447
60	1.3489	0.7414	0.01433	69.7700	0.01933	51.7256	1448.65	28.0064
72	1.4320	0.6983	0.01157	86.4089	0.01657	60.3395	2012.35	33.3504
75	1.4536	0.6879	0.01102	90.7265	0.01602	62.4136	2163.75	34.6679
84	1.5204	0.6577	0.00961	104.0739	0.01461	68.4530	2640.66	38.5763
90	1.5666	0.6383	0.00883	113.3109	0.01383	72.3313	2976.08	41.1451
96	1.6141	0.6195	0.00814	122.8285	0.01314	76.0952	3324.18	43.6845
100	1.6467	0.6073	0.00773	129.3337	0.01273	78.5426	3562.79	45.3613
108	1.7137	0.5835	0.00701	142.7399	0.01201	83.2934	4054.37	48.6758
120	1.8194	0.5496	0.00610	163.8793	0.01110	90.0735	4823.51	53.5508
132	1.9316	0.5177	0.00537	186.3226	0.01037	96.4596	5624.59	58.3103
144	2.0508	0.4876	0.00476	210.1502	0.00976	102.4747	6451.31	62.9551
240	3.3102	0.3021	0.00216	462.0409	0.00716	139.5808	13416	96.1131
360	6.0226	0.1660	0.00100	1004.52	0.00600	166.7916	21403	128.3236
480	10.9575	0.0913	0.00050	1991.49	0.00550	181.7476	27588	151.7949

0.75%			TABLE 3	Discrete Cash Flow: Compound Interest Factors				0.75%
	Single Payments			Uniform Series Payments			Arithmetic Gradients	
	F/P Compound Amount	**P/F** Present Worth	**A/F** Sinking Fund	**F/A** Compound Amount	**A/P** Capital Recovery	**P/A** Present Worth	**P/G** Gradient Present Worth	**A/G** Gradient Uniform Series
n								
1	1.0075	0.9926	1.00000	1.0000	1.00750	0.9926		
2	1.0151	0.9852	0.49813	2.0075	0.50563	1.9777	0.9852	0.4981
3	1.0227	0.9778	0.33085	3.0226	0.33835	2.9556	2.9408	0.9950
4	1.0303	0.9706	0.24721	4.0452	0.25471	3.9261	5.8525	1.4907
5	1.0381	0.9633	0.19702	5.0756	0.20452	4.8894	9.7058	1.9851
6	1.0459	0.9562	0.16357	6.1136	0.17107	5.8456	14.4866	2.4782
7	1.0537	0.9490	0.13967	7.1595	0.14717	6.7946	20.1808	2.9701
8	1.0616	0.9420	0.12176	8.2132	0.12926	7.7366	26.7747	3.4608
9	1.0696	0.9350	0.10782	9.2748	0.11532	8.6716	34.2544	3.9502
10	1.0776	0.9280	0.09667	10.3443	0.10417	9.5996	42.6064	4.4384
11	1.0857	0.9211	0.08755	11.4219	0.09505	10.5207	51.8174	4.9253
12	1.0938	0.9142	0.07995	12.5076	0.08745	11.4349	61.8740	5.4110
13	1.1020	0.9074	0.07352	13.6014	0.08102	12.3423	72.7632	5.8954
14	1.1103	0.9007	0.06801	14.7034	0.07551	13.2430	84.4720	6.3786
15	1.1186	0.8940	0.06324	15.8137	0.07074	14.1370	96.9876	6.8606
16	1.1270	0.8873	0.05906	16.9323	0.06656	15.0243	110.2973	7.3413
17	1.1354	0.8807	0.05537	18.0593	0.06287	15.9050	124.3887	7.8207
18	1.1440	0.8742	0.05210	19.1947	0.05960	16.7792	139.2494	8.2989
19	1.1525	0.8676	0.04917	20.3387	0.05667	17.6468	154.8671	8.7759
20	1.1612	0.8612	0.04653	21.4912	0.05403	18.5080	171.2297	9.2516
21	1.1699	0.8548	0.04415	22.6524	0.05165	19.3628	188.3253	9.7261
22	1.1787	0.8484	0.04198	23.8223	0.04948	20.2112	206.1420	10.1994
23	1.1875	0.8421	0.04000	25.0010	0.04750	21.0533	224.6682	10.6714
24	1.1964	0.8358	0.03818	26.1885	0.04568	21.8891	243.8923	11.1422
25	1.2054	0.8296	0.03652	27.3849	0.04402	22.7188	263.8029	11.6117
26	1.2144	0.8234	0.03498	28.5903	0.04248	23.5422	284.3888	12.0800
27	1.2235	0.8173	0.03355	29.8047	0.04105	24.3595	305.6387	12.5470
28	1.2327	0.8112	0.03223	31.0282	0.03973	25.1707	327.5416	13.0128
29	1.2420	0.8052	0.03100	32.2609	0.03850	25.9759	350.0867	13.4774
30	1.2513	0.7992	0.02985	33.5029	0.03735	26.7751	373.2631	13.9407
36	1.3086	0.7641	0.02430	41.1527	0.03180	31.4468	524.9924	16.6946
40	1.3483	0.7416	0.02153	46.4465	0.02903	34.4469	637.4693	18.5058
48	1.4314	0.6986	0.01739	57.5207	0.02489	40.1848	886.8404	22.0691
50	1.4530	0.6883	0.01656	60.3943	0.02406	41.5664	953.8486	22.9476
52	1.4748	0.6780	0.01580	63.3111	0.02330	42.9276	1022.59	23.8211
55	1.5083	0.6630	0.01476	67.7688	0.02226	44.9316	1128.79	25.1223
60	1.5657	0.6387	0.01326	75.4241	0.02076	48.1734	1313.52	27.2665
72	1.7126	0.5839	0.01053	95.0070	0.01803	55.4768	1791.25	32.2882
75	1.7514	0.5710	0.00998	100.1833	0.01748	57.2027	1917.22	33.5163
84	1.8732	0.5338	0.00859	116.4269	0.01609	62.1540	2308.13	37.1357
90	1.9591	0.5104	0.00782	127.8790	0.01532	65.2746	2578.00	39.4946
96	2.0489	0.4881	0.00715	139.8562	0.01465	68.2584	2853.94	41.8107
100	2.1111	0.4737	0.00675	148.1445	0.01425	70.1746	3040.75	43.3311
108	2.2411	0.4462	0.00604	165.4832	0.01354	73.8394	3419.90	46.3154
120	2.4514	0.4079	0.00517	193.5143	0.01267	78.9417	3998.56	50.6521
132	2.6813	0.3730	0.00446	224.1748	0.01196	83.6064	4583.57	54.8232
144	2.9328	0.3410	0.00388	257.7116	0.01138	87.8711	5169.58	58.8314
240	6.0092	0.1664	0.00150	667.8869	0.00900	111.1450	9494.12	85.4210
360	14.7306	0.0679	0.00055	1830.74	0.00805	124.2819	13312	107.1145
480	36.1099	0.0277	0.00021	4681.32	0.00771	129.6409	15513	119.6620

1%			TABLE 4	Discrete Cash Flow: Compound Interest Factors				1%
	Single Payments		Uniform Series Payments				Arithmetic Gradients	
	F/P Compound Amount	*P/F* Present Worth	*A/F* Sinking Fund	*F/A* Compound Amount	*A/P* Capital Recovery	*P/A* Present Worth	*P/G* Gradient Present Worth	*A/G* Gradient Uniform Series
n								
1	1.0100	0.9901	1.00000	1.0000	1.01000	0.9901		
2	1.0201	0.9803	0.49751	2.0100	0.50751	1.9704	0.9803	0.4975
3	1.0303	0.9706	0.33002	3.0301	0.34002	2.9410	2.9215	0.9934
4	1.0406	0.9610	0.24628	4.0604	0.25628	3.9020	5.8044	1.4876
5	1.0510	0.9515	0.19604	5.1010	0.20604	4.8534	9.6103	1.9801
6	1.0615	0.9420	0.16255	6.1520	0.17255	5.7955	14.3205	2.4710
7	1.0721	0.9327	0.13863	7.2135	0.14863	6.7282	19.9168	2.9602
8	1.0829	0.9235	0.12069	8.2857	0.13069	7.6517	26.3812	3.4478
9	1.0937	0.9143	0.10674	9.3685	0.11674	8.5660	33.6959	3.9337
10	1.1046	0.9053	0.09558	10.4622	0.10558	9.4713	41.8435	4.4179
11	1.1157	0.8963	0.08645	11.5668	0.09645	10.3676	50.8067	4.9005
12	1.1268	0.8874	0.07885	12.6825	0.08885	11.2551	60.5687	5.3815
13	1.1381	0.8787	0.07241	13.8093	0.08241	12.1337	71.1126	5.8607
14	1.1495	0.8700	0.06690	14.9474	0.07690	13.0037	82.4221	6.3384
15	1.1610	0.8613	0.06212	16.0969	0.07212	13.8651	94.4810	6.8143
16	1.1726	0.8528	0.05794	17.2579	0.06794	14.7179	107.2734	7.2886
17	1.1843	0.8444	0.05426	18.4304	0.06426	15.5623	120.7834	7.7613
18	1.1961	0.8360	0.05098	19.6147	0.06098	16.3983	134.9957	8.2323
19	1.2081	0.8277	0.04805	20.8109	0.05805	17.2260	149.8950	8.7017
20	1.2202	0.8195	0.04542	22.0190	0.05542	18.0456	165.4664	9.1694
21	1.2324	0.8114	0.04303	23.2392	0.05303	18.8570	181.6950	9.6354
22	1.2447	0.8034	0.04086	24.4716	0.05086	19.6604	198.5663	10.0998
23	1.2572	0.7954	0.03889	25.7163	0.04889	20.4558	216.0660	10.5626
24	1.2697	0.7876	0.03707	26.9735	0.04707	21.2434	234.1800	11.0237
25	1.2824	0.7798	0.03541	28.2432	0.04541	22.0232	252.8945	11.4831
26	1.2953	0.7720	0.03387	29.5256	0.04387	22.7952	272.1957	11.9409
27	1.3082	0.7644	0.03245	30.8209	0.04245	23.5596	292.0702	12.3971
28	1.3213	0.7568	0.03112	32.1291	0.04112	24.3164	312.5047	12.8516
29	1.3345	0.7493	0.02990	33.4504	0.03990	25.0658	333.4863	13.3044
30	1.3478	0.7419	0.02875	34.7849	0.03875	25.8077	355.0021	13.7557
36	1.4308	0.6989	0.02321	43.0769	0.03321	30.1075	494.6207	16.4285
40	1.4889	0.6717	0.02046	48.8864	0.03046	32.8347	596.8561	18.1776
48	1.6122	0.6203	0.01633	61.2226	0.02633	37.9740	820.1460	21.5976
50	1.6446	0.6080	0.01551	64.4632	0.02551	39.1961	879.4176	22.4363
52	1.6777	0.5961	0.01476	67.7689	0.02476	40.3942	939.9175	23.2686
55	1.7285	0.5785	0.01373	72.8525	0.02373	42.1472	1032.81	24.5049
60	1.8167	0.5504	0.01224	81.6697	0.02224	44.9550	1192.81	26.5333
72	2.0471	0.4885	0.00955	104.7099	0.01955	51.1504	1597.87	31.2386
75	2.1091	0.4741	0.00902	110.9128	0.01902	52.5871	1702.73	32.3793
84	2.3067	0.4335	0.00765	130.6723	0.01765	56.6485	2023.32	35.7170
90	2.4486	0.4084	0.00690	144.8633	0.01690	59.1609	2240.57	37.8724
96	2.5993	0.3847	0.00625	159.9273	0.01625	61.5277	2459.43	39.9727
100	2.7048	0.3697	0.00587	170.4814	0.01587	63.0289	2605.78	41.3426
108	2.9289	0.3414	0.00518	192.8926	0.01518	65.8578	2898.42	44.0103
120	3.3004	0.3030	0.00435	230.0387	0.01435	69.7005	3334.11	47.8349
132	3.7190	0.2689	0.00368	271.8959	0.01368	73.1108	3761.69	51.4520
144	4.1906	0.2386	0.00313	319.0616	0.01313	76.1372	4177.47	54.8676
240	10.8926	0.0918	0.00101	989.2554	0.01101	90.8194	6878.60	75.7393
360	35.9496	0.0278	0.00029	3494.96	0.01029	97.2183	8720.43	89.6995
480	118.6477	0.0084	0.00008	11765	0.01008	99.1572	9511.16	95.9200

1.25%		TABLE **5**	Discrete Cash Flow: Compound Interest Factors				1.25%
	Single Payments		Uniform Series Payments				Arithmetic Gradients	
	F/P Compound Amount	*P/F* Present Worth	*A/F* Sinking Fund	*F/A* Compound Amount	*A/P* Capital Recovery	*P/A* Present Worth	*P/G* Gradient Present Worth	*A/G* Gradient Uniform Series
n								
1	1.0125	0.9877	1.00000	1.0000	1.01250	0.9877		
2	1.0252	0.9755	0.49680	2.0125	0.50939	1.9631	0.9755	0.4969
3	1.0380	0.9634	0.32920	3.0377	0.34170	2.9265	2.9023	0.9917
4	1.0509	0.9515	0.24536	4.0756	0.25786	3.8781	5.7569	1.4845
5	1.0641	0.9398	0.19506	5.1266	0.20756	4.8178	9.5160	1.9752
6	1.0774	0.9282	0.16153	6.1907	0.17403	5.7460	14.1569	2.4638
7	1.0909	0.9167	0.13759	7.2680	0.15009	6.6627	19.6571	2.9503
8	1.1045	0.9054	0.11963	8.3589	0.13213	7.5681	25.9949	3.4348
9	1.1183	0.8942	0.10567	9.4634	0.11817	8.4623	33.1487	3.9172
10	1.1323	0.8832	0.09450	10.5817	0.10700	9.3455	41.0973	4.3975
11	1.1464	0.8723	0.08537	11.7139	0.09787	10.2178	49.8201	4.8758
12	1.1608	0.8615	0.07776	12.8604	0.09026	11.0793	59.2967	5.3520
13	1.1753	0.8509	0.07132	14.0211	0.08382	11.9302	69.5072	5.8262
14	1.1900	0.8404	0.06581	15.1964	0.07831	12.7706	80.4320	6.2982
15	1.2048	0.8300	0.06103	16.3863	0.07353	13.6005	92.0519	6.7682
16	1.2199	0.8197	0.05685	17.5912	0.06935	14.4203	104.3481	7.2362
17	1.2351	0.8096	0.05316	18.8111	0.06566	15.2299	117.3021	7.7021
18	1.2506	0.7996	0.04988	20.0462	0.06238	16.0295	130.8958	8.1659
19	1.2662	0.7898	0.04696	21.2968	0.05946	16.8193	145.1115	8.6277
20	1.2820	0.7800	0.04432	22.5630	0.05682	17.5993	159.9316	9.0874
21	1.2981	0.7704	0.04194	23.8450	0.05444	18.3697	175.3392	9.5450
22	1.3143	0.7609	0.03977	25.1431	0.05227	19.1306	191.3174	10.0006
23	1.3307	0.7515	0.03780	26.4574	0.05030	19.8820	207.8499	10.4542
24	1.3474	0.7422	0.03599	27.7881	0.04849	20.6242	224.9204	10.9056
25	1.3642	0.7330	0.03432	29.1354	0.04682	21.3573	242.5132	11.3551
26	1.3812	0.7240	0.03279	30.4996	0.04529	22.0813	260.6128	11.8024
27	1.3985	0.7150	0.03137	31.8809	0.04387	22.7963	279.2040	12.2478
28	1.4160	0.7062	0.03005	33.2794	0.04255	23.5025	298.2719	12.6911
29	1.4337	0.6975	0.02882	34.6954	0.04132	24.2000	317.8019	13.1323
30	1.4516	0.6889	0.02768	36.1291	0.04018	24.8889	337.7797	13.5715
36	1.5639	0.6394	0.02217	45.1155	0.03467	28.8473	466.2830	16.1639
40	1.6436	0.6084	0.01942	51.4896	0.03192	31.3269	559.2320	17.8515
48	1.8154	0.5509	0.01533	65.2284	0.02783	35.9315	759.2296	21.1299
50	1.8610	0.5373	0.01452	68.8818	0.02702	37.0129	811.6738	21.9295
52	1.9078	0.5242	0.01377	72.6271	0.02627	38.0677	864.9409	22.7211
55	1.9803	0.5050	0.01275	78.4225	0.02525	39.6017	946.2277	23.8936
60	2.1072	0.4746	0.01129	88.5745	0.02379	42.0346	1084.84	25.8083
72	2.4459	0.4088	0.00865	115.6736	0.02115	47.2925	1428.46	30.2047
75	2.5388	0.3939	0.00812	123.1035	0.02062	48.4890	1515.79	31.2605
84	2.8391	0.3522	0.00680	147.1290	0.01930	51.8222	1778.84	34.3258
90	3.0588	0.3269	0.00607	164.7050	0.01857	53.8461	1953.83	36.2855
96	3.2955	0.3034	0.00545	183.6411	0.01795	55.7246	2127.52	38.1793
100	3.4634	0.2887	0.00507	197.0723	0.01757	56.9013	2242.24	39.4058
108	3.8253	0.2614	0.00442	226.0226	0.01692	59.0865	2468.26	41.7737
120	4.4402	0.2252	0.00363	275.2171	0.01613	61.9828	2796.57	45.1184
132	5.1540	0.1940	0.00301	332.3198	0.01551	64.4781	3109.35	48.2234
144	5.9825	0.1672	0.00251	398.6021	0.01501	66.6277	3404.61	51.0990
240	19.7155	0.0507	0.00067	1497.24	0.01317	75.9423	5101.53	67.1764
360	87.5410	0.0114	0.00014	6923.28	0.01264	79.0861	5997.90	75.8401
480	388.7007	0.0026	0.00003	31016	0.01253	79.7942	6284.74	78.7619

	Single Payments		Uniform Series Payments				Arithmetic Gradients	
	F/P	*P/F*	*A/F*	*F/A*	*A/P*	*P/A*	*P/G*	*A/G*
	Compound Amount	Present Worth	Sinking Fund	Compound Amount	Capital Recovery	Present Worth	Gradient Present Worth	Gradient Uniform Series
n								
1	1.0150	0.9852	1.00000	1.0000	1.01500	0.9852		
2	1.0302	0.9707	0.49628	2.0150	0.51128	1.9559	0.9707	0.4963
3	1.0457	0.9563	0.32838	3.0452	0.34338	2.9122	2.8833	0.9901
4	1.0614	0.9422	0.24444	4.0909	0.25944	3.8544	5.7098	1.4814
5	1.0773	0.9283	0.19409	5.1523	0.20909	4.7826	9.4229	1.9702
6	1.0934	0.9145	0.16053	6.2296	0.17553	5.6972	13.9956	2.4566
7	1.1098	0.9010	0.13656	7.3230	0.15156	6.5982	19.4018	2.9405
8	1.1265	0.8877	0.11858	8.4328	0.13358	7.4859	25.6157	3.4219
9	1.1434	0.8746	0.10461	9.5593	0.11961	8.3605	32.6125	3.9008
10	1.1605	0.8617	0.09343	10.7027	0.10843	9.2222	40.3675	4.3772
11	1.1779	0.8489	0.08429	11.8633	0.09929	10.0711	48.8568	4.8512
12	1.1956	0.8364	0.07668	13.0412	0.09168	10.9075	58.0571	5.3227
13	1.2136	0.8240	0.07024	14.2368	0.08524	11.7315	67.9454	5.7917
14	1.2318	0.8118	0.06472	15.4504	0.07972	12.5434	78.4994	6.2582
15	1.2502	0.7999	0.05994	16.6821	0.07494	13.3432	89.6974	6.7223
16	1.2690	0.7880	0.05577	17.9324	0.07077	14.1313	101.5178	7.1839
17	1.2880	0.7764	0.05208	19.2014	0.06708	14.9076	113.9400	7.6431
18	1.3073	0.7649	0.04881	20.4894	0.06381	15.6726	126.9435	8.0997
19	1.3270	0.7536	0.04588	21.7967	0.06088	16.4262	140.5084	8.5539
20	1.3469	0.7425	0.04325	23.1237	0.05825	17.1686	154.6154	9.0057
21	1.3671	0.7315	0.04087	24.4705	0.05587	17.9001	169.2453	9.4550
22	1.3876	0.7207	0.03870	25.8376	0.05370	18.6208	184.3798	9.9018
23	1.4084	0.7100	0.03673	27.2251	0.05173	19.3309	200.0006	10.3462
24	1.4295	0.6995	0.03492	28.6335	0.04992	20.0304	216.0901	10.7881
25	1.4509	0.6892	0.03326	30.0630	0.04826	20.7196	232.6310	11.2276
26	1.4727	0.6790	0.03173	31.5140	0.04673	21.3986	249.6065	11.6646
27	1.4948	0.6690	0.03032	32.9867	0.04532	22.0676	267.0002	12.0992
28	1.5172	0.6591	0.02900	34.4815	0.04400	22.7267	284.7958	12.5313
29	1.5400	0.6494	0.02778	35.9987	0.04278	23.3761	302.9779	12.9610
30	1.5631	0.6398	0.02664	37.5387	0.04164	24.0158	321.5310	13.3883
36	1.7091	0.5851	0.02115	47.2760	0.03615	27.6607	439.8303	15.9009
40	1.8140	0.5513	0.01843	54.2679	0.03343	29.9158	524.3568	17.5277
48	2.0435	0.4894	0.01437	69.5652	0.02937	34.0426	703.5462	20.6667
50	2.1052	0.4750	0.01357	73.6828	0.02857	34.9997	749.9636	21.4277
52	2.1689	0.4611	0.01283	77.9249	0.02783	35.9287	796.8774	22.1794
55	2.2679	0.4409	0.01183	84.5296	0.02683	37.2715	868.0285	23.2894
60	2.4432	0.4093	0.01039	96.2147	0.02539	39.3803	988.1674	25.0930
72	2.9212	0.3423	0.00781	128.0772	0.02281	43.8447	1279.79	29.1893
75	3.0546	0.3274	0.00730	136.9728	0.02230	44.8416	1352.56	30.1631
84	3.4926	0.2863	0.00602	166.1726	0.02102	47.5786	1568.51	32.9668
90	3.8189	0.2619	0.00532	187.9299	0.02032	49.2099	1709.54	34.7399
96	4.1758	0.2395	0.00472	211.7202	0.01972	50.7017	1847.47	36.4381
100	4.4320	0.2256	0.00437	228.8030	0.01937	51.6247	1937.45	37.5295
108	4.9927	0.2003	0.00376	266.1778	0.01876	53.3137	2112.13	39.6171
120	5.9693	0.1675	0.00302	331.2882	0.01802	55.4985	2359.71	42.5185
132	7.1370	0.1401	0.00244	409.1354	0.01744	57.3257	2588.71	45.1579
144	8.5332	0.1172	0.00199	502.2109	0.01699	58.8540	2798.58	47.5512
240	35.6328	0.0281	0.00043	2308.85	0.01543	64.7957	3870.69	59.7368
360	212.7038	0.0047	0.00007	14114	0.01507	66.3532	4310.72	64.9662
480	1269.70	0.0008	0.00001	84580	0.01501	66.6142	4415.74	66.2883

1.5%   **TABLE 6**   Discrete Cash Flow: Compound Interest Factors   **1.5%**

2%   **TABLE 7**   Discrete Cash Flow: Compound Interest Factors   2%

	Single Payments		Uniform Series Payments				Arithmetic Gradients	
	F/P Compound Amount	**P/F** Present Worth	**A/F** Sinking Fund	**F/A** Compound Amount	**A/P** Capital Recovery	**P/A** Present Worth	**P/G** Gradient Present Worth	**A/G** Gradient Uniform Series
n								
1	1.0200	0.9804	1.00000	1.0000	1.02000	0.9804		
2	1.0404	0.9612	0.49505	2.0200	0.51505	1.9416	0.9612	0.4950
3	1.0612	0.9423	0.32675	3.0604	0.34675	2.8839	2.8458	0.9868
4	1.0824	0.9238	0.24262	4.1216	0.26262	3.8077	5.6173	1.4752
5	1.1041	0.9057	0.19216	5.2040	0.21216	4.7135	9.2403	1.9604
6	1.1262	0.8880	0.15853	6.3081	0.17853	5.6014	13.6801	2.4423
7	1.1487	0.8706	0.13451	7.4343	0.15451	6.4720	18.9035	2.9208
8	1.1717	0.8535	0.11651	8.5830	0.13651	7.3255	24.8779	3.3961
9	1.1951	0.8368	0.10252	9.7546	0.12252	8.1622	31.5720	3.8681
10	1.2190	0.8203	0.09133	10.9497	0.11133	8.9826	38.9551	4.3367
11	1.2434	0.8043	0.08218	12.1687	0.10218	9.7868	46.9977	4.8021
12	1.2682	0.7885	0.07456	13.4121	0.09456	10.5753	55.6712	5.2642
13	1.2936	0.7730	0.06812	14.6803	0.08812	11.3484	64.9475	5.7231
14	1.3195	0.7579	0.06260	15.9739	0.08260	12.1062	74.7999	6.1786
15	1.3459	0.7430	0.05783	17.2934	0.07783	12.8493	85.2021	6.6309
16	1.3728	0.7284	0.05365	18.6393	0.07365	13.5777	96.1288	7.0799
17	1.4002	0.7142	0.04997	20.0121	0.06997	14.2919	107.5554	7.5256
18	1.4282	0.7002	0.04670	21.4123	0.06670	14.9920	119.4581	7.9681
19	1.4568	0.6864	0.04378	22.8406	0.06378	15.6785	131.8139	8.4073
20	1.4859	0.6730	0.04116	24.2974	0.06116	16.3514	144.6003	8.8433
21	1.5157	0.6598	0.03878	25.7833	0.05878	17.0112	157.7959	9.2760
22	1.5460	0.6468	0.03663	27.2990	0.05663	17.6580	171.3795	9.7055
23	1.5769	0.6342	0.03467	28.8450	0.05467	18.2922	185.3309	10.1317
24	1.6084	0.6217	0.03287	30.4219	0.05287	18.9139	199.6305	10.5547
25	1.6406	0.6095	0.03122	32.0303	0.05122	19.5235	214.2592	10.9745
26	1.6734	0.5976	0.02970	33.6709	0.04970	20.1210	229.1987	11.3910
27	1.7069	0.5859	0.02829	35.3443	0.04829	20.7069	244.4311	11.8043
28	1.7410	0.5744	0.02699	37.0512	0.04699	21.2813	259.9392	12.2145
29	1.7758	0.5631	0.02578	38.7922	0.04578	21.8444	275.7064	12.6214
30	1.8114	0.5521	0.02465	40.5681	0.04465	22.3965	291.7164	13.0251
36	2.0399	0.4902	0.01923	51.9944	0.03923	25.4888	392.0405	15.3809
40	2.2080	0.4529	0.01656	60.4020	0.03656	27.3555	461.9931	16.8885
48	2.5871	0.3865	0.01260	79.3535	0.03260	30.6731	605.9657	19.7556
50	2.6916	0.3715	0.01182	84.5794	0.03182	31.4236	642.3606	20.4420
52	2.8003	0.3571	0.01111	90.0164	0.03111	32.1449	678.7849	21.1164
55	2.9717	0.3365	0.01014	98.5865	0.03014	33.1748	733.3527	22.1057
60	3.2810	0.3048	0.00877	114.0515	0.02877	34.7609	823.6975	23.6961
72	4.1611	0.2403	0.00633	158.0570	0.02633	37.9841	1034.06	27.2234
75	4.4158	0.2265	0.00586	170.7918	0.02586	38.6771	1084.64	28.0434
84	5.2773	0.1895	0.00468	213.8666	0.02468	40.5255	1230.42	30.3616
90	5.9431	0.1683	0.00405	247.1567	0.02405	41.5869	1322.17	31.7929
96	6.6929	0.1494	0.00351	284.6467	0.02351	42.5294	1409.30	33.1370
100	7.2446	0.1380	0.00320	312.2323	0.02320	43.0984	1464.75	33.9863
108	8.4883	0.1178	0.00267	374.4129	0.02267	44.1095	1569.30	35.5774
120	10.7652	0.0929	0.00205	488.2582	0.02205	45.3554	1710.42	37.7114
132	13.6528	0.0732	0.00158	632.6415	0.02158	46.3378	1833.47	39.5676
144	17.3151	0.0578	0.00123	815.7545	0.02123	47.1123	1939.79	41.1738
240	115.8887	0.0086	0.00017	5744.44	0.02017	49.5686	2374.88	47.9110
360	1247.56	0.0008	0.00002	62328	0.02002	49.9599	2482.57	49.7112
480	13430	0.0001			0.02000	49.9963	2498.03	49.9643

| 3% | | | | TABLE 8 Discrete Cash Flow: Compound Interest Factors | | | | 3% |

	Single Payments		Uniform Series Payments				Arithmetic Gradients	
n	F/P Compound Amount	P/F Present Worth	A/F Sinking Fund	F/A Compound Amount	A/P Capital Recovery	P/A Present Worth	P/G Gradient Present Worth	A/G Gradient Uniform Series
1	1.0300	0.9709	1.00000	1.0000	1.03000	0.9709		
2	1.0609	0.9426	0.49261	2.0300	0.52261	1.9135	0.9426	0.4926
3	1.0927	0.9151	0.32353	3.0909	0.35353	2.8286	2.7729	0.9803
4	1.1255	0.8885	0.23903	4.1836	0.26903	3.7171	5.4383	1.4631
5	1.1593	0.8626	0.18835	5.3091	0.21835	4.5797	8.8888	1.9409
6	1.1941	0.8375	0.15460	6.4684	0.18460	5.4172	13.0762	2.4138
7	1.2299	0.8131	0.13051	7.6625	0.16051	6.2303	17.9547	2.8819
8	1.2668	0.7894	0.11246	8.8923	0.14246	7.0197	23.4806	3.3450
9	1.3048	0.7664	0.09843	10.1591	0.12843	7.7861	29.6119	3.8032
10	1.3439	0.7441	0.08723	11.4639	0.11723	8.5302	36.3088	4.2565
11	1.3842	0.7224	0.07808	12.8078	0.10808	9.2526	43.5330	4.7049
12	1.4258	0.7014	0.07046	14.1920	0.10046	9.9540	51.2482	5.1485
13	1.4685	0.6810	0.06403	15.6178	0.09403	10.6350	59.4196	5.5872
14	1.5126	0.6611	0.05853	17.0863	0.08853	11.2961	68.0141	6.0210
15	1.5580	0.6419	0.05377	18.5989	0.08377	11.9379	77.0002	6.4500
16	1.6047	0.6232	0.04961	20.1569	0.07961	12.5611	86.3477	6.8742
17	1.6528	0.6050	0.04595	21.7616	0.07595	13.1661	96.0280	7.2936
18	1.7024	0.5874	0.04271	23.4144	0.07271	13.7535	106.0137	7.7081
19	1.7535	0.5703	0.03981	25.1169	0.06981	14.3238	116.2788	8.1179
20	1.8061	0.5537	0.03722	26.8704	0.06722	14.8775	126.7987	8.5229
21	1.8603	0.5375	0.03487	28.6765	0.06487	15.4150	137.5496	8.9231
22	1.9161	0.5219	0.03275	30.5368	0.06275	15.9369	148.5094	9.3186
23	1.9736	0.5067	0.03081	32.4529	0.06081	16.4436	159.6566	9.7093
24	2.0328	0.4919	0.02905	34.4265	0.05905	16.9355	170.9711	10.0954
25	2.0938	0.4776	0.02743	36.4593	0.05743	17.4131	182.4336	10.4768
26	2.1566	0.4637	0.02594	38.5530	0.05594	17.8768	194.0260	10.8535
27	2.2213	0.4502	0.02456	40.7096	0.05456	18.3270	205.7309	11.2255
28	2.2879	0.4371	0.02329	42.9309	0.05329	18.7641	217.5320	11.5930
29	2.3566	0.4243	0.02211	45.2189	0.05211	19.1885	229.4137	11.9558
30	2.4273	0.4120	0.02102	47.5754	0.05102	19.6004	241.3613	12.3141
31	2.5001	0.4000	0.02000	50.0027	0.05000	20.0004	253.3609	12.6678
32	2.5751	0.3883	0.01905	52.5028	0.04905	20.3888	265.3993	13.0169
33	2.6523	0.3770	0.01816	55.0778	0.04816	20.7658	277.4642	13.3616
34	2.7319	0.3660	0.01732	57.7302	0.04732	21.1318	289.5437	13.7018
35	2.8139	0.3554	0.01654	60.4621	0.04654	21.4872	301.6267	14.0375
40	3.2620	0.3066	0.01326	75.4013	0.04326	23.1148	361.7499	15.6502
45	3.7816	0.2644	0.01079	92.7199	0.04079	24.5187	420.6325	17.1556
50	4.3839	0.2281	0.00887	112.7969	0.03887	25.7298	477.4803	18.5575
55	5.0821	0.1968	0.00735	136.0716	0.03735	26.7744	531.7411	19.8600
60	5.8916	0.1697	0.00613	163.0534	0.03613	27.6756	583.0526	21.0674
65	6.8300	0.1464	0.00515	194.3328	0.03515	28.4529	631.2010	22.1841
70	7.9178	0.1263	0.00434	230.5941	0.03434	29.1234	676.0869	23.2145
75	9.1789	0.1089	0.00367	272.6309	0.03367	29.7018	717.6978	24.1634
80	10.6409	0.0940	0.00311	321.3630	0.03311	30.2008	756.0865	25.0353
84	11.9764	0.0835	0.00273	365.8805	0.03273	30.5501	784.5434	25.6806
85	12.3357	0.0811	0.00265	377.8570	0.03265	30.6312	791.3529	25.8349
90	14.3005	0.0699	0.00226	443.3489	0.03226	31.0024	823.6302	26.5667
96	17.0755	0.0586	0.00187	535.8502	0.03187	31.3812	858.6377	27.3615
108	24.3456	0.0411	0.00129	778.1863	0.03129	31.9642	917.6013	28.7072
120	34.7110	0.0288	0.00089	1123.70	0.03089	32.3730	963.8635	29.7737

4%			TABLE 9	Discrete Cash Flow: Compound Interest Factors				4%
	Single Payments		Uniform Series Payments				Arithmetic Gradients	
	F/P Compound Amount	P/F Present Worth	A/F Sinking Fund	F/A Compound Amount	A/P Capital Recovery	P/A Present Worth	P/G Gradient Present Worth	A/G Gradient Uniform Series
n								
1	1.0400	0.9615	1.00000	1.0000	1.04000	0.9615		
2	1.0816	0.9246	0.49020	2.0400	0.53020	1.8861	0.9246	0.4902
3	1.1249	0.8890	0.32035	3.1216	0.36035	2.7751	2.7025	0.9739
4	1.1699	0.8548	0.23549	4.2465	0.27549	3.6299	5.2670	1.4510
5	1.2167	0.8219	0.18463	5.4163	0.22463	4.4518	8.5547	1.9216
6	1.2653	0.7903	0.15076	6.6330	0.19076	5.2421	12.5062	2.3857
7	1.3159	0.7599	0.12661	7.8983	0.16661	6.0021	17.0657	2.8433
8	1.3686	0.7307	0.10853	9.2142	0.14853	6.7327	22.1806	3.2944
9	1.4233	0.7026	0.09449	10.5828	0.13449	7.4353	27.8013	3.7391
10	1.4802	0.6756	0.08329	12.0061	0.12329	8.1109	33.8814	4.1773
11	1.5395	0.6496	0.07415	13.4864	0.11415	8.7605	40.3772	4.6090
12	1.6010	0.6246	0.06655	15.0258	0.10655	9.3851	47.2477	5.0343
13	1.6651	0.6006	0.06014	16.6268	0.10014	9.9856	54.4546	5.4533
14	1.7317	0.5775	0.05467	18.2919	0.09467	10.5631	61.9618	5.8659
15	1.8009	0.5553	0.04994	20.0236	0.08994	11.1184	69.7355	6.2721
16	1.8730	0.5339	0.04582	21.8245	0.08582	11.6523	77.7441	6.6720
17	1.9479	0.5134	0.04220	23.6975	0.08220	12.1657	85.9581	7.0656
18	2.0258	0.4936	0.03899	25.6454	0.07899	12.6593	94.3498	7.4530
19	2.1068	0.4746	0.03614	27.6712	0.07614	13.1339	102.8933	7.8342
20	2.1911	0.4564	0.03358	29.7781	0.07358	13.5903	111.5647	8.2091
21	2.2788	0.4388	0.03128	31.9692	0.07128	14.0292	120.3414	8.5779
22	2.3699	0.4220	0.02920	34.2480	0.06920	14.4511	129.2024	8.9407
23	2.4647	0.4057	0.02731	36.6179	0.06731	14.8568	138.1284	9.2973
24	2.5633	0.3901	0.02559	39.0826	0.06559	15.2470	147.1012	9.6479
25	2.6658	0.3751	0.02401	41.6459	0.06401	15.6221	156.1040	9.9925
26	2.7725	0.3607	0.02257	44.3117	0.06257	15.9828	165.1212	10.3312
27	2.8834	0.3468	0.02124	47.0842	0.06124	16.3296	174.1385	10.6640
28	2.9987	0.3335	0.02001	49.9676	0.06001	16.6631	183.1424	10.9909
29	3.1187	0.3207	0.01888	52.9663	0.05888	16.9837	192.1206	11.3120
30	3.2434	0.3083	0.01783	56.0849	0.05783	17.2920	201.0618	11.6274
31	3.3731	0.2965	0.01686	59.3283	0.05686	17.5885	209.9556	11.9371
32	3.5081	0.2851	0.01595	62.7015	0.05595	17.8736	218.7924	12.2411
33	3.6484	0.2741	0.01510	66.2095	0.05510	18.1476	227.5634	12.5396
34	3.7943	0.2636	0.01431	69.8579	0.05431	18.4112	236.2607	12.8324
35	3.9461	0.2534	0.01358	73.6522	0.05358	18.6646	244.8768	13.1198
40	4.8010	0.2083	0.01052	95.0255	0.05052	19.7928	286.5303	14.4765
45	5.8412	0.1712	0.00826	121.0294	0.04826	20.7200	325.4028	15.7047
50	7.1067	0.1407	0.00655	152.6671	0.04655	21.4822	361.1638	16.8122
55	8.6464	0.1157	0.00523	191.1592	0.04523	22.1086	393.6890	17.8070
60	10.5196	0.0951	0.00420	237.9907	0.04420	22.6235	422.9966	18.6972
65	12.7987	0.0781	0.00339	294.9684	0.04339	23.0467	449.2014	19.4909
70	15.5716	0.0642	0.00275	364.2905	0.04275	23.3945	472.4789	20.1961
75	18.9453	0.0528	0.00223	448.6314	0.04223	23.6804	493.0408	20.8206
80	23.0498	0.0434	0.00181	551.2450	0.04181	23.9154	511.1161	21.3718
85	28.0436	0.0357	0.00148	676.0901	0.04148	24.1085	526.9384	21.8569
90	34.1193	0.0293	0.00121	827.9833	0.04121	24.2673	540.7369	22.2826
96	43.1718	0.0232	0.00095	1054.30	0.04095	24.4209	554.9312	22.7236
108	69.1195	0.0145	0.00059	1702.99	0.04059	24.6383	576.8949	23.4146
120	110.6626	0.0090	0.00036	2741.56	0.04036	24.7741	592.2428	23.9057
144	283.6618	0.0035	0.00014	7066.55	0.04014	24.9119	610.1055	24.4906

5% TABLE 10 Discrete Cash Flow: Compound Interest Factors 5%

	Single Payments		Uniform Series Payments				Arithmetic Gradients	
n	F/P Compound Amount	P/F Present Worth	A/F Sinking Fund	F/A Compound Amount	A/P Capital Recovery	P/A Present Worth	P/G Gradient Present Worth	A/G Gradient Uniform Series
1	1.0500	0.9524	1.00000	1.0000	1.05000	0.9524		
2	1.1025	0.9070	0.48780	2.0500	0.53780	1.8594	0.9070	0.4878
3	1.1576	0.8638	0.31721	3.1525	0.36721	2.7232	2.6347	0.9675
4	1.2155	0.8227	0.23201	4.3101	0.28201	3.5460	5.1028	1.4391
5	1.2763	0.7835	0.18097	5.5256	0.23097	4.3295	8.2369	1.9025
6	1.3401	0.7462	0.14702	6.8019	0.19702	5.0757	11.9680	2.3579
7	1.4071	0.7107	0.12282	8.1420	0.17282	5.7864	16.2321	2.8052
8	1.4775	0.6768	0.10472	9.5491	0.15472	6.4632	20.9700	3.2445
9	1.5513	0.6446	0.09069	11.0266	0.14069	7.1078	26.1268	3.6758
10	1.6289	0.6139	0.07950	12.5779	0.12950	7.7217	31.6520	4.0991
11	1.7103	0.5847	0.07039	14.2068	0.12039	8.3064	37.4988	4.5144
12	1.7959	0.5568	0.06283	15.9171	0.11283	8.8633	43.6241	4.9219
13	1.8856	0.5303	0.05646	17.7130	0.10646	9.3936	49.9879	5.3215
14	1.9799	0.5051	0.05102	19.5986	0.10102	9.8986	56.5538	5.7133
15	2.0789	0.4810	0.04634	21.5786	0.09634	10.3797	63.2880	6.0973
16	2.1829	0.4581	0.04227	23.6575	0.09227	10.8378	70.1597	6.4736
17	2.2920	0.4363	0.03870	25.8404	0.08870	11.2741	77.1405	6.8423
18	2.4066	0.4155	0.03555	28.1324	0.08555	11.6896	84.2043	7.2034
19	2.5270	0.3957	0.03275	30.5390	0.08275	12.0853	91.3275	7.5569
20	2.6533	0.3769	0.03024	33.0660	0.08024	12.4622	98.4884	7.9030
21	2.7860	0.3589	0.02800	35.7193	0.07800	12.8212	105.6673	8.2416
22	2.9253	0.3418	0.02597	38.5052	0.07597	13.1630	112.8461	8.5730
23	3.0715	0.3256	0.02414	41.4305	0.07414	13.4886	120.0087	8.8971
24	3.2251	0.3101	0.02247	44.5020	0.07247	13.7986	127.1402	9.2140
25	3.3864	0.2953	0.02095	47.7271	0.07095	14.0939	134.2275	9.5238
26	3.5557	0.2812	0.01956	51.1135	0.06956	14.3752	141.2585	9.8266
27	3.7335	0.2678	0.01829	54.6691	0.06829	14.6430	148.2226	10.1224
28	3.9201	0.2551	0.01712	58.4026	0.06712	14.8981	155.1101	10.4114
29	4.1161	0.2429	0.01605	62.3227	0.06605	15.1411	161.9126	10.6936
30	4.3219	0.2314	0.01505	66.4388	0.06505	15.3725	168.6226	10.9691
31	4.5380	0.2204	0.01413	70.7608	0.06413	15.5928	175.2333	11.2381
32	4.7649	0.2099	0.01328	75.2988	0.06328	15.8027	181.7392	11.5005
33	5.0032	0.1999	0.01249	80.0638	0.06249	16.0025	188.1351	11.7566
34	5.2533	0.1904	0.01176	85.0670	0.06176	16.1929	194.4168	12.0063
35	5.5160	0.1813	0.01107	90.3203	0.06107	16.3742	200.5807	12.2498
40	7.0400	0.1420	0.00828	120.7998	0.05828	17.1591	229.5452	13.3775
45	8.9850	0.1113	0.00626	159.7002	0.05626	17.7741	255.3145	14.3644
50	11.4674	0.0872	0.00478	209.3480	0.05478	18.2559	277.9148	15.2233
55	14.6356	0.0683	0.00367	272.7126	0.05367	18.6335	297.5104	15.9664
60	18.6792	0.0535	0.00283	353.5837	0.05283	18.9293	314.3432	16.6062
65	23.8399	0.0419	0.00219	456.7980	0.05219	19.1611	328.6910	17.1541
70	30.4264	0.0329	0.00170	588.5285	0.05170	19.3427	340.8409	17.6212
75	38.8327	0.0258	0.00132	756.6537	0.05132	19.4850	351.0721	18.0176
80	49.5614	0.0202	0.00103	971.2288	0.05103	19.5965	359.6460	18.3526
85	63.2544	0.0158	0.00080	1245.09	0.05080	19.6838	366.8007	18.6346
90	80.7304	0.0124	0.00063	1594.61	0.05063	19.7523	372.7488	18.8712
95	103.0347	0.0097	0.00049	2040.69	0.05049	19.8059	377.6774	19.0689
96	108.1864	0.0092	0.00047	2143.73	0.05047	19.8151	378.5555	19.1044
98	119.2755	0.0084	0.00042	2365.51	0.05042	19.8323	380.2139	19.1714
100	131.5013	0.0076	0.00038	2610.03	0.05038	19.8479	381.7492	19.2337

6%					TABLE 11	Discrete Cash Flow: Compound Interest Factors			6%

	Single Payments		Uniform Series Payments				Arithmetic Gradients	
	F/P	*P/F*	*A/F*	*F/A*	*A/P*	*P/A*	*P/G*	*A/G*
	Compound	Present	Sinking	Compound	Capital	Present	Gradient	Gradient
n	Amount	Worth	Fund	Amount	Recovery	Worth	Present Worth	Uniform Series
1	1.0600	0.9434	1.00000	1.0000	1.06000	0.9434		
2	1.1236	0.8900	0.48544	2.0600	0.54544	1.8334	0.8900	0.4854
3	1.1910	0.8396	0.31411	3.1836	0.37411	2.6730	2.5692	0.9612
4	1.2625	0.7921	0.22859	4.3746	0.28859	3.4651	4.9455	1.4272
5	1.3382	0.7473	0.17740	5.6371	0.23740	4.2124	7.9345	1.8836
6	1.4185	0.7050	0.14336	6.9753	0.20336	4.9173	11.4594	2.3304
7	1.5036	0.6651	0.11914	8.3938	0.17914	5.5824	15.4497	2.7676
8	1.5938	0.6274	0.10104	9.8975	0.16104	6.2098	19.8416	3.1952
9	1.6895	0.5919	0.08702	11.4913	0.14702	6.8017	24.5768	3.6133
10	1.7908	0.5584	0.07587	13.1808	0.13587	7.3601	29.6023	4.0220
11	1.8983	0.5268	0.06679	14.9716	0.12679	7.8869	34.8702	4.4213
12	2.0122	0.4970	0.05928	16.8699	0.11928	8.3838	40.3369	4.8113
13	2.1329	0.4688	0.05296	18.8821	0.11296	8.8527	45.9629	5.1920
14	2.2609	0.4423	0.04758	21.0151	0.10758	9.2950	51.7128	5.5635
15	2.3966	0.4173	0.04296	23.2760	0.10296	9.7122	57.5546	5.9260
16	2.5404	0.3936	0.03895	25.6725	0.09895	10.1059	63.4592	6.2794
17	2.6928	0.3714	0.03544	28.2129	0.09544	10.4773	69.4011	6.6240
18	2.8543	0.3503	0.03236	30.9057	0.09236	10.8276	75.3569	6.9597
19	3.0256	0.3305	0.02962	33.7600	0.08962	11.1581	81.3062	7.2867
20	3.2071	0.3118	0.02718	36.7856	0.08718	11.4699	87.2304	7.6051
21	3.3996	0.2942	0.02500	39.9927	0.08500	11.7641	93.1136	7.9151
22	3.6035	0.2775	0.02305	43.3923	0.08305	12.0416	98.9412	8.2166
23	3.8197	0.2618	0.02128	46.9958	0.08128	12.3034	104.7007	8.5099
24	4.0489	0.2470	0.01968	50.8156	0.07968	12.5504	110.3812	8.7951
25	4.2919	0.2330	0.01823	54.8645	0.07823	12.7834	115.9732	9.0722
26	4.5494	0.2198	0.01690	59.1564	0.07690	13.0032	121.4684	9.3414
27	4.8223	0.2074	0.01570	63.7058	0.07570	13.2105	126.8600	9.6029
28	5.1117	0.1956	0.01459	68.5281	0.07459	13.4062	132.1420	9.8568
29	5.4184	0.1846	0.01358	73.6398	0.07358	13.5907	137.3096	10.1032
30	5.7435	0.1741	0.01265	79.0582	0.07265	13.7648	142.3588	10.3422
31	6.0881	0.1643	0.01179	84.8017	0.07179	13.9291	147.2864	10.5740
32	6.4534	0.1550	0.01100	90.8898	0.07100	14.0840	152.0901	10.7988
33	6.8406	0.1462	0.01027	97.3432	0.07027	14.2302	156.7681	11.0166
34	7.2510	0.1379	0.00960	104.1838	0.06960	14.3681	161.3192	11.2276
35	7.6861	0.1301	0.00897	111.4348	0.06897	14.4982	165.7427	11.4319
40	10.2857	0.0972	0.00646	154.7620	0.06646	15.0463	185.9568	12.3590
45	13.7646	0.0727	0.00470	212.7435	0.06470	15.4558	203.1096	13.1413
50	18.4202	0.0543	0.00344	290.3359	0.06344	15.7619	217.4574	13.7964
55	24.6503	0.0406	0.00254	394.1720	0.06254	15.9905	229.3222	14.3411
60	32.9877	0.0303	0.00188	533.1282	0.06188	16.1614	239.0428	14.7909
65	44.1450	0.0227	0.00139	719.0829	0.06139	16.2891	246.9450	15.1601
70	59.0759	0.0169	0.00103	967.9322	0.06103	16.3845	253.3271	15.4613
75	79.0569	0.0126	0.00077	1300.95	0.06077	16.4558	258.4527	15.7058
80	105.7960	0.0095	0.00057	1746.60	0.06057	16.5091	262.5493	15.9033
85	141.5789	0.0071	0.00043	2342.98	0.06043	16.5489	265.8096	16.0620
90	189.4645	0.0053	0.00032	3141.08	0.06032	16.5787	268.3946	16.1891
95	253.5463	0.0039	0.00024	4209.10	0.06024	16.6009	270.4375	16.2905
96	268.7590	0.0037	0.00022	4462.65	0.06022	16.6047	270.7909	16.3081
98	301.9776	0.0033	0.00020	5016.29	0.06020	16.6115	271.4491	16.3411
100	339.3021	0.0029	0.00018	5638.37	0.06018	16.6175	272.0471	16.3711

7%				TABLE **12**	Discrete Cash Flow: Compound Interest Factors			7%
	Single Payments		**Uniform Series Payments**				**Arithmetic Gradients**	
	F/P Compound Amount	*P/F* Present Worth	*A/F* Sinking Fund	*F/A* Compound Amount	*A/P* Capital Recovery	*P/A* Present Worth	*P/G* Gradient Present Worth	*A/G* Gradient Uniform Series
n								
1	1.0700	0.9346	1.00000	1.0000	1.07000	0.9346		
2	1.1449	0.8734	0.48309	2.0700	0.55309	1.8080	0.8734	0.4831
3	1.2250	0.8163	0.31105	3.2149	0.38105	2.6243	2.5060	0.9549
4	1.3108	0.7629	0.22523	4.4399	0.29523	3.3872	4.7947	1.4155
5	1.4026	0.7130	0.17389	5.7507	0.24389	4.1002	7.6467	1.8650
6	1.5007	0.6663	0.13980	7.1533	0.20980	4.7665	10.9784	2.3032
7	1.6058	0.6227	0.11555	8.6540	0.18555	5.3893	14.7149	2.7304
8	1.7182	0.5820	0.09747	10.2598	0.16747	5.9713	18.7889	3.1465
9	1.8385	0.5439	0.08349	11.9780	0.15349	6.5152	23.1404	3.5517
10	1.9672	0.5083	0.07238	13.8164	0.14238	7.0236	27.7156	3.9461
11	2.1049	0.4751	0.06336	15.7836	0.13336	7.4987	32.4665	4.3296
12	2.2522	0.4440	0.05590	17.8885	0.12590	7.9427	37.3506	4.7025
13	2.4098	0.4150	0.04965	20.1406	0.11965	8.3577	42.3302	5.0648
14	2.5785	0.3878	0.04434	22.5505	0.11434	8.7455	47.3718	5.4167
15	2.7590	0.3624	0.03979	25.1290	0.10979	9.1079	52.4461	5.7583
16	2.9522	0.3387	0.03586	27.8881	0.10586	9.4466	57.5271	6.0897
17	3.1588	0.3166	0.03243	30.8402	0.10243	9.7632	62.5923	6.4110
18	3.3799	0.2959	0.02941	33.9990	0.09941	10.0591	67.6219	6.7225
19	3.6165	0.2765	0.02675	37.3790	0.09675	10.3356	72.5991	7.0242
20	3.8697	0.2584	0.02439	40.9955	0.09439	10.5940	77.5091	7.3163
21	4.1406	0.2415	0.02229	44.8652	0.09229	10.8355	82.3393	7.5990
22	4.4304	0.2257	0.02041	49.0057	0.09041	11.0612	87.0793	7.8725
23	4.7405	0.2109	0.01871	53.4361	0.08871	11.2722	91.7201	8.1369
24	5.0724	0.1971	0.01719	58.1767	0.08719	11.4693	96.2545	8.3923
25	5.4274	0.1842	0.01581	63.2490	0.08581	11.6536	100.6765	8.6391
26	5.8074	0.1722	0.01456	68.6765	0.08456	11.8258	104.9814	8.8773
27	6.2139	0.1609	0.01343	74.4838	0.08343	11.9867	109.1656	9.1072
28	6.6488	0.1504	0.01239	80.6977	0.08239	12.1371	113.2264	9.3289
29	7.1143	0.1406	0.01145	87.3465	0.08145	12.2777	117.1622	9.5427
30	7.6123	0.1314	0.01059	94.4608	0.08059	12.4090	120.9718	9.7487
31	8.1451	0.1228	0.00980	102.0730	0.07980	12.5318	124.6550	9.9471
32	8.7153	0.1147	0.00907	110.2182	0.07907	12.6466	128.2120	10.1381
33	9.3253	0.1072	0.00841	118.9334	0.07841	12.7538	131.6435	10.3219
34	9.9781	0.1002	0.00780	128.2588	0.07780	12.8540	134.9507	10.4987
35	10.6766	0.0937	0.00723	138.2369	0.07723	12.9477	138.1353	10.6687
40	14.9745	0.0668	0.00501	199.6351	0.07501	13.3317	152.2928	11.4233
45	21.0025	0.0476	0.00350	285.7493	0.07350	13.6055	163.7559	12.0360
50	29.4570	0.0339	0.00246	406.5289	0.07246	13.8007	172.9051	12.5287
55	41.3150	0.0242	0.00174	575.9286	0.07174	13.9399	180.1243	12.9215
60	57.9464	0.0173	0.00123	813.5204	0.07123	14.0392	185.7677	13.2321
65	81.2729	0.0123	0.00087	1146.76	0.07087	14.1099	190.1452	13.4760
70	113.9894	0.0088	0.00062	1614.13	0.07062	14.1604	193.5185	13.6662
75	159.8760	0.0063	0.00044	2269.66	0.07044	14.1964	196.1035	13.8136
80	224.2344	0.0045	0.00031	3189.06	0.07031	14.2220	198.0748	13.9273
85	314.5003	0.0032	0.00022	4478.58	0.07022	14.2403	199.5717	14.0146
90	441.1030	0.0023	0.00016	6287.19	0.07016	14.2533	200.7042	14.0812
95	618.6697	0.0016	0.00011	8823.85	0.07011	14.2626	201.5581	14.1319
96	661.9766	0.0015	0.00011	9442.52	0.07011	14.2641	201.7016	14.1405
98	757.8970	0.0013	0.00009	10813	0.07009	14.2669	201.9651	14.1562
100	867.7163	0.0012	0.00008	12382	0.07008	14.2693	202.2001	14.1703

8%				TABLE 13	Discrete Cash Flow: Compound Interest Factors				8%
	Single Payments		Uniform Series Payments					Arithmetic Gradients	
	F/P Compound Amount	P/F Present Worth	A/F Sinking Fund	F/A Compound Amount	A/P Capital Recovery	P/A Present Worth	P/G Gradient Present Worth	A/G Gradient Uniform Series
n								
1	1.0800	0.9259	1.00000	1.0000	1.08000	0.9259		
2	1.1664	0.8573	0.48077	2.0800	0.56077	1.7833	0.8573	0.4808
3	1.2597	0.7938	0.30803	3.2464	0.38803	2.5771	2.4450	0.9487
4	1.3605	0.7350	0.22192	4.5061	0.30192	3.3121	4.6501	1.4040
5	1.4693	0.6806	0.17046	5.8666	0.25046	3.9927	7.3724	1.8465
6	1.5869	0.6302	0.13632	7.3359	0.21632	4.6229	10.5233	2.2763
7	1.7138	0.5835	0.11207	8.9228	0.19207	5.2064	14.0242	2.6937
8	1.8509	0.5403	0.09401	10.6366	0.17401	5.7466	17.8061	3.0985
9	1.9990	0.5002	0.08008	12.4876	0.16008	6.2469	21.8081	3.4910
10	2.1589	0.4632	0.06903	14.4866	0.14903	6.7101	25.9768	3.8713
11	2.3316	0.4289	0.06008	16.6455	0.14008	7.1390	30.2657	4.2395
12	2.5182	0.3971	0.05270	18.9771	0.13270	7.5361	34.6339	4.5957
13	2.7196	0.3677	0.04652	21.4953	0.12652	7.9038	39.0463	4.9402
14	2.9372	0.3405	0.04130	24.2149	0.12130	8.2442	43.4723	5.2731
15	3.1722	0.3152	0.03683	27.1521	0.11683	8.5595	47.8857	5.5945
16	3.4259	0.2919	0.03298	30.3243	0.11298	8.8514	52.2640	5.9046
17	3.7000	0.2703	0.02963	33.7502	0.10963	9.1216	56.5883	6.2037
18	3.9960	0.2502	0.02670	37.4502	0.10670	9.3719	60.8426	6.4920
19	4.3157	0.2317	0.02413	41.4463	0.10413	9.6036	65.0134	6.7697
20	4.6610	0.2145	0.02185	45.7620	0.10185	9.8181	69.0898	7.0369
21	5.0338	0.1987	0.01983	50.4229	0.09983	10.0168	73.0629	7.2940
22	5.4365	0.1839	0.01803	55.4568	0.09803	10.2007	76.9257	7.5412
23	5.8715	0.1703	0.01642	60.8933	0.09642	10.3711	80.6726	7.7786
24	6.3412	0.1577	0.01498	66.7648	0.09498	10.5288	84.2997	8.0066
25	6.8485	0.1460	0.01368	73.1059	0.09368	10.6748	87.8041	8.2254
26	7.3964	0.1352	0.01251	79.9544	0.09251	10.8100	91.1842	8.4352
27	7.9881	0.1252	0.01145	87.3508	0.09145	10.9352	94.4390	8.6363
28	8.6271	0.1159	0.01049	95.3388	0.09049	11.0511	97.5687	8.8289
29	9.3173	0.1073	0.00962	103.9659	0.08962	11.1584	100.5738	9.0133
30	10.0627	0.0994	0.00883	113.2832	0.08883	11.2578	103.4558	9.1897
31	10.8677	0.0920	0.00811	123.3459	0.08811	11.3498	106.2163	9.3584
32	11.7371	0.0852	0.00745	134.2135	0.08745	11.4350	108.8575	9.5197
33	12.6760	0.0789	0.00685	145.9506	0.08685	11.5139	111.3819	9.6737
34	13.6901	0.0730	0.00630	158.6267	0.08630	11.5869	113.7924	9.8208
35	14.7853	0.0676	0.00580	172.3168	0.08580	11.6546	116.0920	9.9611
40	21.7245	0.0460	0.00386	259.0565	0.08386	11.9246	126.0422	10.5699
45	31.9204	0.0313	0.00259	386.5056	0.08259	12.1084	133.7331	11.0447
50	46.9016	0.0213	0.00174	573.7702	0.08174	12.2335	139.5928	11.4107
55	68.9139	0.0145	0.00118	848.9232	0.08118	12.3186	144.0065	11.6902
60	101.2571	0.0099	0.00080	1253.21	0.08080	12.3766	147.3000	11.9015
65	148.7798	0.0067	0.00054	1847.25	0.08054	12.4160	149.7387	12.0602
70	218.6064	0.0046	0.00037	2720.08	0.08037	12.4428	151.5326	12.1783
75	321.2045	0.0031	0.00025	4002.56	0.08025	12.4611	152.8448	12.2658
80	471.9548	0.0021	0.00017	5886.94	0.08017	12.4735	153.8001	12.3301
85	693.4565	0.0014	0.00012	8655.71	0.08012	12.4820	154.4925	12.3772
90	1018.92	0.0010	0.00008	12724	0.08008	12.4877	154.9925	12.4116
95	1497.12	0.0007	0.00005	18702	0.08005	12.4917	155.3524	12.4365
96	1616.89	0.0006	0.00005	20199	0.08005	12.4923	155.4112	12.4406
98	1885.94	0.0005	0.00004	23562	0.08004	12.4934	155.5176	12.4480
100	2199.76	0.0005	0.00004	27485	0.08004	12.4943	155.6107	12.4545

9%			TABLE **14**	Discrete Cash Flow: Compound Interest Factors				9%
	Single Payments			**Uniform Series Payments**			**Arithmetic Gradients**	
	F/P	**P/F**	**A/F**	**F/A**	**A/P**	**P/A**	**P/G**	**A/G**
	Compound	Present	Sinking	Compound	Capital	Present	Gradient	Gradient
n	Amount	Worth	Fund	Amount	Recovery	Worth	Present Worth	Uniform Series
1	1.0900	0.9174	1.00000	1.0000	1.09000	0.9174		
2	1.1881	0.8417	0.47847	2.0900	0.56847	1.7591	0.8417	0.4785
3	1.2950	0.7722	0.30505	3.2781	0.39505	2.5313	2.3860	0.9426
4	1.4116	0.7084	0.21867	4.5731	0.30867	3.2397	4.5113	1.3925
5	1.5386	0.6499	0.16709	5.9847	0.25709	3.8897	7.1110	1.8282
6	1.6771	0.5963	0.13292	7.5233	0.22292	4.4859	10.0924	2.2498
7	1.8280	0.5470	0.10869	9.2004	0.19869	5.0330	13.3746	2.6574
8	1.9926	0.5019	0.09067	11.0285	0.18067	5.5348	16.8877	3.0512
9	2.1719	0.4604	0.07680	13.0210	0.16680	5.9952	20.5711	3.4312
10	2.3674	0.4224	0.06582	15.1929	0.15582	6.4177	24.3728	3.7978
11	2.5804	0.3875	0.05695	17.5603	0.14695	6.8052	28.2481	4.1510
12	2.8127	0.3555	0.04965	20.1407	0.13965	7.1607	32.1590	4.4910
13	3.0658	0.3262	0.04357	22.9534	0.13357	7.4869	36.0731	4.8182
14	3.3417	0.2992	0.03843	26.0192	0.12843	7.7862	39.9633	5.1326
15	3.6425	0.2745	0.03406	29.3609	0.12406	8.0607	43.8069	5.4346
16	3.9703	0.2519	0.03030	33.0034	0.12030	8.3126	47.5849	5.7245
17	4.3276	0.2311	0.02705	36.9737	0.11705	8.5436	51.2821	6.0024
18	4.7171	0.2120	0.02421	41.3013	0.11421	8.7556	54.8860	6.2687
19	5.1417	0.1945	0.02173	46.0185	0.11173	8.9501	58.3868	6.5236
20	5.6044	0.1784	0.01955	51.1601	0.10955	9.1285	61.7770	6.7674
21	6.1088	0.1637	0.01762	56.7645	0.10762	9.2922	65.0509	7.0006
22	6.6586	0.1502	0.01590	62.8733	0.10590	9.4424	68.2048	7.2232
23	7.2579	0.1378	0.01438	69.5319	0.10438	9.5802	71.2359	7.4357
24	7.9111	0.1264	0.01302	76.7898	0.10302	9.7066	74.1433	7.6384
25	8.6231	0.1160	0.01181	84.7009	0.10181	9.8226	76.9265	7.8316
26	9.3992	0.1064	0.01072	93.3240	0.10072	9.9290	79.5863	8.0156
27	10.2451	0.0976	0.00973	102.7231	0.09973	10.0266	82.1241	8.1906
28	11.1671	0.0895	0.00885	112.9682	0.09885	10.1161	84.5419	8.3571
29	12.1722	0.0822	0.00806	124.1354	0.09806	10.1983	86.8422	8.5154
30	13.2677	0.0754	0.00734	136.3075	0.09734	10.2737	89.0280	8.6657
31	14.4618	0.0691	0.00669	149.5752	0.09669	10.3428	91.1024	8.8083
32	15.7633	0.0634	0.00610	164.0370	0.09610	10.4062	93.0690	8.9436
33	17.1820	0.0582	0.00556	179.8003	0.09556	10.4644	94.9314	9.0718
34	18.7284	0.0534	0.00508	196.9823	0.09508	10.5178	96.6935	9.1933
35	20.4140	0.0490	0.00464	215.7108	0.09464	10.5668	98.3590	9.3083
40	31.4094	0.0318	0.00296	337.8824	0.09296	10.7574	105.3762	9.7957
45	48.3273	0.0207	0.00190	525.8587	0.09190	10.8812	110.5561	10.1603
50	74.3575	0.0134	0.00123	815.0836	0.09123	10.9617	114.3251	10.4295
55	114.4083	0.0087	0.00079	1260.09	0.09079	11.0140	117.0362	10.6261
60	176.0313	0.0057	0.00051	1944.79	0.09051	11.0480	118.9683	10.7683
65	270.8460	0.0037	0.00033	2998.29	0.09033	11.0701	120.3344	10.8702
70	416.7301	0.0024	0.00022	4619.22	0.09022	11.0844	121.2942	10.9427
75	641.1909	0.0016	0.00014	7113.23	0.09014	11.0938	121.9646	10.9940
80	986.5517	0.0010	0.00009	10951	0.09009	11.0998	122.4306	11.0299
85	1517.93	0.0007	0.00006	16855	0.09006	11.1038	122.7533	11.0551
90	2335.53	0.0004	0.00004	25939	0.09004	11.1064	122.9758	11.0726
95	3593.50	0.0003	0.00003	39917	0.09003	11.1080	123.1287	11.0847
96	3916.91	0.0003	0.00002	43510	0.09002	11.1083	123.1529	11.0866
98	4653.68	0.0002	0.00002	51696	0.09002	11.1087	123.1963	11.0900
100	5529.04	0.0002	0.00002	61423	0.09002	11.1091	123.2335	11.0930

10%				TABLE 15	Discrete Cash Flow: Compound Interest Factors				10%
	Single Payments		Uniform Series Payments					Arithmetic Gradients	
	F/P	P/F	A/F	F/A	A/P	P/A	P/G	A/G
	Compound Amount	Present Worth	Sinking Fund	Compound Amount	Capital Recovery	Present Worth	Gradient Present Worth	Gradient Uniform Series
n								
1	1.1000	0.9091	1.00000	1.0000	1.10000	0.9091		
2	1.2100	0.8264	0.47619	2.1000	0.57619	1.7355	0.8264	0.4762
3	1.3310	0.7513	0.30211	3.3100	0.40211	2.4869	2.3291	0.9366
4	1.4641	0.6830	0.21547	4.6410	0.31547	3.1699	4.3781	1.3812
5	1.6105	0.6209	0.16380	6.1051	0.26380	3.7908	6.8618	1.8101
6	1.7716	0.5645	0.12961	7.7156	0.22961	4.3553	9.6842	2.2236
7	1.9487	0.5132	0.10541	9.4872	0.20541	4.8684	12.7631	2.6216
8	2.1436	0.4665	0.08744	11.4359	0.18744	5.3349	16.0287	3.0045
9	2.3579	0.4241	0.07364	13.5795	0.17364	5.7590	19.4215	3.3724
10	2.5937	0.3855	0.06275	15.9374	0.16275	6.1446	22.8913	3.7255
11	2.8531	0.3505	0.05396	18.5312	0.15396	6.4951	26.3963	4.0641
12	3.1384	0.3186	0.04676	21.3843	0.14676	6.8137	29.9012	4.3884
13	3.4523	0.2897	0.04078	24.5227	0.14078	7.1034	33.3772	4.6988
14	3.7975	0.2633	0.03575	27.9750	0.13575	7.3667	36.8005	4.9955
15	4.1772	0.2394	0.03147	31.7725	0.13147	7.6061	40.1520	5.2789
16	4.5950	0.2176	0.02782	35.9497	0.12782	7.8237	43.4164	5.5493
17	5.0545	0.1978	0.02466	40.5447	0.12466	8.0216	46.5819	5.8071
18	5.5599	0.1799	0.02193	45.5992	0.12193	8.2014	49.6395	6.0526
19	6.1159	0.1635	0.01955	51.1591	0.11955	8.3649	52.5827	6.2861
20	6.7275	0.1486	0.01746	57.2750	0.11746	8.5136	55.4069	6.5081
21	7.4002	0.1351	0.01562	64.0025	0.11562	8.6487	58.1095	6.7189
22	8.1403	0.1228	0.01401	71.4027	0.11401	8.7715	60.6893	6.9189
23	8.9543	0.1117	0.01257	79.5430	0.11257	8.8832	63.1462	7.1085
24	9.8497	0.1015	0.01130	88.4973	0.11130	8.9847	65.4813	7.2881
25	10.8347	0.0923	0.01017	98.3471	0.11017	9.0770	67.6964	7.4580
26	11.9182	0.0839	0.00916	109.1818	0.10916	9.1609	69.7940	7.6186
27	13.1100	0.0763	0.00826	121.0999	0.10826	9.2372	71.7773	7.7704
28	14.4210	0.0693	0.00745	134.2099	0.10745	9.3066	73.6495	7.9137
29	15.8631	0.0630	0.00673	148.6309	0.10673	9.3696	75.4146	8.0489
30	17.4494	0.0573	0.00608	164.4940	0.10608	9.4269	77.0766	8.1762
31	19.1943	0.0521	0.00550	181.9434	0.10550	9.4790	78.6395	8.2962
32	21.1138	0.0474	0.00497	201.1378	0.10497	9.5264	80.1078	8.4091
33	23.2252	0.0431	0.00450	222.2515	0.10450	9.5694	81.4856	8.5152
34	25.5477	0.0391	0.00407	245.4767	0.10407	9.6086	82.7773	8.6149
35	28.1024	0.0356	0.00369	271.0244	0.10369	9.6442	83.9872	8.7086
40	45.2593	0.0221	0.00226	442.5926	0.10226	9.7791	88.9525	9.0962
45	72.8905	0.0137	0.00139	718.9048	0.10139	9.8628	92.4544	9.3740
50	117.3909	0.0085	0.00086	1163.91	0.10086	9.9148	94.8889	9.5704
55	189.0591	0.0053	0.00053	1880.59	0.10053	9.9471	96.5619	9.7075
60	304.4816	0.0033	0.00033	3034.82	0.10033	9.9672	97.7010	9.8023
65	490.3707	0.0020	0.00020	4893.71	0.10020	9.9796	98.4705	9.8672
70	789.7470	0.0013	0.00013	7887.47	0.10013	9.9873	98.9870	9.9113
75	1271.90	0.0008	0.00008	12709	0.10008	9.9921	99.3317	9.9410
80	2048.40	0.0005	0.00005	20474	0.10005	9.9951	99.5606	9.9609
85	3298.97	0.0003	0.00003	32980	0.10003	9.9970	99.7120	9.9742
90	5313.02	0.0002	0.00002	53120	0.10002	9.9981	99.8118	9.9831
95	8556.68	0.0001	0.00001	85557	0.10001	9.9988	99.8773	9.9889
96	9412.34	0.0001	0.00001	94113	0.10001	9.9989	99.8874	9.9898
98	11389	0.0001	0.00001		0.10001	9.9991	99.9052	9.9914
100	13781	0.0001	0.00001		0.10001	9.9993	99.9202	9.9927

11%			TABLE 16	Discrete Cash Flow: Compound Interest Factors				11%

	Single Payments		Uniform Series Payments				Arithmetic Gradients	
	F/P Compound Amount	*P/F* Present Worth	*A/F* Sinking Fund	*F/A* Compound Amount	*A/P* Capital Recovery	*P/A* Present Worth	*P/G* Gradient Present Worth	*A/G* Gradient Uniform Series
n								
1	1.1100	0.9009	1.00000	1.0000	1.11000	0.9009		
2	1.2321	0.8116	0.47393	2.1100	0.58393	1.7125	0.8116	0.4739
3	1.3676	0.7312	0.29921	3.3421	0.40921	2.4437	2.2740	0.9306
4	1.5181	0.6587	0.21233	4.7097	0.32233	3.1024	4.2502	1.3700
5	1.6851	0.5935	0.16057	6.2278	0.27057	3.6959	6.6240	1.7923
6	1.8704	0.5346	0.12638	7.9129	0.23638	4.2305	9.2972	2.1976
7	2.0762	0.4817	0.10222	9.7833	0.21222	4.7122	12.1872	2.5863
8	2.3045	0.4339	0.08432	11.8594	0.19432	5.1461	15.2246	2.9585
9	2.5580	0.3909	0.07060	14.1640	0.18060	5.5370	18.3520	3.3144
10	2.8394	0.3522	0.05980	16.7220	0.16980	5.8892	21.5217	3.6544
11	3.1518	0.3173	0.05112	19.5614	0.16112	6.2065	24.6945	3.9788
12	3.4985	0.2858	0.04403	22.7132	0.15403	6.4924	27.8388	4.2879
13	3.8833	0.2575	0.03815	26.2116	0.14815	6.7499	30.9290	4.5822
14	4.3104	0.2320	0.03323	30.0949	0.14323	6.9819	33.9449	4.8619
15	4.7846	0.2090	0.02907	34.4054	0.13907	7.1909	36.8709	5.1275
16	5.3109	0.1883	0.02552	39.1899	0.13552	7.3792	39.6953	5.3794
17	5.8951	0.1696	0.02247	44.5008	0.13247	7.5488	42.4095	5.6180
18	6.5436	0.1528	0.01984	50.3959	0.12984	7.7016	45.0074	5.8439
19	7.2633	0.1377	0.01756	56.9395	0.12756	7.8393	47.4856	6.0574
20	8.0623	0.1240	0.01558	64.2028	0.12558	7.9633	49.8423	6.2590
21	8.9492	0.1117	0.01384	72.2651	0.12384	8.0751	52.0771	6.4491
22	9.9336	0.1007	0.01231	81.2143	0.12231	8.1757	54.1912	6.6283
23	11.0263	0.0907	0.01097	91.1479	0.12097	8.2664	56.1864	6.7969
24	12.2392	0.0817	0.00979	102.1742	0.11979	8.3481	58.0656	6.9555
25	13.5855	0.0736	0.00874	114.4133	0.11874	8.4217	59.8322	7.1045
26	15.0799	0.0663	0.00781	127.9988	0.11781	8.4881	61.4900	7.2443
27	16.7386	0.0597	0.00699	143.0786	0.11699	8.5478	63.0433	7.3754
28	18.5799	0.0538	0.00626	159.8173	0.11626	8.6016	64.4965	7.4982
29	20.6237	0.0485	0.00561	178.3972	0.11561	8.6501	65.8542	7.6131
30	22.8923	0.0437	0.00502	199.0209	0.11502	8.6938	67.1210	7.7206
31	25.4104	0.0394	0.00451	221.9132	0.11451	8.7331	68.3016	7.8210
32	28.2056	0.0355	0.00404	247.3236	0.11404	8.7686	69.4007	7.9147
33	31.3082	0.0319	0.00363	275.5292	0.11363	8.8005	70.4228	8.0021
34	34.7521	0.0288	0.00326	306.8374	0.11326	8.8293	71.3724	8.0836
35	38.5749	0.0259	0.00293	341.5896	0.11293	8.8552	72.2538	8.1594
40	65.0009	0.0154	0.00172	581.8261	0.11172	8.9511	75.7789	8.4659
45	109.5302	0.0091	0.00101	986.6386	0.11101	9.0079	78.1551	8.6763
50	184.5648	0.0054	0.00060	1668.77	0.11060	9.0417	79.7341	8.8185
55	311.0025	0.0032	0.00035	2818.20	0.11035	9.0617	80.7712	8.9135
60	524.0572	0.0019	0.00021	4755.07	0.11021	9.0736	81.4461	8.9762
65	883.0669	0.0011	0.00012	8018.79	0.11012	9.0806	81.8819	9.0172
70	1488.02	0.0007	0.00007	13518	0.11007	9.0848	82.1614	9.0438
75	2507.40	0.0004	0.00004	22785	0.11004	9.0873	82.3397	9.0610
80	4225.11	0.0002	0.00003	38401	0.11003	9.0888	82.4529	9.0720
85	7119.56	0.0001	0.00002	64714	0.11002	9.0896	82.5245	9.0790

TABLE 17 Discrete Cash Flow: Compound Interest Factors

	Single Payments		Uniform Series Payments				Arithmetic Gradients	
	F/P	*P/F*	*A/F*	*F/A*	*A/P*	*P/A*	*P/G*	*A/G*
	Compound Amount	Present Worth	Sinking Fund	Compound Amount	Capital Recovery	Present Worth	Gradient Present Worth	Gradient Uniform Series
n								
1	1.1200	0.8929	1.00000	1.0000	1.12000	0.8929		0.4717
2	1.2544	0.7972	0.47170	2.1200	0.59170	1.6901	0.7972	0.9246
3	1.4049	0.7118	0.29635	3.3744	0.41635	2.4018	2.2208	1.3589
4	1.5735	0.6355	0.20923	4.7793	0.32923	3.0373	4.1273	1.7746
5	1.7623	0.5674	0.15741	6.3528	0.27741	3.6048	6.3970	2.1720
6	1.9738	0.5066	0.12323	8.1152	0.24323	4.1114	8.9302	2.5512
7	2.2107	0.4523	0.09912	10.0890	0.21912	4.5638	11.6443	2.9131
8	2.4760	0.4039	0.08130	12.2997	0.20130	4.9676	14.4714	3.2574
9	2.7731	0.3606	0.06768	14.7757	0.18768	5.3282	17.3563	3.5847
10	3.1058	0.3220	0.05698	17.5487	0.17698	5.6502	20.2541	3.8953
11	3.4785	0.2875	0.04842	20.6546	0.16842	5.9377	23.1288	4.1897
12	3.8960	0.2567	0.04144	24.1331	0.16144	6.1944	25.9523	4.4683
13	4.3635	0.2292	0.03568	28.0291	0.15568	6.4235	28.7024	4.7317
14	4.8871	0.2046	0.03087	32.3926	0.15087	6.6282	31.3624	4.9803
15	5.4736	0.1827	0.02682	37.2797	0.14682	6.8109	33.9202	5.2147
16	6.1304	0.1631	0.02339	42.7533	0.14339	6.9740	36.3670	5.4353
17	6.8660	0.1456	0.02046	48.8837	0.14046	7.1196	38.6973	5.6427
18	7.6900	0.1300	0.01794	55.7497	0.13794	7.2497	40.9080	5.8375
19	8.6128	0.1161	0.01576	63.4397	0.13576	7.3658	42.9979	6.0202
20	9.6463	0.1037	0.01388	72.0524	0.13388	7.4694	44.9676	6.1913
21	10.8038	0.0926	0.01224	81.6987	0.13224	7.5620	46.8188	6.3514
22	12.1003	0.0826	0.01081	92.5026	0.13081	7.6446	48.5543	6.5010
23	13.5523	0.0738	0.00956	104.6029	0.12956	7.7184	50.1776	6.6406
24	15.1786	0.0659	0.00846	118.1552	0.12846	7.7843	51.6929	6.7708
25	17.0001	0.0588	0.00750	133.3339	0.12750	7.8431	53.1046	6.8921
26	19.0401	0.0525	0.00665	150.3339	0.12665	7.8957	54.4177	7.0049
27	21.3249	0.0469	0.00590	169.3740	0.12590	7.9426	55.6369	7.1098
28	23.8839	0.0419	0.00524	190.6989	0.12524	7.9844	56.7674	7.2071
29	26.7499	0.0374	0.00466	214.5828	0.12466	8.0218	57.8141	7.2974
30	29.9599	0.0334	0.00414	241.3327	0.12414	8.0552	58.7821	7.3811
31	33.5551	0.0298	0.00369	271.2926	0.12369	8.0850	59.6761	7.4586
32	37.5817	0.0266	0.00328	304.8477	0.12328	8.1116	60.5010	7.5302
33	42.0915	0.0238	0.00292	342.4294	0.12292	8.1354	61.2612	7.5965
34	47.1425	0.0212	0.00260	384.5210	0.12260	8.1566	61.9612	7.6577
35	52.7996	0.0189	0.00232	431.6635	0.12232	8.1755	62.6052	7.6577
40	93.0510	0.0107	0.00130	767.0914	0.12130	8.2438	65.1159	7.8988
45	163.9876	0.0061	0.0074	1358.23	0.12074	8.2825	66.7342	8.0572
50	289.0022	0.0035	0.00042	2400.02	0.12042	8.3045	67.7624	8.1597
55	509.3206	0.0020	0.00024	4236.01	0.12024	8.3170	68.4082	8.2251
60	897.5969	0.0011	0.00013	7471.64	0.12013	8.3240	68.8100	8.2664
65	1581.87	0.0006	0.00008	13174	0.12008	8.3281	69.0581	8.2922
70	2787.80	0.0004	0.00004	23223	0.12004	8.3303	69.2103	8.3082
75	4913.06	0.0002	0.00002	40934	0.12002	8.3316	69.3031	8.3181
80	8658.48	0.0001	0.00001	72146	0.12001	8.3324	69.3594	8.3241
85	15259	0.0001	0.00001		0.12001	8.3328	69.3935	8.3278

14%				TABLE **18**	Discrete Cash Flow: Compound Interest Factors			14%
	Single Payments		Uniform Series Payments				Arithmetic Gradients	
	F/P Compound Amount	*P/F* Present Worth	*A/F* Sinking Fund	*F/A* Compound Amount	*A/P* Capital Recovery	*P/A* Present Worth	*P/G* Gradient Present Worth	*A/G* Gradient Uniform Series
n								
1	1.1400	0.8772	1.00000	1.0000	1.14000	0.8772		
2	1.2996	0.7695	0.46729	2.1400	0.60729	1.6467	0.7695	0.4673
3	1.4815	0.6750	0.29073	3.4396	0.43073	2.3216	2.1194	0.9129
4	1.6890	0.5921	0.20320	4.9211	0.34320	2.9137	3.8957	1.3370
5	1.9254	0.5194	0.15128	6.6101	0.29128	3.4331	5.9731	1.7399
6	2.1950	0.4556	0.11716	8.5355	0.25716	3.8887	8.2511	2.1218
7	2.5023	0.3996	0.09319	10.7305	0.23319	4.2883	10.6489	2.4832
8	2.8526	0.3506	0.07557	13.2328	0.21557	4.6389	13.1028	2.8246
9	3.2519	0.3075	0.06217	16.0853	0.20217	4.9464	15.5629	3.1463
10	3.7072	0.2697	0.05171	19.3373	0.19171	5.2161	17.9906	3.4490
11	4.2262	0.2366	0.04339	23.0445	0.18339	5.4527	20.3567	3.7333
12	4.8179	0.2076	0.03667	27.2707	0.17667	5.6603	22.6399	3.9998
13	5.4924	0.1821	0.03116	32.0887	0.17116	5.8424	24.8247	4.2491
14	6.2613	0.1597	0.02661	37.5811	0.16661	6.0021	26.9009	4.4819
15	7.1379	0.1401	0.02281	43.8424	0.16281	6.1422	28.8623	4.6990
16	8.1372	0.1229	0.01962	50.9804	0.15962	6.2651	30.7057	4.9011
17	9.2765	0.1078	0.01692	59.1176	0.15692	6.3729	32.4305	5.0888
18	10.5752	0.0946	0.01462	68.3941	0.15462	6.4674	34.0380	5.2630
19	12.0557	0.0829	0.01266	78.9692	0.15266	6.5504	35.5311	5.4243
20	13.7435	0.0728	0.01099	91.0249	0.15099	6.6231	36.9135	5.5734
21	15.6676	0.0638	0.00954	104.7684	0.14954	6.6870	38.1901	5.7111
22	17.8610	0.0560	0.00830	120.4360	0.14830	6.7429	39.3658	5.8381
23	20.3616	0.0491	0.00723	138.2970	0.14723	6.7921	40.4463	5.9549
24	23.2122	0.0431	0.00630	158.6586	0.14630	6.8351	41.4371	6.0624
25	26.4619	0.0378	0.00550	181.8708	0.14550	6.8729	42.3441	6.1610
26	30.1666	0.0331	0.00480	208.3327	0.14480	6.9061	43.1728	6.2514
27	34.3899	0.0291	0.00419	238.4993	0.14419	6.9352	43.9289	6.3342
28	39.2045	0.0255	0.00366	272.8892	0.14366	6.9607	44.6176	6.4100
29	44.6931	0.0224	0.00320	312.0937	0.14320	6.9830	45.2441	6.4791
30	50.9502	0.0196	0.00280	356.7868	0.14280	7.0027	45.8132	6.5423
31	58.0832	0.0172	0.00245	407.7370	0.14245	7.0199	46.3297	6.5998
32	66.2148	0.0151	0.00215	465.8202	0.14215	7.0350	46.7979	6.6522
33	75.4849	0.0132	0.00188	532.0350	0.14188	7.0482	47.2218	6.6998
34	86.0528	0.0116	0.00165	607.5199	0.14165	7.0599	47.6053	6.7431
35	98.1002	0.0102	0.00144	693.5727	0.14144	7.0700	47.9519	6.7824
40	188.8835	0.0053	0.00075	1342.03	0.14075	7.1050	49.2376	6.9300
45	363.6791	0.0027	0.00039	2590.56	0.14039	7.1232	49.9963	7.0188
50	700.2330	0.0014	0.00020	4994.52	0.14020	7.1327	50.4375	7.0714
55	1348.24	0.0007	0.00010	9623.13	0.14010	7.1376	50.6912	7.1020
60	2595.92	0.0004	0.00005	18535	0.14005	7.1401	50.8357	7.1197
65	4998.22	0.0002	0.00003	35694	0.14003	7.1414	50.9173	7.1298
70	9623.64	0.0001	0.00001	68733	0.14001	7.1421	50.9632	7.1356
75	18530	0.0001	0.00001		0.14001	7.1425	50.9887	7.1388
80	35677				0.14000	7.1427	51.0030	7.1406
85	68693				0.14000	7.1428	51.0108	7.1416

15%			TABLE 19	Discrete Cash Flow: Compound Interest Factors				15%

	Single Payments		Uniform Series Payments				Arithmetic Gradients	
	F/P Compound Amount	P/F Present Worth	A/F Sinking Fund	F/A Compound Amount	A/P Capital Recovery	P/A Present Worth	P/G Gradient Present Worth	A/G Gradient Uniform Series
n								
1	1.1500	0.8696	1.00000	1.0000	1.15000	0.8696		
2	1.3225	0.7561	0.46512	2.1500	0.61512	1.6257	0.7561	0.4651
3	1.5209	0.6575	0.28798	3.4725	0.43798	2.2832	2.0712	0.9071
4	1.7490	0.5718	0.20027	4.9934	0.35027	2.8550	3.7864	1.3263
5	2.0114	0.4972	0.14832	6.7424	0.29832	3.3522	5.7751	1.7228
6	2.3131	0.4323	0.11424	8.7537	0.26424	3.7845	7.9368	2.0972
7	2.6600	0.3759	0.09036	11.0668	0.24036	4.1604	10.1924	2.4498
8	3.0590	0.3269	0.07285	13.7268	0.22285	4.4873	12.4807	2.7813
9	3.5179	0.2843	0.05957	16.7858	0.20957	4.7716	14.7548	3.0922
10	4.0456	0.2472	0.04925	20.3037	0.19925	5.0188	16.9795	3.3832
11	4.6524	0.2149	0.04107	24.3493	0.19107	5.2337	19.1289	3.6549
12	5.3503	0.1869	0.03448	29.0017	0.18448	5.4206	21.1849	3.9082
13	6.1528	0.1625	0.02911	34.3519	0.17911	5.5831	23.1352	4.1438
14	7.0757	0.1413	0.02469	40.5047	0.17469	5.7245	24.9725	4.3624
15	8.1371	0.1229	0.02102	47.5804	0.17102	5.8474	26.6930	4.5650
16	9.3576	0.1069	0.01795	55.7175	0.16795	5.9542	28.2960	4.7522
17	10.7613	0.0929	0.01537	65.0751	0.16537	6.0472	29.7828	4.9251
18	12.3755	0.0808	0.01319	75.8364	0.16319	6.1280	31.1565	5.0843
19	14.2318	0.0703	0.01134	88.2118	0.16134	6.1982	32.4213	5.2307
20	16.3665	0.0611	0.00976	102.4436	0.15976	6.2593	33.5822	5.3651
21	18.8215	0.0531	0.00842	118.8101	0.15842	6.3125	34.6448	5.4883
22	21.6447	0.0462	0.00727	137.6316	0.15727	6.3587	35.6150	5.6010
23	24.8915	0.0402	0.00628	159.2764	0.15628	6.3988	36.4988	5.7040
24	28.6252	0.0349	0.00543	184.1678	0.15543	6.4338	37.3023	5.7979
25	32.9190	0.0304	0.00470	212.7930	0.15470	6.4641	38.0314	5.8834
26	37.8568	0.0264	0.00407	245.7120	0.15407	6.4906	38.6918	5.9612
27	43.5353	0.0230	0.00353	283.5688	0.15353	6.5135	39.2890	6.0319
28	50.0656	0.0200	0.00306	327.1041	0.15306	6.5335	39.8283	6.0960
29	57.5755	0.0174	0.00265	377.1697	0.15265	6.5509	40.3146	6.1541
30	66.2118	0.0151	0.00230	434.7451	0.15230	6.5660	40.7526	6.2066
31	76.1435	0.0131	0.00200	500.9569	0.15200	6.5791	41.1466	6.2541
32	87.5651	0.0114	0.00173	577.1005	0.15173	6.5905	41.5006	6.2970
33	100.6998	0.0099	0.00150	664.6655	0.15150	6.6005	41.8184	6.3357
34	115.8048	0.0086	0.00131	765.3654	0.15131	6.6091	42.1033	6.3705
35	133.1755	0.0075	0.00113	881.1702	0.15113	6.6166	42.3586	6.4019
40	267.8635	0.0037	0.00056	1779.09	0.15056	6.6418	43.2830	6.5168
45	538.7693	0.0019	0.00028	3585.13	0.15028	6.6543	43.8051	6.5830
50	1083.66	0.0009	0.00014	7217.72	0.15014	6.6605	44.0958	6.6205
55	2179.62	0.0005	0.00007	14524	0.15007	6.6636	44.2558	6.6414
60	4384.00	0.0002	0.00003	29220	0.15003	6.6651	44.3431	6.6530
65	8817.79	0.0001	0.00002	58779	0.15002	6.6659	44.3903	6.6593
70	17736	0.0001	0.00001		0.15001	6.6663	44.4156	6.6627
75	35673				0.15000	6.6665	44.4292	6.6646
80	71751				0.15000	6.6666	44.4364	6.6656
85					0.15000	6.6666	44.4402	6.6661

16%			TABLE 20	Discrete Cash Flow: Compound Interest Factors				16%
	Single Payments		**Uniform Series Payments**				**Arithmetic Gradients**	
	F/P Compound Amount	*P/F* Present Worth	*A/F* Sinking Fund	*F/A* Compound Amount	*A/P* Capital Recovery	*P/A* Present Worth	*P/G* Gradient Present Worth	*A/G* Gradient Uniform Series
n								
1	1.1600	0.8621	1.00000	1.0000	1.16000	0.8621		
2	1.3456	0.7432	0.46296	2.1600	0.62296	1.6052	0.7432	0.4630
3	1.5609	0.6407	0.28526	3.5056	0.44526	2.2459	2.0245	0.9014
4	1.8106	0.5523	0.19738	5.0665	0.35738	2.7982	3.6814	1.3156
5	2.1003	0.4761	0.14541	6.8771	0.30541	3.2743	5.5858	1.7060
6	2.4364	0.4104	0.11139	8.9775	0.27139	3.6847	7.6380	2.0729
7	2.8262	0.3538	0.08761	11.4139	0.24761	4.0386	9.7610	2.4169
8	3.2784	0.3050	0.07022	14.2401	0.23022	4.3436	11.8962	2.7388
9	3.8030	0.2630	0.05708	17.5185	0.21708	4.6065	13.9998	3.0391
10	4.4114	0.2267	0.04690	21.3215	0.20690	4.8332	16.0399	3.3187
11	5.1173	0.1954	0.03886	25.7329	0.19886	5.0286	17.9941	3.5783
12	5.9360	0.1685	0.03241	30.8502	0.19241	5.1971	19.8472	3.8189
13	6.8858	0.1452	0.02718	36.7862	0.18718	5.3423	21.5899	4.0413
14	7.9875	0.1252	0.02290	43.6720	0.18290	5.4675	23.2175	4.2464
15	9.2655	0.1079	0.01936	51.6595	0.17936	5.5755	24.7284	4.4352
16	10.7480	0.0930	0.01641	60.9250	0.17641	5.6685	26.1241	4.6086
17	12.4677	0.0802	0.01395	71.6730	0.17395	5.7487	27.4074	4.7676
18	14.4625	0.0691	0.01188	84.1407	0.17188	5.8178	28.5828	4.9130
19	16.7765	0.0596	0.01014	98.6032	0.17014	5.8775	29.6557	5.0457
20	19.4608	0.0514	0.00867	115.3797	0.16867	5.9288	30.6321	5.1666
22	26.1864	0.0382	0.00635	157.4150	0.16635	6.0113	32.3200	5.3765
24	35.2364	0.0284	0.00467	213.9776	0.16467	6.0726	33.6970	5.5490
26	47.4141	0.0211	0.00345	290.0883	0.16345	6.1182	34.8114	5.6898
28	63.8004	0.0157	0.00255	392.5028	0.16255	6.1520	35.7073	5.8041
30	85.8499	0.0116	0.00189	530.3117	0.16189	6.1772	36.4234	5.8964
32	115.5196	0.0087	0.00140	715.7475	0.16140	6.1959	36.9930	5.9706
34	155.4432	0.0064	0.00104	965.2698	0.16104	6.2098	37.4441	6.0299
35	180.3141	0.0055	0.00089	1120.71	0.16089	6.2153	37.6327	6.0548
36	209.1643	0.0048	0.00077	1301.03	0.16077	6.2201	37.8000	6.0771
38	281.4515	0.0036	0.00057	1752.82	0.16057	6.2278	38.0799	6.1145
40	378.7212	0.0026	0.00042	2360.76	0.16042	6.2335	38.2992	6.1441
45	795.4438	0.0013	0.00020	4965.27	0.16020	6.2421	38.6598	6.1934
50	1670.70	0.0006	0.00010	10436	0.16010	6.2463	38.8521	6.2201
55	3509.05	0.0003	0.00005	21925	0.16005	6.2482	38.9534	6.2343
60	7370.20	0.0001	0.00002	46058	0.16002	6.2492	39.0063	6.2419

18%			TABLE 21	Discrete Cash Flow: Compound Interest Factors				18%
	Single Payments		**Uniform Series Payments**				**Arithmetic Gradients**	
	F/P Compound Amount	*P/F* Present Worth	*A/F* Sinking Fund	*F/A* Compound Amount	*A/P* Capital Recovery	*P/A* Present Worth	*P/G* Gradient Present Worth	*A/G* Gradient Uniform Series
n								
1	1.1800	0.8475	1.00000	1.0000	1.18000	0.8475		
2	1.3924	0.7182	0.45872	2.1800	0.63872	1.5656	0.7182	0.4587
3	1.6430	0.6086	0.27992	3.5724	0.45992	2.1743	1.9354	0.8902
4	1.9388	0.5158	0.19174	5.2154	0.37174	2.6901	3.4828	1.2947
5	2.2878	0.4371	0.13978	7.1542	0.31978	3.1272	5.2312	1.6728
6	2.6996	0.3704	0.10591	9.4420	0.28591	3.4976	7.0834	2.0252
7	3.1855	0.3139	0.08236	12.1415	0.26236	3.8115	8.9670	2.3526
8	3.7589	0.2660	0.06524	15.3270	0.24524	4.0776	10.8292	2.6558
9	4.4355	0.2255	0.05239	19.0859	0.23239	4.3030	12.6329	2.9358
10	5.2338	0.1911	0.04251	23.5213	0.22251	4.4941	14.3525	3.1936
11	6.1759	0.1619	0.03478	28.7551	0.21478	4.6560	15.9716	3.4303
12	7.2876	0.1372	0.02863	34.9311	0.20863	4.7932	17.4811	3.6470
13	8.5994	0.1163	0.02369	42.2187	0.20369	4.9095	18.8765	3.8449
14	10.1472	0.0985	0.01968	50.8180	0.19968	5.0081	20.1576	4.0250
15	11.9737	0.0835	0.01640	60.9653	0.19640	5.0916	21.3269	4.1887
16	14.1290	0.0708	0.01371	72.9390	0.19371	5.1624	22.3885	4.3369
17	16.6722	0.0600	0.01149	87.0680	0.19149	5.2223	23.3482	4.4708
18	19.6733	0.0508	0.00964	103.7403	0.18964	5.2732	24.2123	4.5916
19	23.2144	0.0431	0.00810	123.4135	0.18810	5.3162	24.9877	4.7003
20	27.3930	0.0365	0.00682	146.6280	0.18682	5.3527	25.6813	4.7978
22	38.1421	0.0262	0.00485	206.3448	0.18485	5.4099	26.8506	4.9632
24	53.1090	0.0188	0.00345	289.4945	0.18345	5.4509	27.7725	5.0950
26	73.9490	0.0135	0.00247	405.2721	0.18247	5.4804	28.4935	5.1991
28	102.9666	0.0097	0.00177	566.4809	0.18177	5.5016	29.0537	5.2810
30	143.3706	0.0070	0.00126	790.9480	0.18126	5.5168	29.4864	5.3448
32	199.6293	0.0050	0.00091	1103.50	0.18091	5.5277	29.8191	5.3945
34	277.9638	0.0036	0.00065	1538.69	0.18065	5.5356	30.0736	5.4328
35	327.9973	0.0030	0.00055	1816.65	0.18055	5.5386	30.1773	5.4485
36	387.0368	0.0026	0.00047	2144.65	0.18047	5.5412	30.2677	5.4623
38	538.9100	0.0019	0.00033	2988.39	0.18033	5.5452	30.4152	5.4849
40	750.3783	0.0013	0.00024	4163.21	0.18024	5.5482	30.5269	5.5022
45	1716.68	0.0006	0.00010	9531.58	0.18010	5.5523	30.7006	5.5293
50	3927.36	0.0003	0.00005	21813	0.18005	5.5541	30.7856	5.5428
55	8984.84	0.0001	0.00002	49910	0.18002	5.5549	30.8268	5.5494
60	20555			114190	0.18001	5.5553	30.8465	5.5526

20%			TABLE 22	Discrete Cash Flow: Compound Interest Factors				20%
	Single Payments		Uniform Series Payments				Arithmetic Gradients	
n	F/P Compound Amount	P/F Present Worth	A/F Sinking Fund	F/A Compound Amount	A/P Capital Recovery	P/A Present Worth	P/G Gradient Present Worth	A/G Gradient Uniform Series
1	1.2000	0.8333	1.00000	1.0000	1.20000	0.8333		
2	1.4400	0.6944	0.45455	2.2000	0.65455	1.5278	0.6944	0.4545
3	1.7280	0.5787	0.27473	3.6400	0.47473	2.1065	1.8519	0.8791
4	2.0736	0.4823	0.18629	5.3680	0.38629	2.5887	3.2986	1.2742
5	2.4883	0.4019	0.13438	7.4416	0.33438	2.9906	4.9061	1.6405
6	2.9860	0.3349	0.10071	9.9299	0.30071	3.3255	6.5806	1.9788
7	3.5832	0.2791	0.07742	12.9159	0.27742	3.6046	8.2551	2.2902
8	4.2998	0.2326	0.06061	16.4991	0.26061	3.8372	9.8831	2.5756
9	5.1598	0.1938	0.04808	20.7989	0.24808	4.0310	11.4335	2.8364
10	6.1917	0.1615	0.03852	25.9587	0.23852	4.1925	12.8871	3.0739
11	7.4301	0.1346	0.03110	32.1504	0.23110	4.3271	14.2330	3.2893
12	8.9161	0.1122	0.02526	39.5805	0.22526	4.4392	15.4667	3.4841
13	10.6993	0.0935	0.02062	48.4966	0.22062	4.5327	16.5883	3.6597
14	12.8392	0.0779	0.01689	59.1959	0.21689	4.6106	17.6008	3.8175
15	15.4070	0.0649	0.01388	72.0351	0.21388	4.6755	18.5095	3.9588
16	18.4884	0.0541	0.01144	87.4421	0.21144	4.7296	19.3208	4.0851
17	22.1861	0.0451	0.00944	105.9306	0.20944	4.7746	20.0419	4.1976
18	26.6233	0.0376	0.00781	128.1167	0.20781	4.8122	20.6805	4.2975
19	31.9480	0.0313	0.00646	154.7400	0.20646	4.8435	21.2439	4.3861
20	38.3376	0.0261	0.00536	186.6880	0.20536	4.8696	21.7395	4.4643
22	55.2061	0.0181	0.00369	271.0307	0.20369	4.9094	22.5546	4.5941
24	79.4968	0.0126	0.00255	392.4842	0.20255	4.9371	23.1760	4.6943
26	114.4755	0.0087	0.00176	567.3773	0.20176	4.9563	23.6460	4.7709
28	164.8447	0.0061	0.00122	819.2233	0.20122	4.9697	23.9991	4.8291
30	237.3763	0.0042	0.00085	1181.88	0.20085	4.9789	24.2628	4.8731
32	341.8219	0.0029	0.00059	1704.11	0.20059	4.9854	24.4588	4.9061
34	492.2235	0.0020	0.00041	2456.12	0.20041	4.9898	24.6038	4.9308
35	590.6682	0.0017	0.00034	2948.34	0.20034	4.9915	24.6614	4.9406
36	708.8019	0.0014	0.00028	3539.01	0.20028	4.9929	24.7108	4.9491
38	1020.67	0.0010	0.00020	5098.37	0.20020	4.9951	24.7894	4.9627
40	1469.77	0.0007	0.00014	7343.86	0.20014	4.9966	24.8469	4.9728
45	3657.26	0.0003	0.00005	18281	0.20005	4.9986	24.9316	4.9877
50	9100.44	0.0001	0.00002	45497	0.20002	4.9995	24.9698	4.9945
55	22645		0.00001		0.20001	4.9998	24.9868	4.9976

22%			TABLE 23	Discrete Cash Flow: Compound Interest Factors				22%
	Single Payments		**Uniform Series Payments**				**Arithmetic Gradients**	
n	*F/P* Compound Amount	*P/F* Present Worth	*A/F* Sinking Fund	*F/A* Compound Amount	*A/P* Capital Recovery	*P/A* Present Worth	*P/G* Gradient Present Worth	*A/G* Gradient Uniform Series
1	1.2200	0.8197	1.00000	1.0000	1.22000	0.8197		
2	1.4884	0.6719	0.45045	2.2200	0.67045	1.4915	0.6719	0.4505
3	1.8158	0.5507	0.26966	3.7084	0.48966	2.0422	1.7733	0.8683
4	2.2153	0.4514	0.18102	5.5242	0.40102	2.4936	3.1275	1.2542
5	2.7027	0.3700	0.12921	7.7396	0.34921	2.8636	4.6075	1.6090
6	3.2973	0.3033	0.09576	10.4423	0.31576	3.1669	6.1239	1.9337
7	4.0227	0.2486	0.07278	13.7396	0.29278	3.4155	7.6154	2.2297
8	4.9077	0.2038	0.05630	17.7623	0.27630	3.6193	9.0417	2.4982
9	5.9874	0.1670	0.04411	22.6700	0.26411	3.7863	10.3779	2.7409
10	7.3046	0.1369	0.03489	28.6574	0.25489	3.9232	11.6100	2.9593
11	8.9117	0.1122	0.02781	35.9620	0.24781	4.0354	12.7321	3.1551
12	10.8722	0.0920	0.02228	44.8737	0.24228	4.1274	13.7438	3.3299
13	13.2641	0.0754	0.01794	55.7459	0.23794	4.2028	14.6485	3.4855
14	16.1822	0.0618	0.01449	69.0100	0.23449	4.2646	15.4519	3.6233
15	19.7423	0.0507	0.01174	85.1922	0.23174	4.3152	16.1610	3.7451
16	24.0856	0.0415	0.00953	104.9345	0.22953	4.3567	16.7838	3.8524
17	29.3844	0.0340	0.00775	129.0201	0.22775	4.3908	17.3283	3.9465
18	35.8490	0.0279	0.00631	158.4045	0.22631	4.4187	17.8025	4.0289
19	43.7358	0.0229	0.00515	194.2535	0.22515	4.4415	18.2141	4.1009
20	53.3576	0.0187	0.00420	237.9893	0.22420	4.4603	18.5702	4.1635
22	79.4175	0.0126	0.00281	356.4432	0.22281	4.4882	19.1418	4.2649
24	118.2050	0.0085	0.00188	532.7501	0.22188	4.5070	19.5635	4.3407
26	175.9364	0.0057	0.00126	795.1653	0.22126	4.5196	19.8720	4.3968
28	261.8637	0.0038	0.00084	1185.74	0.22084	4.5281	20.0962	4.4381
30	389.7579	0.0026	0.00057	1767.08	0.22057	4.5338	20.2583	4.4683
32	580.1156	0.0017	0.00038	2632.34	0.22038	4.5376	20.3748	4.4902
34	863.4441	0.0012	0.00026	3920.20	0.22026	4.5402	20.4582	4.5060
35	1053.40	0.0009	0.00021	4783.64	0.22021	4.5411	20.4905	4.5122
36	1285.15	0.0008	0.00017	5837.05	0.22017	4.5419	20.5178	4.5174
38	1912.82	0.0005	0.00012	8690.08	0.22012	4.5431	20.5601	4.5256
40	2847.04	0.0004	0.00008	12937	0.22008	4.5439	20.5900	4.5314
45	7694.71	0.0001	0.00003	34971	0.22003	4.5449	20.6319	4.5396
50	20797		0.00001	94525	0.22001	4.5452	20.6492	4.5431
55	56207				0.22000	4.5454	20.6563	4.5445

TABLE 24 Discrete Cash Flow: Compound Interest Factors

	Single Payments		Uniform Series Payments				Arithmetic Gradients	
	F/P Compound Amount	**P/F** Present Worth	**A/F** Sinking Fund	**F/A** Compound Amount	**A/P** Capital Recovery	**P/A** Present Worth	**P/G** Gradient Present Worth	**A/G** Gradient Uniform Series
n								
1	1.2400	0.8065	1.00000	1.0000	1.24000	0.8065		
2	1.5376	0.6504	0.44643	2.2400	0.68643	1.4568	0.6504	0.4464
3	1.9066	0.5245	0.26472	3.7776	0.50472	1.9813	1.6993	0.8577
4	2.3642	0.4230	0.17593	5.6842	0.41593	2.4043	2.9683	1.2346
5	2.9316	0.3411	0.12425	8.0484	0.36425	2.7454	4.3327	1.5782
6	3.6352	0.2751	0.09107	10.9801	0.33107	3.0205	5.7081	1.8898
7	4.5077	0.2218	0.06842	14.6153	0.30842	3.2423	7.0392	2.1710
8	5.5895	0.1789	0.05229	19.1229	0.29229	3.4212	8.2915	2.4236
9	6.9310	0.1443	0.04047	24.7125	0.28047	3.5655	9.4458	2.6492
10	8.5944	0.1164	0.03160	31.6434	0.27160	3.6819	10.4930	2.8499
11	10.6571	0.0938	0.02485	40.2379	0.26485	3.7757	11.4313	3.0276
12	13.2148	0.0757	0.01965	50.8950	0.25965	3.8514	12.2637	3.1843
13	16.3863	0.0610	0.01560	64.1097	0.25560	3.9124	12.9960	3.3218
14	20.3191	0.0492	0.01242	80.4961	0.25242	3.9616	13.6358	3.4420
15	25.1956	0.0397	0.00992	100.8151	0.24992	4.0013	14.1915	3.5467
16	31.2426	0.0320	0.00794	126.0108	0.24794	4.0333	14.6716	3.6376
17	38.7408	0.0258	0.00636	157.2534	0.24636	4.0591	15.0846	3.7162
18	48.0386	0.0208	0.00510	195.9942	0.24510	4.0799	15.4385	3.7840
19	59.5679	0.0168	0.00410	244.0328	0.24410	4.0967	15.7406	3.8423
20	73.8641	0.0135	0.00329	303.6006	0.24329	4.1103	15.9979	3.8922
22	113.5735	0.0088	0.00213	469.0563	0.24213	4.1300	16.4011	3.9712
24	174.6306	0.0057	0.00138	723.4610	0.24138	4.1428	16.6891	4.0284
26	268.5121	0.0037	0.00090	1114.63	0.24090	4.1511	16.8930	4.0695
28	412.8642	0.0024	0.00058	1716.10	0.24058	4.1566	17.0365	4.0987
30	634.8199	0.0016	0.00038	2640.92	0.24038	4.1601	17.1369	4.1193
32	976.0991	0.0010	0.00025	4062.91	0.24025	4.1624	17.2067	4.1338
34	1500.85	0.0007	0.00016	6249.38	0.24016	4.1639	17.2552	4.1440
35	1861.05	0.0005	0.00013	7750.23	0.24013	4.1664	17.2734	4.1479
36	2307.71	0.0004	0.00010	9611.28	0.24010	4.1649	17.2886	4.1511
38	3548.33	0.0003	0.00007	14781	0.24007	4.1655	17.3116	4.1560
40	5455.91	0.0002	0.00004	22729	0.24004	4.1659	17.3274	4.1593
45	15995	0.0001	0.00002	66640	0.24002	4.1664	17.3483	4.1639
50	46890		0.00001		0.24001	4.1666	17.3563	4.1653
55					0.24000	4.1666	17.3593	4.1663

| 25% | TABLE 25 | | | Discrete Cash Flow: Compound Interest Factors | | | | 25% |

	Single Payments		Uniform Series Payments				Arithmetic Gradients	
	F/P Compound Amount	*P/F* Present Worth	*A/F* Sinking Fund	*F/A* Compound Amount	*A/P* Capital Recovery	*P/A* Present Worth	*P/G* Gradient Present Worth	*A/G* Gradient Uniform Series
n								
1	1.2500	0.8000	1.00000	1.0000	1.25000	0.8000		
2	1.5625	0.6400	0.44444	2.2500	0.69444	1.4400	0.6400	0.4444
3	1.9531	0.5120	0.26230	3.8125	0.51230	1.9520	1.6640	0.8525
4	2.4414	0.4096	0.17344	5.7656	0.42344	2.3616	2.8928	1.2249
5	3.0518	0.3277	0.12185	8.2070	0.37185	2.6893	4.2035	1.5631
6	3.8147	0.2621	0.08882	11.2588	0.33882	2.9514	5.5142	1.8683
7	4.7684	0.2097	0.06634	15.0735	0.31634	3.1611	6.7725	2.1424
8	5.9605	0.1678	0.05040	19.8419	0.30040	3.3289	7.9469	2.3872
9	7.4506	0.1342	0.03876	25.8023	0.28876	3.4631	9.0207	2.6048
10	9.3132	0.1074	0.03007	33.2529	0.28007	3.5705	9.9870	2.7971
11	11.6415	0.0859	0.02349	42.5661	0.27349	3.6564	10.8460	2.9663
12	14.5519	0.0687	0.01845	54.2077	0.26845	3.7251	11.6020	3.1145
13	18.1899	0.0550	0.01454	68.7596	0.26454	3.7801	12.2617	3.2437
14	22.7374	0.0440	0.01150	86.9495	0.26150	3.8241	12.8334	3.3559
15	28.4217	0.0352	0.00912	109.6868	0.25912	3.8593	13.3260	3.4530
16	35.5271	0.0281	0.00724	138.1085	0.25724	3.8874	13.7482	3.5366
17	44.4089	0.0225	0.00576	173.6357	0.25576	3.9099	14.1085	3.6084
18	55.5112	0.0180	0.00459	218.0446	0.25459	3.9279	14.4147	3.6698
19	69.3889	0.0144	0.00366	273.5558	0.25366	3.9424	14.6741	3.7222
20	86.7362	0.0115	0.00292	342.9447	0.25292	3.9539	14.8932	3.7667
22	135.5253	0.0074	0.00186	538.1011	0.25186	3.9705	15.2326	3.8365
24	211.7582	0.0047	0.00119	843.0329	0.25119	3.9811	15.4711	3.8861
26	330.8722	0.0030	0.00076	1319.49	0.25076	3.9879	15.6373	3.9212
28	516.9879	0.0019	0.00048	2063.95	0.25048	3.9923	15.7524	3.9457
30	807.7936	0.0012	0.00031	3227.17	0.25031	3.9950	15.8316	3.9628
32	1262.18	0.0008	0.00020	5044.71	0.25020	3.9968	15.8859	3.9746
34	1972.15	0.0005	0.00013	7884.61	0.25013	3.9980	15.9229	3.9828
35	2465.19	0.0004	0.00010	9856.76	.025010	3.9984	15.9367	3.9858
36	3081.49	0.0003	0.00008	12322	0.25008	3.9987	15.9481	3.9883
38	4814.82	0.0002	0.00005	19255	0.25005	3.9992	15.9651	3.9921
40	7523.16	0.0001	0.00003	30089	0.25003	3.9995	15.9766	3.9947
45	22959		0.00001	91831	0.25001	3.9998	15.9915	3.9980
50	70065				0.25000	3.9999	15.9969	3.9993
55					0.25000	4.0000	15.9989	3.9997

30%   TABLE 26   Discrete Cash Flow: Compound Interest Factors   30%

	Single Payments		Uniform Series Payments				Arithmetic Gradients	
n	F/P Compound Amount	P/F Present Worth	A/F Sinking Fund	F/A Compound Amount	A/P Capital Recovery	P/A Present Worth	P/G Gradient Present Worth	A/G Gradient Uniform Series
1	1.3000	0.7692	1.00000	1.0000	1.30000	0.7692		
2	1.6900	0.5917	0.43478	2.3000	0.73478	1.3609	0.5917	0.4348
3	2.1970	0.4552	0.25063	3.9900	0.55063	1.8161	1.5020	0.8271
4	2.8561	0.3501	0.16163	6.1870	0.46163	2.1662	2.5524	1.1783
5	3.7129	0.2693	0.11058	9.0431	0.41058	2.4356	3.6297	1.4903
6	4.8268	0.2072	0.07839	12.7560	0.37839	2.6427	4.6656	1.7654
7	6.2749	0.1594	0.05687	17.5828	0.35687	2.8021	5.6218	2.0063
8	8.1573	0.1226	0.04192	23.8577	0.34192	2.9247	6.4800	2.2156
9	10.6045	0.0943	0.03124	32.0150	0.33124	3.0190	7.2343	2.3963
10	13.7858	0.0725	0.02346	42.6195	0.32346	3.0915	7.8872	2.5512
11	17.9216	0.0558	0.01773	56.4053	0.31773	3.1473	8.4452	2.6833
12	23.2981	0.0429	0.01345	74.3270	0.31345	3.1903	8.9173	2.7952
13	30.2875	0.0330	0.01024	97.6250	0.31024	3.2233	9.3135	2.8895
14	39.3738	0.0254	0.00782	127.9125	0.30782	3.2487	9.6437	2.9685
15	51.1859	0.0195	0.00598	167.2863	0.30598	3.2682	9.9172	3.0344
16	66.5417	0.0150	0.00458	218.4722	0.30458	3.2832	10.1426	3.0892
17	86.5042	0.0116	0.00351	285.0139	0.30351	3.2948	10.3276	3.1345
18	112.4554	0.0089	0.00269	371.5180	0.30269	3.3037	10.4788	3.1718
19	146.1920	0.0068	0.00207	483.9734	0.30207	3.3105	10.6019	3.2025
20	190.0496	0.0053	0.00159	630.1655	0.30159	3.3158	10.7019	3.2275
22	321.1839	0.0031	0.00094	1067.28	0.30094	3.3230	10.8482	3.2646
24	542.8008	0.0018	0.00055	1806.00	0.30055	3.3272	10.9433	3.2890
25	705.6410	0.0014	0.00043	2348.80	0.30043	3.3286	10.9773	3.2979
26	917.3333	0.0011	0.00033	3054.44	0.30033	3.3297	11.0045	3.3050
28	1550.29	0.0006	0.00019	5164.31	0.30019	3.3312	11.0437	3.3153
30	2620.00	0.0004	0.00011	8729.99	0.30011	3.3321	11.0687	3.3219
32	4427.79	0.0002	0.00007	14756	0.30007	3.3326	11.0845	3.3261
34	7482.97	0.0001	0.00004	24940	0.30004	3.3329	11.0945	3.3288
35	9727.86	0.0001	0.00003	32423	0.30003	3.3330	11.0980	3.3297

Index

Glossary of Terms and Symbols

Term	Symbol	Description
Annual amount or worth	A or AW	Equivalent uniform annual worth of all cash inflows and outflows over estimated life.
Annual operating cost	AOC	Estimated annual costs to maintain and support an alternative.
Average	$\overline{X}$	Measure of central tendency; average of sample values.
Benefit/cost ratio	B/C	Ratio of a project's benefits to costs expressed in PW, AW, or FW terms.
Bond dividend	I	Dividend (interest) paid periodically on a bond.
Book value	BV	Remaining capital investment in an asset after depreciation is accounted for.
Breakeven point	Q_{BE}	Quantity at which revenues and costs are equal, or two alternatives are equivalent.
Capital recovery	CR	Equivalent annual cost of owning an asset plus the required return on the initial investment.
Capitalized cost	CC or P	Present worth of an alternative that will last forever (or a long time).
Cash flow	CF	Actual cash amounts which are receipts (inflow) and disbursements (outflow).
Cash flow before or after taxes	CFBT or CFAT	Cash flow amount before relevant taxes or after taxes are applied.
Composite rate of return	i'	Unique rate of return when a reinvestment rate c is applied to a multiple-rate cash flow series.
Compounding frequency	m	Number of times interest is compounded per period (year).
Cost estimating relationships	C_2 or C_T	Relations that use design variables and changing costs over time to estimate current and future costs.
Cost of capital	WACC	Interest rate paid for the use of capital funds; includes both debt and equity funds. For debt and equity considered, it is weighted average cost of capital.
Debt-equity mix	D-E	Percentages of debt and equity investment capital used by a corporation to fund projects.
Depreciation	D	Reduction in the value of assets using specific models and rules; there are book and tax depreciation methods.
Depreciation rate	d_t	Annual rate for reducing the value of assets using depreciation.
Economic service life	ESL or n	Number of years at which the AW of costs is a minimum.
Expenses	E	All corporate costs incurred in transacting business.
First cost	P	Total initial cost—purchase, construction, setup, etc.
Future amount or worth	F or FW	Amount at some future date considering time value of money.

(Continued)